The implementation and constructive use of misspecification tests in econometrics

The implementation and constructive use of misspecification tests in econometrics

edited by L. G. Godfrey

Manchester University Press

Manchester and New York

Distributed exclusively in the USA and Canada by St. Martin's Press

Published by Manchester University Press
Oxford Road, Manchester M13 9PL, UK
and Room 400, 175 Fifth Avenue, New York, NY 10010, USA

Distributed exclusively in the USA and Canada by St. Martin's Press, Inc., 175 Fifth Avenue, NY 10010, USA

British Library Cataloguing-in-Publication Data
A catalogue record for this book is available from the British Library

Library of Congress Cataloging-in-Publication Data
The implementation and constructive use of misspecification tests in
econometrics / edited by L. G. Godfrey
p. cm.
Includes index.
ISBN 0–7190–3274–1 (hardback)
1. Econometrics. 2. Mathematical statistics. I. Godfrey, L. G.
HB139.I554 1992
330'.01'5195—dc20 91–42764
CIP

ISBN 0 7190 3274 1 hardback

Printed in Great Britain
by Biddles Ltd, Guidford and King's Lynn

Contents

Preface

This book contains a collection of papers coauthored by members of the Department of Economics and Related Studies and the Institute for Research in the Social Sciences in the University of York. The derivation and evaluation of tests for misspecification has been a major area for research by econometricians at York since the late 1970s and recent work has been devoted to two general questions of considerable interest. These questions, which concern the implementation and constructive use of misspecification tests, are as follows:

(i) which methods for calculating asymptotically valid tests can be recommended for use with samples of the size available in empirical economics?

and

(ii) what is the scope for using test statistics to determine the nature of specification errors and to provide suitable corrections to estimates of parameters?

The contents of this book address these two questions in the context of a wide variety of estimators and models. Several of the contributors had the good fortune to be taught by Professor J. Durbin and/or Professor J. D. Sargan in postgraduate courses at the LSE and there is, not surprisingly, some emphasis on testing for serial correlation and instrumental variable methods. However, models with limited or qualitative dependent variables also receive considerable attention.

The first two chapters contain discussions of the implementation of checks of the adequacy of autocorrelation models for observable variables and unobservable disturbance terms. Chapter 1 by Godfrey and Tremayne deals with various types of univariate time series models, and includes findings and Monte Carlo results that provide useful reference points for analyses in subsequent chapters. In terms of the standard Box-Jenkins acronym, Godfrey and Tremayne concentrate on tests related to the choice of p and q in an ARIMA(p,d,q) specification, rather than on the degree of differencing d. Chapter 1 does, however, include a section on testing for unit roots and cointegration. The literature concerning these two related topics is vast and still growing rapidly. Godfrey and Tremayne outline the main ideas and results, and also give references to recent more comprehensive surveys. In the second chapter, Burke and Godfrey provide results on testing for serial correlation after estimation by instrumental variable/two stage least squares type procedures. This topic is important because instrumental variable estimates in applied studies are often accompanied by invalid checks for serial correlation, e.g. the Durbin-Watson statistic or the Box-Pierce portmanteau test. Burke and Godfrey derive new tests and compare them to existing procedures.

The next two chapters, both written by Eastwood and Godfrey, focus on the second general question given above, i.e. the interpretation and constructive value of test statistics. Many computer packages for the estimation of multiple regression models provide a number of tests for individual misspecifications, presumably in the hope that these separate tests can be used to identify and isolate specification errors. Theoretical results relating to the value of misspecification tests as guides to model reformulation are given in Chapter 3 while Chapter 4 contains Monte Carlo results that permit a reappraisal of the practical value of a multiple comparison procedure for regression equations.

Two points stressed by Eastwood and Godfrey are that researchers sometimes have only vague information about the nature of specification errors and that it seems useful to avoid tests that require normality of model disturbances even for large sample validity. Thus, in some cases, there will be a need for information parsimonious general tests that are asymptotically robust to nonnormality. In Chapter 5, Burke, Godfrey and McAleer provide a discussion of such tests in the context of checking for incorrect functional form and/or omitted variables in linear regression models.

Orme extends the range of models under consideration by discussing several important types of microeconometric specifications in the last two chapters. Chapter 6 contains an examination of the constructive value of test procedures in which Orme evaluates the performance of estimator corrections that can be obtained as a by-product of testing model adequacy. One general check of model adequacy that is widely used in microeconometric analysis is the Information Matrix test. Various algorithms for calculating the Information Matrix test statistic have been proposed. These alternative methods yield statistics that are asymptotically equivalent, but which can have quite different finite sample performances. The final chapter of this collection contains important evidence on the reliability of variants of the Information Matrix test. Orme's work will assist applied researchers to choose reliable checks in the absence of precise ideas about the nature of specification errors.

Acknowledgements

The authors are grateful to seminar participants at the Australian National University, Bristol University, CORE, Hitotsubashi University, London Business School, London School of Economics, Manchester University, University of Melbourne, Osaka University, Southampton University, Yokohama National University and University of York for helpful comments and suggestions. Burke, Eastwood, Godfrey and Tremayne wish to acknowledge the financial support of the Economic and Social Research

Council (UK), reference numbers B00232137 and R000231190. McAleer wishes to acknowledge the financial support of the Australian Research Council and a Japanese Government Foreign Research Fellowship at Kyoto University. Chapter 1 is a revised and extended version of a paper published by Godfrey and Tremayne in Econometric Reviews (Volume 7, Number 1, 1-42, 1988). These authors are grateful to Marcel Dekker Inc. (New York) for permission to include this work. The editor wishes to thank Francis Brooke and Katherine Reeve of Manchester University Press for all their help and encouragement.

Contributors

S. P. Burke *University of Reading*
Alison Eastwood *University of York*
L. G. Godfrey *University of York*
Michael McAleer *University of Western Australia*
C. D. Orme *University of York*
A. R. Tremayne *University of York*

L. G. GODFREY AND A. R. TREMAYNE

1. MISSPECIFICATION TESTS FOR UNIVARIATE TIME SERIES MODELS AND THEIR APPLICATIONS IN ECONOMETRICS

1. Introduction

Econometrics as a discipline clearly has its own subject matter and raises specific problems, but an important shared area of interest with statistical theory and practice is that of time series analysis. Indeed, in recent years, the use of time series models and techniques has become widespread in econometrics. Many of the points discussed by Hendry and Richard (1982), in particular some of their ideas relating to the data coherency of models, are linked to ideas that are employed in time series analysis. As further evidence of the importance of exploiting the connections between time series and econometrics, the contention of Pagan (1985, p.200) that "the interpretation and formulation of dynamic specifications is inextricably bound up with the nature of the time series used in the modelling exercise" seems pertinent. It is thus our belief, in common with that of Hendry and Richard (1983, p.129), that it is sensible to regard time series analysis and econometrics as being complementary, rather than competing.

In applied studies, the degree of reliance on time series methods is extremely variable; for a general discussion of applications, see Granger and Watson (1984, Section 7). At one end of the spectrum, for example in the work of Baillie, Lippens and McMahon (1983) and Cox (1985), such methods are of central importance. There are also many occasions when a model may be specified with a good deal of reliance on economic theory and little on patterns of serial correlation where nevertheless time series techniques can be combined with econometric methods. A prime example of this situation would be in certain models involving expectations, or surprise variables. In such cases, linear time series models are often used to construct regressors; see Pagan (1984) for a review of some of the issues involved. An early example of work addressing problems of expectation formation, together with the use of moving average disturbances in regression analysis (which is by no means commonplace even now), and of diagnostic checks for time series models is provided by Trivedi (1973). More recent papers in empirical economics which utilise portmanteau statistics of time series model adequacy as part of the report of the modelling exercise are Jones and Uri (1987) and Mills and Stephenson (1987), among many others. Finally, mention might be made of Fama (1975) and Fama and Gibbons (1982) who report individual estimated residual autocorrelations as summary statistics.

It is, however, important that any time series method used in econometrics should be adequate and appropriate for its intended purpose. The purpose of this chapter is to review and evaluate a number of time series techniques that may be of value to econometricians. In order to constrain the discussion to be of a manageable length, we shall restrict our attention to the problem of testing for specification errors in the context of univariate autoregressive-moving average models and transfer function (rational distributed lag) models of the type discussed by Box and Jenkins (1976) in their influential book. Details of corresponding tests for multiple time series models are given by Chitturi (1974), Hosking (1981) and Poskitt and Tremayne (1982, 1984).

There are obviously major subjects of relevance to both econometrics and time series analysis which are outside the scope of this chapter. Particular mention might be made of areas such as model determination and selection. This topic is discussed at some length in the econometrics literature by Chow (1981), whilst the work of Hannan (1980) is of importance in the pure time series framework. Further, in what follows we do not address the subject matter of Bayesian time series analysis and econometrics, but a useful summary and an extensive set of references are provided by Zellner (1985). More general topics, such as estimation and forecasting, are covered by Granger and Watson (1984) and Hendry and Richard (1983).

Section 2 contains a review of tests of the adequacy of a time series model with the procedures being classified according to the general approach used in their construction. The relationships between the different types of tests are explored and, where possible, Monte Carlo evidence on small sample behaviour is summarised. In Section 3, the application of such tests in econometric modelling is discussed. In particular, we examine tests of the autocorrelation structure of the disturbance term of a regression model. It is shown that the implementation of several tests only requires the least squares estimation of an auxiliary regression equation.

Econometricians sometimes include lagged values of the dependent variable in the regressors and this important case is also discussed in Section 3. There are, however, arguments against estimating models in this form, with a rational distributed lag specification being preferred in some situations. These arguments, along with asymptotic and small sample results for relevant test procedures, will be considered in Section 4.

Section 5 contains some results from a Monte Carlo study designed to illustrate how the ability to detect autocorrelation varies between alternative tests of the assumption of serial independence. These simulation results also provide information on the effects of using estimated residuals in place of the unobservable disturbance terms of a regression model, and hence on the potential usefulness of time series tests in econometric analysis.

An important feature of the models to be considered in Sections 2-5 is that they are stationary under the null hypothesis. There has recently been a great deal of interest in the case in which the null hypothesis is that the model has an autoregressive component with a unit root and so is nonstationary. This work builds upon the results of Dickey and Fuller (1979, 1981), Fuller (1976) and Evans and Savin (1981, 1984), and test statistics derived from the theoretical advances made have found application in the work of Perron and Phillips (1987), for example. The study of tests for unit roots has been part of a larger body of research concerned with nonstationarity of economic variables. Progress has been made in understanding the inadequacies of classical results on estimation and inference in the presence of nonstationarity, and in obtaining results that are asymptotically valid; see Park and Phillips (1988, 1989) and the references provided by these authors. The modelling of cointegrated variables has also received considerable attention; see the influential paper by Engle and Granger (1987), the collection of papers edited by Hendry (1986), and the recent surveys by Dolado, Jenkinson and Sosvilla-Rivero (1990) and Pagan and Wickens (1989). A full discussion of the large and rapidly growing literature on nonstationary processes is outside the scope of this chapter, but Section 6 contains a summary of some results related to testing. References to recent surveys and articles are also provided. Some concluding remarks are given in Section 7.

2. Tests for univariate time series models

The model which is tentatively assumed to be adequate for describing the generation of the random variable u will be taken to be a mixed autoregressive-moving average process of order (p, q) [ARMA(p, q)]. This model will be referred to as the null model or null specification, and will be written as

$$\phi(L)u_t = \theta(L)e_t \tag{1}$$

where

$$\phi(L) = 1 - \phi_1 L - \phi_2 L^2 - \dots - \phi_p L^p,$$

$$\theta(L) = 1 - \theta_1 L - \theta_2 L^2 - \dots - \theta_q L^q,$$

L being the lag operator defined by $L^j u_t = u_{t-j}$, and the e_t are independently distributed $N(0, \sigma_e^2)$ variates. The assumption of normality is adopted for simplicity and the tests discussed below retain their asymptotic validity under much weaker distributional assumptions.

It is also assumed that standard stationarity and invertibility conditions are satisfied with the classical asymptotic theory of maximum likelihood estimation, as exposited by Box and Jenkins (1976, Chapter 7), being applicable. It is worth pointing out that the application of classical likelihood-based inference involves taking the number of parameters to be finite under both null and alternative hypotheses. Thus, for example, the values of p and q in (1) are fixed as $n \rightarrow \infty$. It could be argued that, as more and more data became available, it would be natural to improve the approximation provided by a statistical model by increasing the number of parameters. From a theoretical perspective, therefore, an attractive alternative approach to classical methods might be to have $p = p(n)$ and $q = q(n)$ with $p(n) \rightarrow \infty$, $q(n) \rightarrow \infty$, $p(n)/n \rightarrow 0$ and $q(n)/n \rightarrow 0$ as $n \rightarrow \infty$; see Geweke (1984, Section 5) for a discussion along these lines.

In this chapter, we adopt a fairly pragmatic stance. From a practical point of view, we are trying to adopt a set of assumptions that lead to nondegenerate asymptotic distributions and for which the orders of magnitude involved correspond more or less to the orders of magnitude typically observed in the situations being considered. Given the moderate number of observations available in many econometric applications, we think that classical tests involving "small" numbers of parameters are likely to be more useful than those involving "large" numbers and derived by allowing the number of parameters to increase with n. This preference is based upon practical considerations, e.g. evidence on the relative merits of classical tests and portmanteau tests in finite samples, and certainly does not reflect a view that alternative approaches have limited theoretical interest.

Three general approaches to testing the adequacy of the null model (1) have been discussed in the literature. These approaches can be differentiated by their treatment of the alternative hypothesis and its relationship to the null specification. Before providing a detailed discussion, it will be useful to summarise their principal features as follows.

Pure significance tests

A pure signficance test is derived without the explicit formulation of an alternative hypothesis and so is appropriate when there is little information about the likely nature of any model misspecification. Let $T = T(u_1, u_2, \dots, u_n)$ be a statistic for such a test of the ARMA(p, q) model. Cox and Hinkley (1974, Section 3.2) argue that T should satisfy the following criteria: the distribution of T under the hypothesis of correct specification should be known (at least asymptotically) and should not depend upon the values of the parameters of the null model; and the larger the sample value of T, the stronger the evidence of departure from the null specification of the type it is required to detect.

Nested hypothesis tests

An alternative approach to evaluating a null specification is to embed it in some more general model in such a way that the former is obtained from the latter by imposing a set of parametric restrictions. The adequacy of the null model can then be assessed by testing these restrictions. Provided regularity conditions are satisfied, the likelihood ratio (LR), Wald (W) and score or Lagrange multiplier (LM) principles provide asymptotically equivalent tests of such restrictions. The LM principle is, however, particularly attractive when constructing misspecification checks because it does not require the estimation of the more complex model; see Engle (1984) for a general discussion of the LM, LR and W tests.

Nonnested hypothesis tests

Tests of the adequacy of the ARMA(p, q) model of (1) can also be obtained by specifying an alternative scheme which is neither more general than the null model nor a special case of it. In such a situation, the models are said to be nonnested or separate. For example, the ARMA(p, q) model and a pure autoregression of order (p + s), s > 0, are nonnested. The classical likelihood-based tests such as LR are not applicable when the hypotheses are nonnested, but asymptotic tests can be derived using the results of Cox (1961, 1962).

Each of these three general approaches will be examined below. It will be shown that the relationships between tests derived from different approaches are such that distinctions are often blurred. In addition to describing the theoretical relationships between the various procedures, we shall also discuss Monte Carlo evidence on their small sample behaviour. This evidence is clearly of considerable importance given the limited number of observations sometimes available when studying economic time series.

2.1 Pure significance tests

Suppose that the maximum likelihood estimates (MLE) of the null model are denoted by

$$\hat{\sigma}_e^2,\ \hat{\phi} = (\hat{\phi}_1, \ldots, \hat{\phi}_p)',\ \hat{\theta} = (\hat{\theta}_1, \ \ldots \ , \hat{\theta}_q)',$$

and the associated residuals by $\hat{e}_t$, t = 1, ... , n. Many different statistical criteria could be constructed using these estimates so as to provide a pure significance test. Since the purpose of time series analysis is to obtain a model of the general form (1) such that $e_t = [\phi(L)/\theta(L)]u_t$ is white noise, it is natural to check for evidence of significant serial correlation in the residuals $\hat{e}_t$. Thus a test of model adequacy based upon the estimated autocorrelations

$$\hat{r}_j = \Sigma_{j+1}^n \hat{e}_t\hat{e}_{t-j} \,/\, \Sigma_1^n\hat{e}_t^2 \ , \ j = 1, 2, \ldots , \qquad (2)$$

merits consideration.

The asymptotic distribution of the vector of autocorrelation estimates $\hat{r} = (\hat{r}_1, \hat{r}_2, \ldots, \hat{r}_m)'$ under the null hypothesis of correct specification has been derived by Box and Pierce (1970) and McLeod (1978). This distribution is multivariate normal with zero mean vector and a covariance matrix that is approximately idempotent with rank equal to (m - p - q) as m, the number of autocorrelations, increases at the rate of $n^{1/2}$; see Box and Pierce (1970, p. 1513). Box and Pierce exploit the latter feature and propose the well-known portmanteau test based upon the statistic

$$Q = n \Sigma_1^m \hat{r}_j^2. \tag{3}$$

If the ARMA(p, q) specification is adequate, then the test statistic Q is asymptotically distributed as $\chi^2(m - p - q)$ if m is $O(n^{1/2})$. Significantly large values of Q indicate some sort of model misspecification.

The portmanteau test is asymptotically valid under its own particular set of assumptions which include the special requirements that m be large and m/n be small ($m \to \infty$ and $m/n \to 0$ as $n \to \infty$). As will be discussed below, satisfying these requirements may cause problems when the sample size is moderate. If m were fixed, then Q would not be asymptotically distributed as $\chi^2(m - p - q)$ under the null hypothesis of correct specification. Further, a "naive" test of the type discussed by Durbin (1970) involving comparison of Q with critical values of the $\chi^2(m)$ distribution would also be invalid for fixed m. The portmanteau test is, however, not an invalid "naive" test because the number of estimated autocorrelations is regarded as a function of n that tends to infinity, but at a slower rate than n.

The Box-Pierce Q-test has the two properties recommended by Cox and Hinkley (1974) which were mentioned above and it is widely used, being calculated by many programs as a matter of routine. The practical value of this test is, however, open to question. It has been reported that significant values of Q are not often observed in empirical work and the portmanteau test has proved to be of limited value in choosing between models; e.g., see Prothero and Wallis (1976). The problem seems to lie in the discrepancy between the actual finite sample distribution of Q and the χ^2 distribution predicted by asymptotic theory. Davies, Triggs and Newbold (1977) find that, even for quite large samples, the true significance levels of the Q-test can be much smaller than the asymptotically valid nominal values.

The quality of the asymptotic approximation to the actual null distribution appears to be improved if Q of (3) is replaced by the modified portmanteau statistic

$$Q' = n(n + 2) \Sigma_1^m[(n - j)^{-1}\hat{r}_j^2] \qquad (4)$$

which is proposed by Ljung and Box (1978). There is, however, evidence that the power of this modified portmanteau test can be quite low, even in the presence of severe misspecification; see Davies and Newbold (1979). Accordingly it must be concluded that it is not safe to rely upon general checks such as Q and Q′ to detect misspecified models with high probability.

Part of the difficulty may stem from the choice of the number of terms $\hat{r}_j$ to be tested. The values of m used in empirical work and simulation experiments are often moderately large, e.g. Davies et al. (1977) use m = 20. Large values of m support the use of the approximation that the covariance matrix of $\hat{r}$ is idempotent, but probably require very large values of n to justify appeal to asymptotic theory since m/n should be small. Sample sizes in econometrics are sometimes only of the order of 80 or smaller and so it seems worthwhile to consider tests based upon a few (<<20) estimated autocorrelations. Ljung (1986) has used McLeod's (1978) results to derive a test of the significance of a small number of the $\hat{r}_j$, but such tests can be more easily implemented using equivalence results linking them to LM tests derived by nesting the null in a more general alternative. The latter class of tests will now be discussed.

2.2 Nested hypotheses and LM tests

The null specification of ARMA(p, q) is clearly a special case of an ARMA(p + f, q + g) model written as

$$\rho(L)\phi(L)u_t = \mu(L)\theta(L)e_t \tag{5}$$

where

$$\rho(L) = 1 - \rho_1 L - \dots - \rho_f L^f,\ f > 0,$$

and

$$\mu(L) = 1 + \mu_1 L + \dots + \mu_g L^g,\ g > 0.$$

It will be convenient to define the following parameter vectors: $\phi' = (\phi_1, \phi_2, \dots, \phi_p)$, $\theta' = (\theta_1, \theta_2, \dots, \theta_q)$, $\psi_1{}' = (\phi', \theta')$, $\rho' = (\rho_1, \rho_2, \dots, \rho_f)$, $\mu' = (\mu_1, \mu_2, \dots, \mu_g)$, $\psi_2{}' = (\rho', \mu')$ and $\psi' = (\psi_1{}', \psi_2{}')$. Thus imposing the (f + g) parametric restrictions of $\psi_2 = 0$ in (5) yields the null model. Further the MLE of the null model (1), denoted by $\hat{\psi}_1{}' = (\hat{\phi}', \hat{\theta}')$, can be regarded as being obtained by solving the constrained optimisation problem

$$\textit{maximise}\ \ell(\psi) = -1/2\ \ln\{\Sigma_1^n\ [e_t(\psi)]^2\}$$
$$\textit{subject to}\ \psi_2 = 0,$$

where $\ell(\psi)$ is, apart from an irrelevant constant, the concentrated log-likelihood for (5) divided by n, and

$$e_t(\psi) = \{[\rho(L)\phi(L)]\ /\ [\mu(L)\theta(L)]\}u_t.$$

A suitable Lagrangian for this problem is

$$\Lambda(\psi,\lambda) = \ell(\psi) + \lambda'\psi_2, \tag{6}$$

where λ is the (f + g)-dimensional vector of multipliers.

As suggested above, the LM principle is especially attractive when deriving misspecification tests. The LM procedure proposed by Silvey (1959) (and independently by Rao (1948) as the score test) involves testing the joint significance of the estimated multipliers denoted by $\hat{\lambda}$. Equation (6) implies that these multipliers are given by

$$\hat{\lambda} = \partial\ell(\hat{\psi})/\partial\psi_2, \tag{7}$$

where $\hat{\psi}$ is the constrained MLE, i.e. $\hat{\psi}' = (\hat{\psi}_1', 0')$. This vector has as its elements

$$\partial\ell(\hat{\psi})/\partial\rho_i = \Sigma\, \hat{e}_t\hat{e}_{t-i} \;/\; \Sigma\, \hat{e}_t^2, \; i = 1, \dots, f,$$

and

$$\partial\ell(\hat{\psi})/\partial\mu_i = \Sigma\, \hat{e}_t\hat{e}_{t-i} \;/\; \Sigma\, \hat{e}_t^2, \; i = 1, \dots, g,$$

so that

$$\hat{\lambda}' = (\hat{r}_1, \hat{r}_2, \dots, \hat{r}_f, \hat{r}_1, \hat{r}_2, \dots, \hat{r}_g). \tag{8}$$

Equation (8) reveals that an LM test against a nesting hypothesis is equivalent to a test based upon residual autocorrelations. This result was obtained by Newbold (1980) using a different parameterization of the alternative model.

There are clearly redundant elements in $\hat{\lambda}$ and so this vector must have a singular asymptotic null distribution. This difficulty is easily overcome. All that needs to be done is to check only the significance of the $s = \max(f, g)$ distinct elements of $\hat{\lambda}$, viz. $\hat{r}_1$, $\hat{r}_2, \dots, \hat{r}_s$. This strategy is equivalent to imposing $\min(f, g)$ untested parametric restrictions in (5) to achieve identifiability under (1), as the information matrix for an unrestricted ARMA(p + f, q + g) model is singular when the ARMA(p, q) model is valid; see Hannan (1970, pp.388-9). Consequently, a direct test of the significance of $\hat{r}_1, \hat{r}_2, \dots, \hat{r}_s$ is equivalent to Silvey's (1959) modified LM procedure involving a generalised inverse (g-inverse) of the information matrix $-n\,E[\partial^2\ell(\psi)/\partial\psi\partial\psi']$; see Poskitt and Tremayne (1980, Section 2).

There are several valid ways in which to delete redundant elements of $\hat{\lambda}$, e.g. either the first or the (f + 1)th element can be removed to avoid duplication of $\hat{r}_1$. Each of these ways corresponds to a different set of identifying restrictions on ψ and hence to a different g-inverse; see Silvey (1959) for the form of such g-inverses. The value of the test statistic will, however, be unaltered, provided a common consistent estimate of the covariance matrix of $\hat{r}_1, \hat{r}_2, \ldots, \hat{r}_s$ is used. If the choice of covariance matrix estimate varies with the position of the deleted elements of $\hat{\lambda}$, then the corresponding test statistics will differ by asymptotically negligible terms.

Since the LM tests derived from the nested hypothesis approach are equivalent to tests of the significance of residual autocorrelations, they are similar to the pure significance tests discussed above. The distinction between the two types of procedure is further blurred by a certain lack of sensitivity of the LM test to the precise formulation of the alternative hypothesis. Once s = max(f, g) is fixed, the same LM statistic is appropriate for all restricted ARMA(p + f,q + g) models that are identified under the null ARMA(p, q) specification. In particular, by setting f = g = s, we deduce that the same LM test is valid when the null is ARMA(p, q) and the alternative is either ARMA(p + s, q) or ARMA(p, q + s). The latter models are, in the terminology of Godfrey (1981), locally equivalent alternatives (LEAs) with respect to the ARMA(p, q) specification, and are essentially the same for small departures from the null hypothesis.

Despite the similarities between LM checks and portmanteau tests of residual autocorrelations, it should be stressed that there is an important difference between them. The number of autocorrelations being considered, i.e. s, is fixed and finite in the LM approach. Consequently the covariance matrix of the estimators $\hat{r}_1, \ldots, \hat{r}_s$ cannot be taken to be idempotent with rank equal to (s - p - q) and easily computed forms like those of Q and Q′ are unavailable. Fortunately there is a convenient method for implementing LM tests.

Implementation of LM tests

Godfrey (1979) points out that the LM statistic can be calculated by means of an auxiliary regression equation. This device can be illustrated by considering the ARMA(p + s, q) model which is locally equivalent to the ARMA(p + f, q + g), s = max(f, g), scheme of (5). The former model can be written as

$$e_t(\psi) = [\rho(L)\phi(L)/\theta(L)]u_t, \tag{9}$$

where $\rho(L) = 1 - \rho_1 L - \dots - \rho_s L^s$, and ψ is now to be interpreted as the (p + q + s)-dimensional vector with elements ρ_i (i = 1, ... , s), ϕ_i (i = 1, ... , p) and θ_i (i = 1, ... , q). The LM statistic can then be calculated as n times R^2, the uncentred coefficient of determination, for the least squares regression of the residuals $\hat{e}_t$ on the partial derivatives $\partial e_t(\psi)/\partial\psi$ evaluated at $\psi = \hat{\psi}$, the constrained MLE. Sample values of the nR^2 statistic should be compared to critical points of the $\chi^2(s)$ distribution with significantly large values indicating model inadequacy. It should be noted that the degrees of freedom parameter for the LM test equals the number of (distinct) autocorrelations under scrutiny, whereas these quantities are unequal for portmanteau tests unless p = q = 0.

Equation (9) represents a multiplicative extension of the null model. Several articles in the literature use a different parameterization of the alternative model by specifying an additive extension of the $\phi(L)$ polynomial with ρ_i corresponding to ϕ_{p+i}, i = 1, ... , s. An advantage of this parameterization is that regressors of the auxiliary model, i.e. the elements of $\partial e_t(\hat{\psi})/\partial\psi$, can be obtained by simple recurrence relationships; see Godfrey (1979, p.69). The value of the LM test statistic is, however, invariant to the choice of parameterization for the alternative model. These points can be illustrated by considering the problem of testing an AR(1) null against an AR(2) alternative.

The multiplicative form of the AR(2) model corresponding to (9) is

$$e_t(\psi_1, \psi_2) = (1 - \psi_1 L)(1 - \psi_2 L)u_t,$$

while the additive version is

$$e_t(\psi_1, \psi_2) = (1 - \psi_1 L - \psi_2 L^2)u_t.$$

The auxiliary regression for the multiplicative version is, apart from irrelevant changes of sign, given by

$$e_t(\hat{\psi}_1, 0) = u_{t-1}b_1^m + (u_{t-1} - \hat{\psi}_1 u_{t-2})b_2^m + \text{residual},$$

say, and the corresponding equation for the additive extension is

$$e_t(\hat{\psi}_1, 0) = u_{t-1}b_1^a + u_{t-2}b_2^a + \text{residual}.$$

Standard least squares theory implies that the least squares coefficient of determination is the same for both auxiliary equations. More generally, the R^2 statistic for the regression of $\hat{e}_t$ on $\partial e_t(\hat{\psi})/\partial\psi$ will be the same whichever of the two parameterizations of the alternative ARMA(p + s, q) model is adopted. The multiplicative version is useful in deriving theoretical results and the additive version offers the practical advantage mentioned above.

The R^2 statistic used in Godfrey's (1979) χ^2 form of the LM test of the ARMA(p, q) model against the ARMA(p + s, q) alternative can also be used to derive an asymptotically valid procedure involving the comparison of the "F statistic"

$$F = [R^2/(1 - R^2)]/[(n - p - q - s)/s]$$

to critical points in the right-hand tail of the F(s, n - p - q - s) distribution. Despite the form of the first term in square brackets, this F statistic provides a test of the significance of only a subset of the regressors $\partial e_t(\hat{\psi})/\partial\psi$ of the auxiliary model. This point will be considered below in the discussion of C(α) tests.

The LM principle provides a very flexible and easily implemented approach to the construction of diagnostic checks for time series models, and is easily modified, e.g. to allow for gaps reflecting seasonal patterns; see Newbold (1983). In order to be of value to applied workers, however, the asymptotically valid LM tests should behave well in finite samples. Exact finite sample results are rarely available and it is necessary to rely upon Monte Carlo experiments.

Monte Carlo evidence

Godfrey (1979) carries out a number of simulation experiments, including some similar to those of Davies et al. (1977). He finds that the actual significance levels of LM statistics are much closer to the nominal values than those of the widely used Q-test. Godfrey also provides power comparisons for LM procedures and the modified portmanteau test of Ljung and Box (1978). The LM tests outperform the Ljung-Box test, and appear to be quite powerful even in some cases in which they are based upon inappropriate alternatives. Such results on the behaviour of LM tests are obviously encouraging.

It might be thought that since an LM check requires only the specification of a class of LEAs determined by the value of s, the number of testable restrictions, its small sample performance might be inferior to an asymptotically equivalent LR or W procedure derived using full knowledge of the alternative hypothesis. This issue has also been examined using Monte Carlo methods. The available results suggest that, provided the set of LEAs is identified correctly, there are no substantial gains to be achieved by using knowledge about which member of this set generated the data; see Bera and McKenzie (1986) and Godfrey (1981).

C(α) tests and inverse autocorrelations

The LM test is a special case of Neyman's (1959) C(α)

procedure. An interesting feature of Neyman's procedure in the context of testing the ARMA(p, q) specification against the ARMA(p + s, q) model, or some LEA, is that, in the terminology of Box and Jenkins (1976), it is possible to move from identification to diagnostic testing without the intermediate step of estimation. More precisely, a $C(\alpha)$ test can be derived that is asymptotically equivalent to the LM, LR and W tests, but requires neither the null nor the alternative to be estimated by ML methods. This $C(\alpha)$ test can be calculated using a regression-based algorithm proposed by Breusch and Pagan (1980, Section 4), and requires the calculation of the autocorrelation function (ACF) and inverse autocorrelation function (IACF) of the series u. The inverse autocorrelations of an ARMA(p, q) process are the autocorrelations of the dual ARMA(q, p) model obtained by interchanging the lag polynomials $\phi(L)$ and $\theta(L)$; see Cleveland (1972) and Chatfield (1979) for useful discussions. For example, consider the ARMA(p, 0) process

$$u_t = \phi_1 u_{t-1} + \dots + \phi_p u_{t-p} + e_t.$$

The IACF for this process is the ACF for the ARMA(0, p) scheme

$$u_t = e_t - \phi_1 e_{t-1} - \dots - \phi_p e_{t-p},$$

and is, therefore, given by

$$ir^u_j = 1 \text{ for } j = 0$$

$$= (-\phi_j + \sum_{l=1}^{p-j} \phi_l \phi_{l+j})/(1 + \sum_{l=1}^{p} \phi_l^2) \text{ for } j = \pm 1, \dots, \pm p$$

$$= 0 \text{ for } |j| > p.$$

In order to outline the $C(\alpha)$ test, we shall first describe the general form of the test and then explain how the ACF and IACF are to be used. Let $\tilde{\phi}$ and $\tilde{\theta}$ be estimators such that $(\tilde{\phi} - \phi)$ and

$(\tilde{\theta} - \theta)$ are both $O_p(n^{-1/2})$ when the null model is valid. Thus these estimators are $n^{1/2}$-consistent, but not necessarily asymptotically efficient, under the assumption of correct specification. A restricted consistent estimator of ψ, the parameter vector of the alternative model of (9), is therefore given by $\tilde{\psi}' = (\tilde{\psi}_1', \tilde{\psi}_2')$, where $\tilde{\psi}_1' = (\tilde{\phi}', \tilde{\theta}')$ and $\tilde{\psi}_2$ is an s-dimensional vector with every element equal to zero. This restricted estimator can be used to evaluate residuals $\tilde{e}_t = e_t(\tilde{\psi})$ and estimated partial derivatives $\partial e_t(\tilde{\psi})/\partial\psi$, $t = 1, \ldots, n$. The $C(\alpha)$ test of $\psi_2 = 0$ can then be implemented as follows.

(i) Regress $\tilde{e}_t$ on $\partial e_t(\tilde{\psi})/\partial\psi_1$, $t = 1, \ldots, n$ to obtain a coefficient of determination R_0^2.

(ii) Regress $\tilde{e}_t$ on $\partial e_t(\tilde{\psi})/\partial\psi_1$ and $\partial e_t(\tilde{\psi})/\partial\psi_2$, $t = 1, \ldots, n$, to obtain a coefficient of determination R_1^2.

(iii) Treat $n(R_1^2 - R_0^2)$ as a $\chi^2(s)$ variate with large values of this statistic indicating model misspecification.

The $C(\alpha)$ test is simply a check of the significance of the regressors $\partial e_t(\tilde{\psi})/\partial\psi_2 = \left\{\partial e_t(\tilde{\psi})/\partial\rho_i\right\}$ in the auxiliary regression of $\tilde{e}_t$ on $\partial e_t(\tilde{\psi})/\partial\psi$, and the standard F-test based upon

$$[(R_1^2 - R_0^2)/(1 - R_1^2)][(n - p - q - s)/s]$$

is asymptotically valid and simple to implement.

The LM test is obtained from the general $C(\alpha)$ form by using the restricted MLE $\hat{\psi}_1$ for $\tilde{\psi}_1$. In this special case, R_1^2 equals the R^2 statistic given by Godfrey (1979) and the first-order conditions for $\hat{\psi}_1$ imply that R_0^2 equals zero. Substituting these values in the expression for the F-version of the $C(\alpha)$ statistic yields the LM criterion

$$[R^2/(1 - R^2)][(n - p - q - s)/s]$$

that was given above.

In order to complete the outline of the C(α) test, it only remains to describe how the estimates $\tilde{\phi}$ and $\tilde{\theta}$ can be obtained from the ACF and IACF. Let the autocorrelations of the u series be denoted by

$$r_j^u = \Sigma_{j+1}^n u_t u_{t-j} / \Sigma_1^n u_t^2, \; j = 1, 2, \ldots . \tag{10}$$

These autocorrelations can be used to construct the p-dimensional vector

$$r^u = (r_{q+1}^u, \ldots, r_{q+p}^u)' \tag{11}$$

and the p by p matrix

$$R^u = \left\{ r_{q+i-j}^u \right\}, \; i, j = 1, \ldots, p. \tag{12}$$

It is well-known that

$$\tilde{\phi} = [R^u]^{-1} r^u \tag{13}$$

is consistent for ϕ under the assumption of correct specification; see Box and Jenkins (1976, p.202). Next, if ir_j^u, as before, denotes a typical inverse autocorrelation, ir^u is the q-dimensional vector $(ir_{p+1}^u, \ldots, ir_{p+q}^u)'$, and IR^u is the q by q matrix with typical element (i, j) equal to ir_{p+i-j}^u, then

$$\tilde{\theta} = [IR^u]^{-1} ir^u \tag{14}$$

is consistent for θ under the null, provided the inverse autocorrelations are estimated consistently. Results on the consistency of estimators of the ir_j^u are provided by Bhansali (1980).

2.3 Nonnested hypotheses

The evaluation of a null model using a nonnesting hypothesis has not received a great deal of attention in the time series literature. Walker (1967) has applied the theory developed by Cox (1961, 1962) to the problem in which the u_t are generated by a pure AR(p) process under one hypothesis and by a pure MA(q) process under the other; the special case with $p = q = 1$ being considered in detail. The tests derived by Walker are, however, relatively complicated and they have not been widely adopted. In a recent paper, McAleer, McKenzie and Hall (1986) have developed tests for nonnested time series models based upon auxiliary regressions, and these procedures may prove to be useful in empirical analysis.

An alternative approach to the problem is to embed the two competing separate models in a comprehensive model. The two models of interest are then both tested against the more general formulation. For example, the simple AR(1) and MA(1) schemes considered by Walker (1967, Section 3) are both special cases of the ARMA(1, 1) model

$$u_t - \phi_1 u_{t-1} = e_t - \theta_1 e_{t-1}.$$

Consequently the adequacy of the AR(1) and MA(1) hypotheses can be investigated by testing $\theta_1 = 0$ and $\phi_1 = 0$ respectively. Note that the results on local equivalence imply that the LM check of the AR(1) specification against the ARMA(1, 1) alternative is also appropriate for the AR(2) alternative and so can be calculated by testing the significance of u_{t-2} in the autoregression

$$u_t = \phi_1 u_{t-1} + \phi_2 u_{t-2} + e_t.$$

More generally, the AR(p) and the MA(q) hypotheses can be embedded in the ARMA(p, q) specification and, for example, the adequacy of the pure MA process checked by testing $\phi_1 = \phi_2 = \dots = \phi_p = 0$ using an appropriate LM statistic.

Pure significance tests may also be of value. Consider Walker's special case in which the nonnested models under scrutiny are the AR(1) and MA(1) schemes. If the AR specification is correct, then the autocorrelation estimates for the u_t satisfy

$$\text{plim } [r_j^u - (r_1^u)^j] = 0, \; j = 2, 3, \ldots \, . \tag{15}$$

It follows that the adequacy of the AR(1) scheme can be assessed by calculating terms like $[r_j^u - (r_1^u)^j]$ and testing their significance. It is interesting to note that if the probability limits of (15) are written in the mathematically equivalent form

$$\text{plim } [r_j^u - r_1^u r_{j-1}^u] = 0, \; j = 2, 3, \ldots ,$$

then the corresponding pure significance test is equivalent to an LM test against a higher-order autoregression. This equivalence can be demonstrated as follows.

The least squares estimate of the parameter of the AR(1) null model is

$$\hat{\phi}_1 = \Sigma_2^n u_t u_{t-1} / \Sigma_2^n u_{t-1}^2$$

which differs from the first-order autocorrelation estimate

$$r_1^u = \Sigma_2^n u_t u_{t-1} / \Sigma_1^n u_t^2$$

by asymptotically negligible terms. Thus a typical term to be tested is equivalent to

$$\begin{aligned} r_j^u - \hat{\phi}_1 r_{j-1}^u &= [\Sigma_{j+1}^n u_t u_{t-j} - \hat{\phi}_1 \Sigma_j^n u_t u_{t-j+1}] / \Sigma_1^n u_t^2 \\ &\approx [\Sigma_{j+1}^n u_t u_{t-j} - \hat{\phi}_1 \Sigma_{j+1}^n u_{t-1} u_{t-j}] / \Sigma_1^n u_t^2 \\ &= [\Sigma_{j+1}^n (u_t - \hat{\phi}_1 u_{t-1}) u_{t-j}] / \Sigma_1^n u_t^2 \\ &= \Sigma_{j+1}^n \hat{e}_t u_{t-j} / \Sigma_1^n u_t^2 \, , \; j = 2, 3, \ldots \end{aligned}$$

which is proportional to the LM criterion (score) for testing the AR(1) scheme against

$$u_t - \phi_1 u_{t-1} - \phi_j u_{t-j} = e_t.$$

A joint test of, say, (h - 1) terms is, therefore, equivalent to testing H_0: $\phi_2 = \phi_3 = \ldots = \phi_h = 0$ in the general AR(h) alternative

$$u_t - \phi_1 u_{t-1} - \ldots - \phi_h u_{t-h} = e_t.$$

A standard F-test is easy to calculate and asymptotically valid.

Turning to the MA(1) hypothesis, it seems natural to base a pure significance test on the result that, under this assumption, the autocorrelations of the u_t satisfy

$$\text{plim } r_j^u = 0, \; j = 2, 3, \ldots . \qquad (16)$$

Thus, for example, a test can be based upon r_2^u with significant values leading to the rejection of the MA(1) hypothesis. Bartlett's (1946) results imply that an asymptotically valid test of the significance of r_2^u can be derived by treating

$$n^{1/2}\{r_2^u/[1 + 2(r_1^u)^2]^{1/2}\}$$

as an N(0, 1) variate. Under the alternative of an AR(1) process, plim $r_2^u = \phi_1^2 > 0$ and so researchers may wish to use a one-sided test in this particular context.

3. On the application of tests for univariate time series models in econometrics

Many useful applications of the tests described in the previous section can be obtained by viewing the variable u_t of (1) as the unobservable disturbance term of a regression equation. Let such a regression model be written as

$$y_t = \Sigma_1^k x_{ti}\beta_i + u_t , \; t = 1, \dots , n, \tag{17}$$

where the regressors are strictly exogenous and covariance-stationary, and the errors u_t are generated by the ARMA(p, q) scheme of (1). In order to economise on notation, the MLE for the null model (which now comprises (1) and (17)) will again be denoted by a circumflex, e.g. $\hat{\beta}$ stands for the estimate of the regression parameter β. Pierce (1971a) discusses the asymptotic distribution of the MLE $(\hat{\beta}, \hat{\phi}, \hat{\theta})$. He shows that, under standard regularity conditions, these estimators are consistent and asymptotically normally distributed with $\hat{\beta}$ being asymptotically independent of $(\hat{\phi}, \hat{\theta})$. This independence results from the asymptotic information matrix being block-diagonal between regression parameters and autocorrelation parameters. Pierce (1971b) uses this block-diagonality to demonstrate that the autocorrelation estimates based upon the residuals

$$\hat{e}_t = [\hat{\phi}(L)/\hat{\theta}(L)][y_t - \Sigma_1^k x_{ti}\hat{\beta}_i] \tag{18}$$

are asymptotically equivalent to those calculated using the true value of β, provided the model is specified correctly. Hence, if the residuals of (18) are employed to calculate autocorrelation estimators

$$\hat{r}_j = \Sigma_{j+1}^n \hat{e}_t\hat{e}_{t-j}/\Sigma_1^n \hat{e}_t^2, \; j = 1, 2, \dots ,$$

then the asymptotic distribution of $\hat{r} = (\hat{r}_1, \dots , \hat{r}_m)'$ under the regression model with ARMA errors is the same as for the pure time series model of (1). Consequently the Q and Q′ tests are asymptotically valid and can be applied to least squares residuals as general misspecification checks without modification. Earlier comments on the small sample performance of these tests do not, however, provide strong support for their use in regression models.

The irrelevance of the substitution of $\hat{\beta}$ for β when examining the asymptotic behaviour of the autocorrelation estimates $\hat{r}_j$ also has implications for LM tests. It will be recalled that LM tests are equivalent to tests of the estimates $\hat{r}_j$ and so the nR^2 statistics of Godfrey (1979) can be applied to regression models with autocorrelated errors with no explicit account being taken of the presence of regressors, provided that these regressors are strictly exogenous. Alternative (and asymptotically equivalent) nR^2 statistics which do involve the regressor values x_{ti} can be obtained as special cases of the formulae provided by Godfrey (1978b).

In addition to validating portmanteau and LM tests, the block-diagonality of the information matrix also justifies the application of a very simple procedure when the null hypothesis of AR(p) disturbances is to be tested against the AR(p + s) alternative or some locally equivalent form such as ARMA(p, s). The need for such tests is stressed by Granger and Newbold (1977, p.12). This procedure does not even require ML estimation of the null model and can be implemented using the ordinary least squares (OLS) estimate of β. Let the OLS estimate for (17) be denoted by $\dot{\beta}$ and the associated residuals by $\dot{u}_t$, $t = 1, \dots, n$. Godfrey (1987) shows that the AR(p) error model can be tested against the AR(p + s) or ARMA(p, s) alternative by checking the significance of $\dot{u}_{t-p-1}, \dots, \dot{u}_{t-p-s}$ in the autoregression

$$\dot{u}_t = \Sigma_1^{p+s} \dot{u}_{t-j}\alpha_j + v_t.$$

The conventional F-test is simple to compute and is asymptotically equivalent to the corresponding LM procedure.

The asymptotic equivalence of Godfrey's (1987) procedure and the more familiar LM, LR and W tests can be established by noting that the OLS estimates for the regression of $\dot{u}_t$ on $\dot{u}_{t-1}, \dots, \dot{u}_{t-p-s}$ maximize the likelihood for the regression model with AR(p + s) errors conditionally upon $\beta = \dot{\beta}$. Since $\dot{\beta}$ is consistent with $(\dot{\beta} - \beta) = O_p(n^{-1/2})$ and the information matrix is block-diagonal between autocorrelation parameters and regression parameters, these conditional maximizers are equivalent to the corresponding full MLE and the required relationships between test procedures follow. In fact Godfrey's procedure can be interpreted as a W-type test based upon a two-step estimator of the parameters of the AR(p + s) error model.

Testing the assumption of independent errors

The OLS estimate $\dot{\beta}$ is, of course, the MLE for the special case (p = 0, q = 0) in which the disturbances are taken to be independent. The assumption of serial independence is often made in applied econometric work and the Durbin-Watson test is probably the most commonly used check for serial correlation. This test is well-known to be asymptotically equivalent to a test based upon the first-order autocorrelation estimate $\hat{r}_1$, but it is dangerous to rely solely upon such a check, especially when the data exhibit seasonality. Wallis (1972) derives a quarterly analogue of the Durbin-Watson test which is asymptotically equivalent to a check of the significance of $\hat{r}_4$; also see Vinod (1973). It is, however, important to look at the correlogram of the OLS residuals and not just one element; see Granger and Newbold (1977).

The autocorrelation estimates calculated from OLS residuals, viz. $\hat{r}_j$, are asymptotically distributed as NID(0, n^{-1}) variates under the null hypothesis of correct specification with independent errors. Hence the LM test for general sth-order autocorrelation can be implemented by comparing sample values of

$$LM' = n \, \Sigma_1^s \, \hat{r}_j^2 \tag{19}$$

to right-hand-tail critical points of the $\chi^2(s)$ distribution. (LM′ is asymptotically equivalent to the standard LM test; see Godfrey (1978a).) Equation (19) indicates that, in this case, the Box-Pierce Q-test is asymptotically valid even though the number of autocorrelations being tested is fixed and finite. Thus there is no need to use large values of m in (3) (or expression (4) for Q′) when testing the assumption of serial independence. Indeed small sample performance may be much improved if large values of m are avoided.

The impact of varying the autocorrelations under test can be studied by considering an LM-type statistic which is asymptotically equivalent to the criterion of (19). Breusch and Godfrey (1981) point out that it is asymptotically valid to test against either the AR(s) or MA(s) alternative by testing the exclusion restrictions $\rho_1 = \rho_2 = \ldots = \rho_s = 0$ in the augmented regression model

$$y_t = \Sigma_1^k x_{ti}\beta_i + \Sigma_1^s \rho_i \dot{u}_{t-i} + w_t, \qquad (20)$$

where presample values of the lagged OLS residuals are set equal to zero. Thus the adequacy of the null model is checked by adding lagged OLS residuals as a set of proxy variables and assessing their joint significance. This is a situation of the type considered by Pagan and Hall (1983) and Godfrey's (1983, pp.230-1) comments are pertinent. More precisely, under a suitable sequence of local alternatives, overspecifying the order of the true autocorrelation scheme will lead to a test which is less powerful than that based upon the correct value of s, whereas underspecification can lead to a more powerful test. As suggested above, however, choosing very small values of s can be dangerous. For example, a test against AR(1) errors with s = 1 will not even be consistent when the errors are generated by a simple AR(4) model of the form

$$u_t = \phi_4 u_{t-4} + e_t.$$

In view of the likely uncertainty about the precise nature of possible autocorrelation, it is worth examining the effects of using different sets of test variables in expanded models like (20). Some Monte Carlo results on these effects and the small sample impact of replacing β by its MLE will be provided in Section 5.

Tests of nonnested hypotheses

Like the pure significance and nested hypothesis tests, the tests for nonnested models derived by Walker (1967) are asymptotically valid when applied to regression residuals, as opposed to observed time series variables, provided the regressors are strictly exogenous and satisfy classical regularity conditions. King and McAleer (1987) report some simulation evidence for the case of testing the AR(1) disturbance model against the MA(1) alternative in the context of the regression model (17). The Cox-type test performs very badly with a true null model being rejected far too frequently. In contrast the LM test based upon the comprehensive ARMA(1, 1) model discussed in Section 2 performs well and is quite powerful, despite the fact that it is based upon an inappropriate (overspecified) alternative hypothesis. Since the Cox-type tests are poorly behaved in finite samples and the LM tests are based upon an inappropriate alternative hypothesis, it seems worthwhile to consider the application of a pure significance test of the type outlined in Section 2.

Burke, Godfrey and Tremayne (1990) examine a test based upon (15) with j = 2 and autocorrelation estimates calculated using OLS residuals. Their procedure, called the TS test, is designed to have the property that King (1983, p.37) suggests that any "good" test for this problem should possess. This property is that the size of the test for $|\phi_1| > 0.5$ should tend to zero as n tends to infinity. This restriction on the asymptotic behaviour of tests is based upon the results that ϕ_1 equals the first-order autocorrelation of an AR(1) process and that the absolute value of this autocorrelation cannot exceed 0.5 for an MA(1) model.

Simulation experiments are used by Burke, Godfrey and Tremayne (1990) to study the finite sample performance of the TS test. One set of experiments is obtained by employing a design used by King and McAleer (1987) with the data generation process being

$$y_t = \beta_1 + \beta_2 x_{t2} + \beta_3 x_{t3} + u_t, \; t = 1, \ldots, n,$$

with

$$u_t = \phi_1 u_{t-1} + e_t + \theta_1 e_{t-1},\ e_t \text{ NID}(0, \sigma_e^2),$$

where x_{t2} and x_{t3} are the price and income variables from the data set of Durbin and Watson (1951, p.159). The variance parameter σ_e^2 is set equal to 1 and $\beta_1 = \beta_2 = \beta_3 = 0$, but these restrictions involve no loss of generality. Experiments with $\theta_1 = 0$ give estimates of significance levels and those with $\theta_1 \neq 0$ provide estimates of rejection probabilities for false null hypotheses. The nominal significance level is 5 per cent in all cases and estimates of rejection frequencies are based upon 6000 replications.

Table 1

Testing the AR(1) error hypothesis: some estimates of rejection frequencies

	n = 60		
(ϕ_1, θ_1)	LM	g(0.75,0.75)	TS
(0.9, 0.0)	0.05	0.00*	0.01
(0.6, 0.0)	0.04	0.03	0.04
(0.3, 0.0)	0.04	0.05	0.04
(0.0, 0.0)	0.03	0.02	0.04
(0.0, 0.2)	0.07	0.06	0.07
(0.0, 0.5)	0.35	0.43	0.39
(0.0, 0.7)	0.65	0.89	0.68
(0.0, 0.9)	0.79	0.99	0.81

*The actual figure corresponding to this entry is 0.003.

Table 1 contains some results for n = 60 that are taken from Burke, Godfrey and Tremayne (1990). This table also includes results for the LM test and the point optimal g(0.75, 0.75) test discussed by King and McAleer (1987). As can be seen from Table 1, the estimated significance levels of TS are in quite good agreement with the predictions of asymptotic theory and, under MA schemes, the performance of TS is similar to that of the LM test. A more detailed discussion of Monte Carlo evidence on the behaviour of the TS test is provided by Burke, Godfrey and Tremayne (1990).

There are other interesting possibilities for applying tests of nonnested hypotheses involving time series models. For example, it seems reasonable to argue that if an econometric model is adequate then it should provide evidence against a (nonnested) univariate time series model for the dependent variable. On the other hand, if a structural model is revealed to be data inconsistent when a pure time series model serves as the alternative, then the econometric relationship must be judged to be misspecified. If, as is often likely to be the case, interest focuses on evaluating a structural model with $p = q = 0$, and a pure AR(h) scheme for y_t is employed as the alternative so as to avoid the computational cost of estimating MA parameters, then modified Cox statistics of the type discussed by Godfrey and Pesaran (1983) may be useful. Alternatively, in such cases, the adequacy of (17) could be assessed by testing the restrictions $\alpha_1 = \alpha_2 = \dots = \alpha_h = 0$ in the comprehensive model

$$y_t = \Sigma_1^k x_{ti}\beta_i + \Sigma_1^h y_{t-i}\alpha_i + u_t, \tag{21}$$

where the errors u_t are taken to be NID(0, σ_u^2).

Models with lagged dependent variables

Equations like (21) which include lagged values of the dependent variable as regressors are quite common in econometrics. The inclusion of lagged dependent variables has important implications for inference. The information matrix for a model consisting of (21) and (1) is not block-diagonal between regression parameters (β, α) and autocorrelation parameters

(θ, ϕ). Consequently the autocorrelation estimates $\hat{r}_j$ for such a model are not asymptotically equivalent to those for the pure ARMA scheme; see Durbin (1970) for a general discussion of this type of problem.

The univariate time series checks described in Section 2 are, therefore, not valid when the regressors include lagged dependent variables and, for example, the portmanteau statistics Q and Q′ are inappropriate even though they are often provided routinely by regression packages. Fortunately the LM procedures are easily modified and several tests for error autocorrelation structure of regression models with lagged dependent variables are given by Breusch (1978) and Godfrey (1978b). These tests, like those of Godfrey (1979) for the time series models, can be calculated as n times the R^2 statistic for the regression of the residual on a set of estimated partial derivatives.

The important difference between static and dynamic regression models is that, in the latter case, it is no longer valid to omit terms corresponding to partial differentiation of the residual with respect to regression parameters when constructing the set of variables upon which the residual is regressed; see Godfrey (1978b) for details. For example, if the null specification is one of independent errors, then the LM check against general sth-order autocorrelation can be obtained by applying OLS to

$$\dot{u}_t = \Sigma_1^k x_{ti}\beta_i + \Sigma_1^h y_{t-i}\alpha_i + \Sigma_1^s \dot{u}_{t-i}\rho_i + \varepsilon_t,$$

and then computing n times the R^2 statistic, where $\dot{u}_t$ is a typical residual from the OLS estimation of (21).

The LM procedures given by Godfrey (1978b) could be generalised by obtaining the corresponding $C(\alpha)$ tests. Appropriate $C(\alpha)$ statistics could be calculated using a simple modification of the algorithm for time series models given in Section 2.2. The parameter vector ψ need only be interpreted as having three parts: $\psi' = (\psi_0', \psi_1', \psi_2')$, where ψ_0 contains a set of regression parameters that yield the disturbances u_t when combined with data, and ψ_1 and ψ_2 have their usual meanings. The C(α) test can then be carried out as a test of the joint significance of the terms $\partial e_t(\tilde{\psi})/\partial\psi_2$ in the auxiliary regression of $\tilde{e}_t$ on $\partial e_t(\tilde{\psi})/\partial\psi_0$, $\partial e_t(\tilde{\psi})/\partial\psi_1$ and $\partial e_t(\tilde{\psi})/\partial\psi_2$, $\tilde{\psi}$ being an estimator that is $n^{1/2}$-consistent when $\psi_2 = 0$. Indeed all the regression-based tests described above could be derived as special cases of this general algorithm by restricting the choice of $\tilde{\psi}$ and/or assuming that the regressors are strongly, rather than weakly, exogenous. (In some applications to observed time series, the term ψ_0 will not be required, while in others it might represent an unknown mean value.) The details of the approximations that yield the various special cases are straightforward to obtain, but lengthy. Moreover, these details do not provide much additional insight and, therefore, they will not be provided.

Finally, it should be noted that there are arguments against basing estimation and inference on models like (21). These arguments are used to justify a preference for eliminating the lagged dependent variables and working with

$$y_t = \Sigma_1^k\ [\beta_i/\alpha(L)]x_{ti} + [u_t/\alpha(L)], \tag{22}$$

where $\alpha(L) = 1 - \alpha_1 L - \dots - \alpha_h L^h$. Equation (22) is a special case of the transfer function (TF) model

$$y_t = \Sigma_1^k\ [\beta_i(L)/\alpha_i(L)]x_{ti} + [\theta(L)/\phi(L)]e_t, \tag{23}$$

where e_t is white noise and there are no redundant factors in the polynomial ratios. The general arguments in favour of the use of TF models, along with results on asymptotic misspecification tests and their small sample behaviour, will be considered in Section 4.

Simultaneous equation models

We have so far restricted attention to cases in which the regressors do not include endogenous variables. If the equation being estimated is, however, part of a simultaneous system, then least squares methods are inappropriate, as are the associated autocorrelation tests. Estimates of the parameters of individual structural equations from the model are often obtained using an instrumental variable (IV) technique, e.g. two stage least squares. Applying the time series checks described in Section 2 to IV residuals will not, in general, yield asymptotically valid tests. It is interesting to note that, in contrast to Durbin's (1970, p.414) general results on "naive" likelihood-based tests, invalid checks which treat IV residuals as observed time series variables can lead to the significance of sample outcomes being overestimated; see Godfrey (1978c, p.227). It is of course possible to derive a test of the joint significance of a set of autocorrelation estimates based upon IV residuals, and Breusch and Godfrey (1981, Appendix B) provide a suitable statistic. Monte Carlo evidence on the small sample behaviour of a test of the first-order autocorrelation calculated from IV residuals is provided in Chapter 2 by Burke and Godfrey who also consider other asymptotically valid tests.

4. ARMAX and transfer function models

It has been pointed out in the previous section that difficulties arise in the application of time series checks in regression models like (21) which include lagged dependent variables. As noted by Hendry, Pagan and Sargan (1984), equation (21), along with many other individual equations used in econometric analysis, is a special case of the scalar ARMAX model

$$\alpha(L)y_t = \Sigma_1^k \beta_i(L)x_{ti} + \theta(L)e_t, \qquad (24)$$

where $\beta_i(L)$ is a polynomial of degree g_i in the lag operator, $i = 1, \ldots, k$, and $\alpha(L)$ and $\theta(L)$ have their previous meanings. It is assumed that there are no polynomial factors common to all the polynomials of (24). A number of problems can occur in constructing diagnostic tests for such a model. These difficulties essentially stem from the absence of the block-diagonality of the information matrix. Further, if an alternative model is obtained by allowing for an autoregressive component in the error scheme, then it is only identified because rational distributed lag considerations have been excluded.

An alternative representation of the process of interest y_t is provided by (23) which has been termed a rational distributed lag model (Dhrymes, 1971) with autocorrelated errors, a transfer function-noise (TF) model (Box and Jenkins, 1976), or a dynamic disturbance model (Pierce, 1972a). Poskitt and Tremayne (1981b) introduce the acronym ARMAT for this model. ARMAX and TF models are simply alternative parameterizations of the same class of process, but Pierce (1983) has argued that ARMAX models are best fitted using TF specifications. The essential feature which may make it desirable to work with the TF model (23) is that its information matrix is block-diagonal between the parameters of the distributed lag component and those of the error autocorrelation scheme under the assumption that the regressors are all strongly exogenous.

One consequence of this block-diagonality is that, under the null hypothesis of correct specification, tests of the adequacy of the polynomial ratios $\beta_i(L)/\alpha_i(L)$ are asymptotically independent of tests of $\theta(L)/\phi(L)$. Accordingly it is valid to carry out separate tests and then to sum the two asymptotic χ^2 criteria to obtain a joint test against misspecification of the x-y dynamics and the disturbance autocorrelation model. Evidence on the specificity of the two separate tests will be considered below. The block-diagonality of the information matrix can also be exploited to simplify estimation since it permits a degree of separation; see Pesaran (1981) for a discussion of ML estimation. It is also important to note that the asymptotic distribution of the residual autocorrelations derived for a correctly specified TF model is the same as that for a univariate time series model, i.e. the effects of estimating the parameters of the polynomial ratios $\beta_i(L)/\alpha_i(L)$ are asymptotically negligible when testing error dynamics.

The validity of these useful results, however, depends

crucially upon the assumptions made about the regressors x_{ti}. In most of the early studies concerned with TF modelling, see, for example, Pierce (1971a, 1972b), little consideration is given to the nature of these variables other than to describe them as constants or exogenous. Hendry et al. (1984) have emphasised that the results are not valid if any of the regressors are Granger-caused by y in the model information set. It follows that if any regressor variable is weakly, but not strongly, exogenous then the potentially useful features of TF models are not available and joint modelling of the distributed lag and error components is vital. With regressors that are only weakly exogenous, the information matrix is no longer block-diagonal, implying that the two test statistics are not asymptotically independent and cannot be added together to obtain a valid joint test. Moreover misspecification of, for example, the error autocorrelation model will have spill-over effects and lead to the inconsistency of the estimators of the distributed lag model, as well as causing inconsistency of the estimators $\hat{\phi}_i$ and $\hat{\theta}_j$, $i = 1, \ldots, p$ and $j = 1, \ldots, q$. Newbold (1983) describes how an LM test for unidirectional causality can be implemented.

The TF model of (23) is, of course, a special case of a vector ARMA model; see Tiao and Box (1981). In the case of a single input variable, the dynamic disturbance model reduces to a special case of a bivariate ARMA model and one could seek to test the restrictions implicit in the TF model via this more general framework. However, as Tsay (1985) has indicated, the investigation of the specification of TF models is justified in its own right since the resulting vector ARMA model is not parsimonious.

Tests for models with strongly exogenous regressors

We shall follow most other researchers in this field by considering a simple TF model with a single regressor (input variable). This model, which is adopted to simplify exposition, will be written as

$$y_t = [\beta(L)/\alpha(L)]x_t + [\theta(L)/\phi(L)]e_t, \tag{25}$$

where x_t is strongly exogenous, the e_t are NID(0, σ_e^2) variates, and the degrees of the lag polynomials $\beta(L)$, $\alpha(L)$, $\phi(L)$ and $\theta(L)$ are g, h, p and q respectively. Following Poskitt and Tremayne (1981a, b), model (25) will be denoted ARMAT(g, h, p, q). The methods which will be discussed below can be generalised to deal with more inputs; see Pesaran (1981). There are, however, difficulties in the many-inputs case which have led some authors, e.g. Liu and Hanssens (1982), to turn from hypothesis testing procedures to identification methods such as the corner method of Beguin, Gourieroux and Monfort (1983).

Although there is clearly a greater potential for devising nonnested models when the ARMA model of (1) is expanded to the general TF form (23), we know of no results on testing nonnested TF relationships. Consequently the procedures to be discussed are either portmanteau tests or LM tests.

Portmanteau tests for TF models

The basic building blocks for diagnostic tests of model adequacy of the portmanteau type in TF models are the estimated residual autocorrelations $\hat{r}_j$ given by (2) and the estimated residual cross-correlations

$$\hat{r}^*_j = \Sigma^n_{j+1}\, \hat{\pi}_{t-j}\, \hat{e}_t/(\Sigma^n_1\, \hat{\pi}^2_t \Sigma^n_1\, \hat{e}^2_t)^{1/2}, \tag{26}$$

where $\hat{e}_t$ denotes a typical residual associated with the ML estimation of the TF model (25) and $\hat{\pi}_t$ denotes the estimated prewhitened input. Pierce (1972b) then suggests that an overall check of the adequacy of the distributed lag component $\beta(L)/\alpha(L)$ can be based upon

$$Q_T = n\Sigma^s_0\, \hat{r}^{*2}_j. \tag{27}$$

The statistic Q_T can be compared to an appropriate upper percentage point of the $\chi^2(s - g - h)$ distribution.

As noted above, the estimated residual autocorrelations of the ARMAT(g, h, p, q) model have the same asymptotic distribution as those for the univariate ARMA(p, q) model, under the assumption of correct specification. It follows that the Box-Pierce statistic Q can be used as a general check for misspecification of the error dynamics component of the null model (25). In the univariate ARMA model case, the question of the adequacy of the asymptotic approximation used in deriving the distribution of Q for use in finite samples prompted modification of Q to the statistic Q′ given in (4). Since similar approximations are also required in deriving the distribution of Q_T, a natural modification to this statistic, suggested by Ljung and Box (1978), is to compute

$$Q'_T = n^2\Sigma_0^s (n - j)^{-1}\hat{r}_j^{*2}, \tag{28}$$

with a view to obtaining a statistic whose empirical size properties better approach the nominal size.

A general test against misspecification of both the distributed lag and error dynamics components can be obtained by summing the portmanteau statistics Q′ and Q'_T. The block-diagonality of the information matrix implies that it is asymptotically valid to compare the resulting statistic $(Q' + Q'_T)$ to critical values of the $\chi^2(m + s - g - h - p - q)$ distribution.

LM tests for TF models

The score or LM principle has been suggested as a basis for assessing the specification of TF models by Pesaran (1981) and Poskitt and Tremayne (1981a). A contrast between LM and portmanteau tests is that here, as in the case of ARMA models, the asymptotic validity of the latter tests requires the number of residual auto- and cross-correlations upon which they are based to increase with, though at a slower rate than, sample size. Another difference is that it is not necessary to apply a prewhitening filter to the input variables in order to obtain the quantities required for the calculation of the LM statistic.

Suppose then that an ARMAT(g, h, p, q) model has been fitted to a realization of data. As in the case of ARMA models, it is not possible to entertain a simultaneous increase in the order of both polynomials in either the distributed lag part $\beta(L)/\alpha(L)$ or in the error dynamics component $\theta(L)/\phi(L)$, since identifiability problems arise under the null hypothesis. By virtue of the block-diagonality of the information matrix, it is, however, possible to test the adequacy of the two parts of the model separately. If two separate tests are calculated, then the usual invariance results apply. Thus, if the null model is ARMAT(g, h, p, q), the ARMAT(g + s, h, p, q) and ARMAT(g, h + s, p, q) schemes are locally equivalent alternatives, as are the ARMAT(g, h, p + r, q) and ARMAT(g, h, p, q + r) models. For a joint test of distributed lag and error dynamics, the following alternatives lead to the same LM statistic: ARMAT(g + s, h, p + r, q), ARMAT(g + s, h, p, q + r), ARMAT(g, h + s, p + r, q), and ARMAT(g, h + s, p, q + r).

These points concerning locally equivalent alternatives suggest that LM tests in TF models will have some of the characteristics of pure significance tests. Nevertheless, there is an element of specificity in their use in that separate tests of the two parts of the model can be conducted. This feature is not present, for example, in scalar ARMAX models for which Poskitt and Tremayne (1984) point out that the LM test should be regarded as a general purpose misspecification test on the transfer functions that specify the model.

As with ARMA models, it is possible to implement the LM test via an n times R^2 formulation as discussed in Section 2. The statistic for testing ARMAT(g, h, p, q) against either ARMAT(g, h, p + r, q) or ARMAT (g, h, p, q + r) is exactly as described in that earlier section of the paper. In the case of the rational distributed lag component of the model, the necessary auxiliary regression for a single x variable requires a regression of $\hat{e}_t$ on derivative processes pertaining to the parameters of the ratio of lag polynomials in the alternative model used to construct the test statistic. A test of model adequacy of ARMAT(g, h, p, q) against either ARMAT(g + s, h, p, q) or

ARMAT(g, h + s, p, q) - or any alternative which is, in the terminology of Poskitt and Tremayne (1980), admissible - is thereby obtained. Detailed expressions for the various derivative processes are provided by Poskitt and Tremayne (1981a).

In view of the fact that the principal difference between LM and portmanteau tests lies in the determination of the degrees of freedom parameter, it seems useful to provide a brief review of Monte Carlo results available for TF models.

Monte Carlo evidence

Poskitt and Tremayne (1981a, b) report the results of some simulation experiments relating to TF models with a single input variable. They consider the empirical size and power of both LM and portmanteau tests for both the rational distributed lag and disturbance components of the model. The general empirical size properties of the LM and modified portmanteau statistics seem acceptable in moderate sample sizes, though it would appear that some care must be exercised in the choice of the number of residual cross- and autocorrelations used to structure the latter. A value of m = 20, say, is not to be recommended in samples of 50 or 100; see Poskitt and Tremayne (1981b) for further discussion.

The results for empirical power indicate that the LM procedure is generally a more powerful diagnostic tool than any variant of the portmanteau procedures and that this conclusion holds true for both components of the TF model. This may again reflect the different ways in which the degrees of freedom indices for the two types of test are chosen. Moreover, the pure significance test aspect of the LM test appears evident in some of the results reported in Poskitt and Tremayne (1981a) where an LM test derived from an underspecified alternative is more powerful than one based upon the correct alternative hypothesis. Limited evidence is also provided by these authors (1981a, pp.984-5) that a more powerful test can be derived by summing the separate LM tests of the two components of the TF model and using the resulting value as an overall check for misspecification. The disadvantage of such a joint test is that it cannot point to one inadequate component if the other is specified correctly. Poskitt and Tremayne report that, in several cases, the two separate tests are quite robust to misspecification of the unconsidered component, implying a degree of specificity that might be useful in guiding respecification. The constructive use of the two separate tests is discussed by Box and Jenkins (1976, Section 11.3.4).

Finally, an advantage of LM tests over portmanteau tests, quite apart from their seemingly superior power properties, which may make them more attractive to practitioners is that there is no need to prewhiten input (regressor) variables in order to use them. Many authors, for example Liu and Hanssens (1982), have confessed difficulties in applying prewhitening methods, particularly in the case of multiple input models. Moreover, although the use of linear filters to reduce series to white noise is commonplace in time series, it is not a widely used technique in applied econometrics.

5. Testing for serial correlation: some Monte Carlo results

The simulation results reported and discussed in this section are intended to provide information about the usefulness of time series checks in least squares regression. The null hypothesis is that the disturbances are serially independent, and the results indicate how the behaviour of a test can be altered by the choice of the number and orders of the autocorrelation estimates used. The results also provide some evidence on the consequences of working with estimated residuals rather than observed time series data.

The data generation process employed in the experiments is

$$y_t = \beta_1 + \beta_2 x_t + u_t, \; t = 1, \dots , n,$$

with

$$u_t - \Sigma_1^6 \phi_i u_{t-i} = e_t, \; e_t \; \text{NID}(0, \sigma_e^2),$$

the regressor x being the quarterly US GNP variable given by

Maddala and Rao (1973). The values of β_1, β_2 and σ_e^2 are irrelevant, so that no generality is lost by using ($\beta_1 = 0$, $\beta_2 = 0$, $\sigma_e^2 = 1$) in all experiments; see Breusch (1980). A number of experiments can, however, be obtained by varying the values of the six coefficients ϕ_i, e.g. exclusion restrictions yield lower-order schemes. The sample size is also varied with n = 40 and n = 60 for each selected AR error process.

For every combination of coefficients ϕ_i and sample size n, a sequence of (n + 50) values of u is generated with the initial values set equal to zero. The first fifty observations are then deleted to reduce the impact of the choice of initial values, and the remaining terms, denoted by u, are regressed on an intercept and x_t to derive the corresponding sequence of residuals u_t, t = 1, 2, ... , n. Four tests of H_0: u_t NID(0, σ_u^2) are then applied as follows:

$$T_1 = n\,\hat{r}_1^2 \underset{a}{\sim} \chi^2(1) \text{ on } H_0;$$

$$T_2 = n\,\hat{r}_4^2 \underset{a}{\sim} \chi^2(1) \text{ on } H_0;$$

$$T_3 = n\,(\hat{r}_1^2 + \hat{r}_4^2) \underset{a}{\sim} \chi(2) \text{ on } H_0; \text{ and}$$

$$T_4 = n\,(r_1^2 + \dots + \hat{r}_6^2) \underset{a}{\sim} \chi^2(6) \text{ on } H_0.$$

Thus we have in mind a situation in which a quarterly econometric model has been estimated and the tests available include procedures asymptotically equivalent to the DW test and Wallis's test, and a portmanteau-type test using a relatively low value of m. The statistic $T_3 = T_1 + T_2$ is less familiar than the other three criteria. Under the null and local alternatives, T_3 provides a valid and asymptotically optimal test against the alternatives

$$(1 - \phi_1 L - \phi_4 L^4)u_t = e_t$$

and

$$(1 - \phi_1 L)(1 - \phi_4 L^4)u_t = e_t,$$

the latter scheme being singled out by Wallis for consideration in his work on testing for autocorrelation in quarterly models. At first sight, T_3 might appear to be inappropriate for the second alternative because it takes no account of the fifth-order term $\phi_1\phi_4 L^5$. However, under a sequence of local alternatives with $\phi_1 = O(n^{-1/2})$ and $\phi_4 = O(n^{-1/2})$, the term $\phi_1\phi_4$ is $O(n^{-1})$ and so is asymptotically negligible.

A sample of results is reported in Table 2. The parameter values used to obtain these results lead to fairly low degrees of autocorrelation and are selected to avoid situations in which all or most tests have power very close to one. All rejection frequencies are given as percentages. Estimates of significance levels are based upon 2000 replications, while those of powers are calculated using 500 replications. The results can be summarized as follows.

Case (i): $\phi(L) = 1$, i.e. H_0 true

As the estimated significance levels are based upon 2000 replications, only the estimate for T_2 using the actual u_t with n = 40 differs significantly from the nominal value of 5 per cent. Time series checks based upon the disturbances u_t reject the true null hypothesis less frequently than tests calculated using the OLS residuals with the latter tests performing well for both sample sizes. The results confirm our conjecture about the usefulness of employing a relatively small value of m when using the Box-Pierce test in the (p = 0, q = 0) case.

These results on small sample significance levels are encouraging, but the ability to detect autocorrelation is clearly important. We, therefore, turn to some estimates of power.

Case (ii): $\phi(L) = 1 - \phi_1 L$, $\phi_1 = 0.3$

The T_1 test is designed for the correct alternative, viz. the AR(1) process, and not surprisingly outperforms the other procedures in all cases. Under a sequence of local alternatives, T_1, T_3 and T_4 have noncentral χ^2 distributions with the same noncentrality parameter and degrees of freedom parameters 1, 2 and

Table 2

Simulation results for five per cent significance level

Test	n = 40 true u_t	n = 40 residual	n = 60 true u_t	n = 60 residual
	(i) Independence			
T_1	4	5	4	6
T_2	3	4	5	5
T_3	4	5	4	5
T_4	4	5	4	5
	(ii) $\phi(L) = 1 - 0.3L$			
T_1	41	25	62	48
T_2	6	6	5	6
T_3	29	18	50	38
T_4	19	13	32	26
	(iii) $\phi(L) = 1 - 0.3L^4$			
T_1	5	8	6	7
T_2	38	23	59	47
T_3	27	20	47	35
T_4	19	19	32	29
	(iv) $\phi(L) = 1 - 0.3L - 0.3L^4$			
T_1	50	18	70	43
T_2	43	16	65	32
T_3	59	22	82	52
T_4	49	18	71	41
	(v) $\phi(L) = (1 - 0.3L)(1 - 0.3L^4)$			
T_1	45	20	63	42
T_2	37	15	59	37
T_3	57	22	80	56
T_4	42	18	61	40

6 respectively. Ranking by asymptotic local power, therefore, coincides with the ranking by estimated power (T_4 is outperformed by T_3 which is in turn outperformed by T_1). It should be noted that plim $\hat{r}_4 = \phi_1^4 \neq 0$, so that T_2 yields a consistent test, but the estimated power is close to nominal size for n = 40 and n = 60. Finally the substitution of the OLS residuals for the disturbances u_t always leads to a reduction in power and the magnitude of this reduction is not altered greatly by increasing the sample size from 40 to 60.

Case (iii): $\phi(L) = 1 - \phi_4 L^4$, $\phi_4 = 0.3$

As in the previous case, the test based upon the correct alternative, viz. T_2, is the most powerful procedure and power is lost when the u_t are unobserved and OLS residuals must be used to calculate the estimates $\hat{r}_i$. The T_1 test for AR(1) errors is not consistent in the presence of simple fourth-order autocorrelation and its estimated power is close to its significance level, indicating the dangers of relying upon the DW test when fitting quarterly models. Turning to the other tests, under a sequence of local alternatives,

$$T_2 \underset{a}{\sim} \chi^2(1, \delta^2),\ T_3 \underset{a}{\sim} \chi^2(2, \delta^2),\ T_4 \underset{a}{\sim} \chi^2(6, \delta^2),$$

so that the results of Das Gupta and Perlman (1974) suggest that T_2 should be more powerful than T_3 and T_4, and T_3 should outperform T_4. These qualitative predictions are consistent with the results of Table 2. The differences between the asymptotically equivalent tests based upon the disturbances u_t and the OLS residuals make it clear that asymptotic theory is not a completely accurate guide to finite sample behaviour.

Case (iv): $\phi(L) = 1 - \phi_1 L - \phi_4 L^4$, $\phi_1 = \phi_4 = 0.3$

The results for this case reinforce previous findings on the value of correct specification of the alternative and the impact of using regression residuals rather than the actual disturbances. One interesting feature is that the Box-Pierce test which involves our "irrelevant" terms compares well with the tests T_1 and T_2, each of which omits one term. The differences between tests calculated using the unobservable disturbances and the least squares residuals are substantial even when the correct alternative is chosen.

Case (v): $\phi(L) = (1 - \phi_1 L)(1 - \phi_4 L^4)$, $\phi_1 = \phi_4 = 0.3$

As anticipated, T_3 is the best test. The rankings of the tests are very similar to those of the previous (locally equivalent) alternative model, as are the corresponding pairs of rejection frequencies of T_3 which is the LM test for both (iv) and (v). All tests suffer a considerable loss of power when disturbances u_t are replaced by least squares residuals. This feature of the results will cause concern to econometricians who wish to employ time series methods in empirical research.

Summary

While time series checks may have significance levels close to the nominal value when applied to least squares residuals, power estimates derived from studies of time series models may substantially overstate the sensitivity of such checks when they are employed in regression analysis. In view of this problem, it is all the more important not to rely upon testing a single autocorrelation. If a quarterly model is being estimated and the estimation program only provides the DW statistic d and Wallis's (1972) statistic d_4, then these can be combined to obtain

$$T_3' = n[(1 - d/2)^2 + (1 - d_4/2)^2]$$

which is asymptotically equivalent to T_3. If the Box-Pierce procedure is available, then this should be employed using a relatively small value of m, e.g. 6 rather than 24. In experiments not reported in Table 2, it was found that increasing the value of m reduced the power of the Box-Pierce test. This finding is consistent with Hendry's (1977) calculations of the approximate power of the portmanteau test; also see Ljung (1986).

6. Integrated variables and tests for unit roots

An observable time series variable u_t is said to be integrated of order d, denoted by $u_t \sim I(d)$, if $\Delta^d u_t$ is a stationary and invertible ARMA(p, q) process, where Δ is the first difference operator, i.e. $\Delta = (1 - L)$. The acronym ARIMA(p, d, q) is used for such processes by Box and Jenkins (1976). In Sections 2-5, it has been assumed that u_t itself is a stationary and invertible ARMA process, so that $u_t \sim I(0)$. However, it has been argued that this assumption is inappropriate for some economic variables and that these variables are better modelled as I(1) processes; see Nelson and Plosser (1982).

The simplest model for an integrated I(1) variable is the random walk

$$u_t = u_{t-1} + e_t, \tag{29}$$

with e_t being NID(0, σ_e^2) or, less restrictively, iid(0, σ_e^2). It is sometimes useful to generalise (29) to allow for "drift" by adding an intercept term to obtain the model

$$u_t = \mu + u_{t-1} + e_t. \tag{30}$$

The properties of I(1) variables are quite unlike those of I(0) variables. For example, an I(0) process has a constant variance determined by the parameters of (1), but, assuming that the starting value u_0 equals zero, the variance of u_t in (29) is $t\sigma_e^2$ and so grows with t; see Engle and Granger (1987) and Pagan and Wickens (1989, Section 1.2.1) for useful discussions of the differences between I(0) and I(1) variables.

If it is certain that a variable is I(1), then first differences can be taken and the results described in earlier sections will apply to Δu_t, rather than to u_t. The assumption that u_t is I(1) can, however, be viewed as a testable hypothesis by regarding (29) as the restricted version of the AR(1) model

$$u_t = \alpha u_{t-1} + e_t$$

that is obtained by imposing H_α: $\alpha = 1$. Since conventional asymptotic tests require that $|\alpha| < 1$, they cannot be applied to testing H_α and, more generally, alternative approaches are needed for testing for a unit root in the AR operator of a time series model.

One family of tests for unit roots has its origins in the work of Dickey and Fuller (1979, 1981) and Fuller (1976). The simplest form of the test is obtained by checking the adequacy of (29). The alternative model is taken to be

$$\Delta u_t = (\alpha - 1)u_{t-1} + e_t, \qquad (31)$$

in which the e_t are NID(0, σ_e^2). Since $(\alpha - 1) = 0$ on H_α, the assumption of a unit root can be tested by investigating the significance of u_{t-1} after OLS estimation of (31). As noted above, this testing problem is nonstandard and the classical t-test is inappropriate, even in large samples. Appropriate tests can be based upon either the OLS coefficient estimate $(\hat{\alpha} - 1)$ or the associated t-statistic. Conventional asymptotic theory cannot be applied, but critical values for $n(\hat{\alpha} - 1)$ have been obtained by Monte Carlo methods and are provided by Fuller (1976, p. 371, Table 8.5.1). The use of the norming factor n, rather than $n^{1/2}$, reflects one of the nonstandard features of the analysis.

The null and alternative models of the simplest form of the Dickey-Fuller (DF) test have been generalised in various ways. Models with intercepts and time trends have been considered; see, for example, Dickey and Fuller (1981). The generalisations of (31) that include either an intercept or an intercept and a time trend can be written as

$$\Delta u_t = (\alpha - 1)u_{t-1} + \mu_u + e_t, \qquad (31a)$$

and

$$\Delta u_t = (\alpha - 1)u_{t-1} + \mu_u + \tau_u[t - (n + 1)/2] + e_t, \tag{31b}$$

respectively. Some care is required when testing H_α: $\alpha = 1$ in these two generalisations because the limiting distributions of test statistics under H_α depend upon the values of nuisance parameters in (31a) and (31b).

If $\mu_u = 0$ and $\tau_u = 0$, conventional asymptotic theory cannot be applied to either (31a) or (31b) when testing the significance of the normed OLS estimate $n(\dot{\alpha} - 1)$. Moreover the critical values estimated for (31) are inappropriate and separate sets must be calculated for both generalisations. Estimates of quantiles that permit the application of tests to (31a) when $\mu_u = 0$ and to (31b) when $\tau_u = 0$ are contained in Fuller's table for the empirical cumulative distribution of $n(\dot{\alpha} - 1)$ under $\alpha = 1$; see Fuller (1976, p. 371, Table 8.5.1). Fuller also provides a table relevant to the estimated t-ratio of u_{t-1} in (31a) and (31b).

Fuller's tables and n-norming of $(\dot{\alpha} - 1)$ are not, however, appropriate for all values of the nuisance parameters in (31a) and (31b). If $\mu_u \neq 0$, i.e. there is drift in the process, a standard t-test of H_α is asymptotically valid in (31a); see Pagan and Wickens (1989, pp. 968-9). This conventional test will also be asymptotically valid in (31b) if $\tau_u \neq 0$; see Nankervis and Savin (1987, Table 1, p. 392). Reference to Fuller's tables should, however, be made if $\mu_u \neq 0$ and $\tau_u = 0$ in (31b).

The inclusion of a time trend in models such as (31b) is important because nonstationary time series are sometimes modelled as polynomial trends with covariance-stationary errors. West (1987) argues that the DF tests will be inconsistent if the process under scrutiny is stationary about a trend and the time trend is not included in the regression used to generate the test statistic.

Another important generalisation of the basic DF framework of equation (31) is to allow the series of first differences Δu_t to be autocorrelated. If an AR(p) scheme is used to model this autocorrelation, an augmented Dickey-Fuller (ADF) test is implemented by estimating

$$\Delta u_t = \mu + \Sigma_1^p \phi_j \Delta u_{t-j} + (\alpha - 1)u_{t-1} + e_t, \qquad (32)$$

by OLS. A test of the significance of $n(\dot{\alpha} - 1)$ is then carried out using the table provided by Fuller (1976). Two points should be made concerning the model (32) that is used to construct the ADF test statistic.

First, given the arguments of West (1987), it may be useful to expand the regressor set by including a time trend term in order to avoid the risk of low power. The inclusion of a time trend, however, does not guarantee that ADF tests for unit roots will be powerful in every application. Perron (1989) shows that such tests cannot reject the unit root hypothesis when the true data process is a stationary error about a trend function with a one-time break.

Perron (1989) also points out that, in finite samples, a trend-stationary process is nearly observationally equivalent to a unit root process with Δu_t being ARMA with a root of the MA operator close to -1. If p, the number of terms Δu_{t-j} in the regressors of (32), is regarded as fixed, then the associated ADF test may produce quite misleading results when the true process is such that Δu_t has a general ARMA structure; see Schwert (1987). Concern about the effects of misspecification of the model for Δu_t under the unit root hypothesis leads to the second suggestion relating to the regressors of (32). This suggestion is that the number of lagged first differences should increase with n at a rate that allows for the approximation of general autocorrelation.

Said and Dickey (1984) show that, if Δu_t is ARMA of unknown order, the ADF test is asymptotically valid when p in (32) increases at a rate less than $n^{1/3}$, i.e. p is $o(n^{1/3})$. In a situation such as that mentioned by Perron (1989), the near noninvertibility of the MA component will presumably imply that a relatively large value of p will be required to obtain an adequate AR approximation in finite samples. The effects of moving average components on the behaviour of tests for unit roots are investigated in a Monte Carlo study by Schwert (1989). In another study, Schwert (1987) gives several interesting empirical applications of tests for unit roots and also suggests rules for selecting values of p that are $O(n^{1/4})$. Further comments on the practical use of DF and ADF tests (and many useful references) are provided by Dickey, Bell and Miller (1986).

The DF and ADF tests have been used in many studies, but other tests have recently become available. These new tests are described and examined in papers by Perron and Phillips (1987), Phillips (1987), and Phillips and Perron (1988). In these papers, corrections are derived for DF-type criteria so that the modified test statistics are asymptotically valid for quite general time series models, allowing for fairly general sorts of weak dependence and heterogeneity. The modified statistics, which are often denoted by Z with some appropriate subscript, have been obtained for the basic model and for models fitted with a time trend and/or an intercept; see Phillips and Perron (1988). It has been established that the asymptotic distribution of a Z-statistic is invariant over a wide range of weakly dependent and heterogeneous data processes, and that this distribution is the same as that of the corresponding DF statistic under the much stronger distributional assumptions of Dickey and Fuller (1979). Hence the critical values tabulated by Fuller (1976) are appropriate for assessing the significance of Z-statistics.

The type of correction introduced by Phillips (1987) can be illustrated by considering the DF statistic $n(\dot{\alpha} - 1)$ in the context of the basic model (31). The modified statistic for this case is given by

$$Z_\alpha = n(\dot{\alpha} - 1) - 0.5(s^2_{n\ell} - s^2_e)/(n^{-2}\Sigma_t\, u^2_{t-1}), \tag{33}$$

in which $s^2_{n\ell}$ is a consistent estimator of $\lim n^{-1}[E(\Sigma_t e_t)^2]$ and s^2_e

is a consistent estimator of $\lim n^{-1}\Sigma E(e_t^2)$; see Phillips (1987, Section 5). The difference between Z_α and $n(\alpha - 1)$ in the special case in which Δu_t is MA(1) can be deduced from an example provided by Phillips (1987, p. 283).

The published critical values for DF tests and the corresponding Z-variants have been estimated under the assumption of normality. It is of interest to investigate whether or not these critical values are sensitive to this assumption because there is no reason to believe it will always be appropriate in practice. Also Perron (1989, p. 1389) makes the observation that it will be important that unit root procedures be valid with disturbances with nonnormal tail behaviour. In order to study some of the effects of nonnormal observations, we ran simulation experiments for unit root tests appropriate in models (31a) and (31b), as well as (31), but the results of varying the innovation distribution were similar in each case, so only the results for the basic model are presented here.

The computations carried out were based upon maintaining the assumption of iid innovations, but generating the necessary random variables variously from the following distributions: t-distribution with 5 degrees of freedom, t(5); chi-squared with 2 degrees of freedom, $\chi^2(2)$; uniform; log-normal distribution (denoted by log-n in Table 3); and Cauchy. These choices of distributions reflect differing moment properties with regard to existence and skewness and kurtosis relative to the normal distribution. Where necessary, adjustment is made to give a zero mean.

Table 3 provides some results pertaining to the performance of the t-statistic version of the DF test, denoted t_{DF}, and the corresponding Z-variant, denoted Z_t, for sample sizes n = 50, 250. The numbers in brackets following Z_t represent the value of the truncation lag ℓ used to obtain the term $s_{n\ell}^2$ which apears in (33). The entries are based on 10000 replications, so the standard errors derived from these entries are not greater than about 0.3 per cent.

Table 3

Empirical cumulative distribution of t_{DF} and $Z_t(.)$ for $\alpha = 1$

			Probability of a smaller value under normality					
n	dbn	test	1	5	10	90	95	99
50	t(5)	t_{DF}	1.0	4.6	9.7	90.3	95.1	99.1
50	$\chi^2(2)$	t_{DF}	0.8	4.7	9.4	88.8	93.7	98.0
50	uniform	t_{DF}	1.1	5.2	10.3	90.1	94.9	99.0
50	log-n	t_{DF}	0.5	3.8	8.3	85.3	90.3	95.6
50	Cauchy	t_{DF}	0.7	3.1	6.2	92.3	97.2	99.6
250	t(5)	t_{DF}	1.1	5.0	10.2	89.8	95.2	99.1
250	$\chi^2(2)$	t_{DF}	1.0	4.6	9.3	89.6	94.5	98.8
250	uniform	t_{DF}	0.9	5.0	9.9	90.2	95.2	99.1
250	log-n	t_{DF}	0.8	4.6	9.2	88.0	93.2	97.5
250	Cauchy	t_{DF}	0.8	2.9	5.9	93.2	97.1	99.6
50	t(5)	$Z_t(3)$	0.1	1.7	5.3	90.2	95.9	99.5
50	$\chi^2(2)$	$Z_t(3)$	0.0	1.3	4.8	88.3	94.5	99.6
50	uniform	$Z_t(3)$	0.1	1.9	5.5	89.3	95.5	99.4
50	log-n	$Z_t(3)$	0.1	1.1	4.4	86.1	92.7	99.6
50	Cauchy	$Z_t(3)$	0.1	1.0	3.6	93.1	97.7	99.7
250	t(5)	$Z_t(6)$	0.4	3.4	7.8	89.8	95.1	99.2
250	$\chi^2(2)$	$Z_t(6)$	0.3	3.3	7.7	89.5	95.0	99.1
250	uniform	$Z_t(6)$	0.3	3.5	7.8	90.2	95.3	99.3
250	log-n	$Z_t(6)$	0.3	3.0	7.4	88.2	93.6	98.4
250	Cauchy	$Z_t(6)$	0.2	2.0	4.7	93.3	97.4	99.6

Notes: n denotes the sample size, dbn denotes the distribution of the innovation e_t in (31), and all estimated rejection frequencies are given in percentages.

Consider first the behaviour of the t_{DF} procedure. For a sample size of 50, the rejection frequencies under t(5), $\chi^2(2)$ and uniform are close to those expected under normality. This feature of the results is consistent with asymptotic theory. When the innovations are log-normally distributed, all of the rejection frequencies of t_{DF} fall below the probabilities that Fuller's critical values are designed to give. The usefulness of these critical values is even more limited when the innovations come from the Cauchy distribution. This is, however, not surprising because the Cauchy distribution does not satisfy the regularity conditions on the existence of moments, e.g. Phillips and Perron (1988, p. 336). The results for n = 250 show that Fuller's critical values provide a good approximation for all the distributions considered, except the Cauchy distribution for which the large sample theory is inappropriate.

Turning to the Z_t test, the estimates of Table 3 indicate that, for both values of n, the left-hand tail of the null distribution is rather thinner than that of the distribution underlying Fuller's table. The left-hand-tail rejection frequencies move closer to the probabilities for normal innovations as n increases from 50 to 250, but there is not close agreement even at the larger sample size. If H_α: $\alpha = 1$ is tested against the alternative that $|\alpha| < 1$, i.e. the data process is stationary, then a one-sided test involving a comparison of sample values of Z_t with Fuller's critical values may well be undersized when the innovations are nonnormal. Under the Cauchy distribution, the Z_t test, like t_{DF}, is inappropriate and the use of Fuller's tables is not justified by large sample theory.

To sum up, the general conclusions that are suggested by the estimates presented in Table 3 are that: if H_α is to be rejected on the basis of small values of Z_t, the test is likely to be undersized; and the tables of Fuller (1976) for t_{DF} are fairly robust to a range of departures from normality, but that some care should be exercised in using them.

The DF, ADF and Z tests are often applied to individual

economic time series. In addition, these tests are now becoming quite widely used in the econometric analysis of multiple regression equations. Consider again the regression model of Section 3 in which

$$y_t = \Sigma\, x_{ti}\beta_i + u_t,\ t = 1, \dots , n. \tag{17}$$

In order to appeal to the standard asymptotic theory for MLE and OLS under some ARMA(p, q) specification for the disturbance term u_t, it is assumed in Section 3 that all variables of (17) are I(0). The restriction that all regression variables are I(0) is necessary because the standard large sample theory uses assumptions of covariance stationarity and ergodicity in the derivation of results concerning estimators and tests. These assumptions are obviously inappropriate when nonstationary, nonergodic processes are being studied, and so cases in which the regression model involves I(1) variables should not be analysed by means of classical asymptotic results.

The order of integration of the disturbance u_t in (17) is of crucial importance. If, like y_t and some of the x_{ti}, u_t is I(1), then first-differencing can be used to obtain

$$\Delta y_t = \Sigma\, \Delta x_{ti}\beta_i + \Delta u_t,$$

in which all variables are stationary and Δu_t has a stationary and invertible ARMA representation. Standard asymptotic theory can therefore be applied to the differenced version of the regression model. The only complication is minor: if (17) contains an intercept, this coefficient will be lost in the transformation.

The case in which u_t is I(0), but y_t and some of the x_{ti} are I(1), has much more serious implications for the usefulness of standard theory. Methods developed for the classical (nonintegrated regression variables) model can provide very misleading results in this case and asymptotic normality cannot be assumed. For example, the correct asymptotic theory for a linear model such as (17) when the regressors are generated by random walks and u_t is I(0) is provided as a special case of Theorem 4.1 of Phillips and Durlauf (1986, pp. 481-482). An extensive literature on inference for time series models with unit roots has developed since the pioneering work by Phillips (1986) and we shall not attempt to summarize it here. The interested reader should consult Park and Phillips (1988, 1989) and Sims, Stock and Watson (1990), and the many references given in these articles.

When u_t is I(0) despite the fact that y_t and some x_{ti} are I(1), there must be a linear combination of the integrated regressand and regressors that is stationary. The regression variables are, in the terminology of Engle and Granger (1987), cointegrated. Much research has been carried out on cointegration and related topics, such as error correction models; see Engle and Granger (1987), the collection of papers edited by Hendry (1986), the recent surveys by Dickey, Jansen and Thornton (1991), Dolado, Jenkinson and Sosvilla-Rivero (1990), and Pagan and Wickens (1989), and the article by Johansen and Juselius (1990). (The paper by Dickey, Jansen and Thornton (1991) includes a useful step-by-step exposition of the application of alternative approaches to cointegration testing.) We shall concentrate on testing for cointegration and, for simplicity of exposition, will assume that the model under consideration is

$$y_t = \beta_1 + \beta_2 x_t + u_t, \tag{34}$$

in which y_t and x_t are both I(1) variates. (In practice, ADF or Z statistics would usually be employed to investigate whether or not the data are consistent with the assumption that $y_t \sim I(1)$ and $x_t \sim I(1)$.)

Tests for cointegration can be carried out by applying a unit root test to the residuals obtained by OLS estimation of (34) - this is termed the cointegrating regression. If the residual-based unit root test leads to the acceptance of the hypothesis that $u_t \sim I(1)$, then this is evidence against cointegration. Significant values of the unit root test statistic may indicate that the u_t are I(0) and that x_t and y_t are cointegrated. Thus the application of a unit root test to the OLS residuals from (34) can be viewed as a test of the null hypothesis of no cointegration. At first sight, it might appear more attractive to use the assumption of cointegration as the null hypothesis, but major difficulties are associated with this approach; see Phillips and Ouliaris (1990, Section 6).

Various tests for cointegration based upon the OLS residuals $\dot{u}_t$ have been suggested; see Engle and Granger (1987) and Phillips and Ouliaris (1990). A procedure proposed by Sargan and Bhargava (1983) can be used under the fairly restrictive assumption that

$$u_t = \alpha u_{t-1} + e_t, \ -1 < \alpha \leq 1,$$

with the e_t being NID(0, σ_e^2) variates. In the context of testing for cointegration, this procedure is based upon the Durbin-Watson statistic from the cointegrating regression and the u_t are taken to be I(0) if this statistic is significantly greater than zero. Bounds for critical values that can be used for the cointegrating regression Durbin-Watson statistic are provided by Sargan and Bhargava (1983), but these authors recommend the use of the Imhof routine to obtain exact results because of the size of the inconclusive region. There are other problems associated with the use of the Sargan-Bhargava test. It is derived under relatively restrictive assumptions - the first differences are taken to be NID(0, σ_e^2) under the null hypothesis - and may, therefore, be misleading in more general situations. Further, on the basis of its performance in Monte Carlo experiments, Engle and Granger (1987) argue that the Sargan-Bhargava test is inferior to the ADF test calculated using the OLS residuals $\dot{u}_t$.

The research on tests for unit roots in observed time series suggests that, in addition to the ADF procedure, it may be useful to consider the application of a Z-test to the residuals $\dot{u}_t$; see Phillips and Perron (1988) for some simulation evidence on the finite sample performance of the Z-test. Phillips and Ouliaris (1990) argue that a Z-test derived by correcting the DF-type criterion $n(\dot{\alpha} - 1)$, where $\dot{\alpha} = \Sigma_2^n \dot{u}_t \dot{u}_{t-1} / \Sigma_2^n \dot{u}_{t-1}^2$, should have relatively good power in large samples. The results derived by Phillips and Ouliaris are only asymptotically valid and, as noted above, large sample theory does not always provide a good approximation to finite sample behaviour. We, therefore, share these authors' view that it would be useful to carry out simulation experiments in order to gain more information about the performance and relative merits of residual-based tests for cointegration. The results reported in Section 5 concerning the effects of replacing unobservable disturbances by estimated residuals indicate that regression estimation effects may be important for moderately large sample sizes.

7. Conclusions

We have discussed various general approaches to testing the adequacy of a univariate autoregressive-moving average time series model. It has been shown that LM tests are, like portmanteau tests, simply tests of the significance of estimated residual autocorrelations; see Newbold (1980). Indeed, it can be argued that an LM test has some of the qualities of a pure significance test because it is asymptotically optimal against a class of (locally equivalent) alternative hypotheses and will, in general, be consistent against a wide range of misspecifications. The LM tests do, however, appear to have markedly better small sample performance than portmanteau tests, even when they are based upon an incorrect alternative. In addition, LM statistics can be easily calculated by means of auxiliary least squares regressions and we feel, therefore, that their use is to be recommended.

As yet, we have no evidence on the finite sample behaviour of the $C(\alpha)$ test which used inverse autocorrelations and allows the researcher to move from identification to diagnostic testing without having to estimate the null model by ML methods. The testing of transfer function models has also been considered and again LM procedures seem to be more useful than portmanteau tests.

The application of time series checks in econometrics has

also been examined. Diagnostic checks for misspecification must play an important part in time series econometrics if misleading inferences are to be avoided. If LM procedures (or any other time series tests) are to be employed in econometric analysis, they should be (at least asymptotically) valid for the model under scrutiny and powerful. It is not always the case that time series tests can be applied to the residuals of a fitted econometric model without modification. A simple algorithm for calculating LM statistics involving only the OLS estimation of auxiliary models is available for regression models with strongly exogenous regressors, as is a different one for those with weakly exogenous regressors. More complicated methods are required if the equation being evaluated is contained in a simultaneous system and has endogenous regressors. Some of these methods are discussed by Burke and Godfrey in Chapter 2.

For the important special case of testing the assumption that the errors of a regression model are independent, it appears that the actual significance levels of LM tests are reasonably close to their nominal values. In terms of achieving power, we would stress the need to take account of the seasonality of the data and advise researchers not to rely upon a test based upon a single autocorrelation. We would not, however, recommend using a test involving a large number of additional terms. As always, the data inconsistency of a null model need not imply the data consistency of the alternative, and there is some evidence that powerful tests can be derived from underspecified alternative models.

Even if the alternative is identified correctly and all regressors are strongly exogenous, so that no modification of time series tests is required, there remains the problem of the small sample effects of treating estimated residuals as if they were observed time series variables. We have provided some Monte Carlo evidence that there is sometimes considerable loss of power relative to tests based upon the corresponding unobservable disturbance terms. Such estimation effects, when combined with the relatively short length of many economic time series, suggest that the application of time series tests in econometrics, while extremely important, will not be without difficulty.

References

Baillie, R.T., Lippens, R.E. and McMahon, P.C., (1983). Testing rational expectations and efficiency in the foreign exchange market. *Econometrica*, 50, 553-563.

Bartlett, M.S., (1946). On the theoretical specification and sampling properties of autocorrelated time series. *Journal of the Royal Statistical Society* (Supplement), 8, 27-41. 41.

Beguin, J.M., Gourieroux, C. and Monfort, A., (1980). Identification of a mixed autoregressive-moving average process: the corner method, in *Time Series*, (O.D. Anderson, Ed.) Amsterdam: North-Holland, 423-435.

Bera, A.K. and McKenzie, C.R., (1986). Alternative forms and properties of the score test. *Journal of Applied Statistics*, 13, 13-25.

Bhansali, R.J., (1980). Autoregressive and window estimates of the inverse correlation function. *Biometrika*, 67, 551-565.

Box, G.E.P. and Jenkins, G.M., (1976). *Time Series Analysis, Forecasting and Control.* San Francisco: Holden-Day, revised edition.

Box, G.E.P. and Pierce, D.A., (1970). Distribution of residual auto-correlations in autoregressive integrated moving average time series models. *Journal of the American Statistical Association*, 65, 1509-1526.

Breusch, T.S., (1978). Testing for autocorrelation in dynamic linear models. *Australian Economic Papers*, 17, 334-355.

Breusch, T.S., (1980). Useful invariance results for generalized regression models. *Journal of Econometrics*, 13, 327-340.

Breusch, T.S. and Godfrey, L.G., (1981). A review of recent work on testing for autocorrelation in dynamic simultaneous equation models, in *Macroeconomic Analysis.* (D.A. Currie, A.R. Nobay and D. Peel, Eds.) London: Croom Helm.

Breusch, T.S. and Pagan, A.R., (1980). The Lagrange multiplier test and its applications to model specification in econometrics. *Review of Economic Studies*, 47, 239-253.

Burke, S.P., Godfrey, L.G. and Tremayne, A.R. (1990). Testing AR(1) against MA(1) disturbances in the linear regression model: an alternative procedure. *Review of Economic Studies*, 57, 135-145.

Chatfield, C., (1979). Inverse autocorrelations. *Journal of the Royal Statistical Society*, Series A, 142, 363-377.

Chitturi, R.V., (1974). Distribution of residual autocorrelations in multiple autoregressive schemes. *Journal of the American Statistical Association*, 69, 928-934.

Chow, G.C., (1981). Selection of econometric models by the information criteria, in *Proceedings of the Econometric Society European Meeting, 1979.* (E.G. Charatsis, Ed.) Amsterdam: North-Holland.

Cleveland, W.S., (1972). The inverse autocorrelations of a time series and their applications. *Technometrics*, 14, 277-298.

Cox, D.R., (1961). Tests of separate families of hypotheses, in *Proceedings of the Fourth Berkeley Symposium on Mathematical Statistics and Probability*, 1, 105-123. Berkeley: University of California Press.

Cox, D.R., (1962). Further results on tests of separate families of hypotheses. *Journal of the Royal Statistical Society*, Series B, 24, 406-424.

Cox, D.R. and Hinkley, D.V., (1974). *Theoretical Statistics.* London: Chapman and Hall.

Cox, W.M., (1985). The behaviour of Treasury securities monthly, 1942-84. *Journal of Monetary Economics*, 16, 227-250.

Das Gupta, S. and Perlman, M.D., (1974). Power of the noncentral F-test: effect of additional variates on Hotelling's T^2-test. *Journal of the American Statistical Association*, 69, 174-180.

Davies, N. and Newbold, P., (1979). Some power studies of a portmanteau test of time series model specification. *Biometrika*, 66, 153-155.

Davies, N., Triggs, C.M. and Newbold, P., (1977). Significance levels of the Box-Pierce portmanteau statistic in finite samples. *Biometrika*, 64, 517-522.

Dhrymes, P.J., (1971). *Distributed Lags: Problems of Estimation and Formulation.* San Francisco: Holden Day.

Dickey, D.A., Bell, W.R. and Miller, R.B., (1986). Unit roots series model: tests and implications. *American Statistician*, 40, 12-26.

Dickey, D.A. and Fuller, W.A., (1979). Distribution of the estimators for autoregressive time series with a unit root. *Journal of the American Statistical Association*, 74, 427-431.

Dickey, D.A. and Fuller, W.A., (1981). Likelihood ratio statistics for autoregressive time series with a unit root. *Econometrica*, 49, 1057-1072.

Dickey, D.A., Jansen, D.W. and Thornton, D.L., (1991). A primer on cointegration with an application to money and income. *Federal Reserve Bank of St. Louis Quarterly Review*, 58-78.

Dolado, J.J, Jenkinson, T. and Sosvilla-Rivero, S., (1990). Cointegration and unit roots. *Journal of Economic Surveys*, 4, 249-274.

Durbin, J., (1970). Testing for serial correlation in least squares regression when some of the regressors are lagged dependent variables. *Econometrica*, 38, 410-421.

Durbin, J. and Watson, G.S., (1951). Testing for serial correlation in least squares regression II. *Biometrika*, 38, 159-178.

Engle, R.F., (1984). Wald, likelihood ratio, and Lagrange multiplier tests in econometrics, in *Handbook of Econometrics, Vol. 2.* (Z. Griliches and M.D. Intriligator, Eds.). Amsterdam: North-Holland.

Engle, R.F. and Granger, C.W.J., (1987). Co-integration: representation, estimation and testing, *Econometrica*, 55, 251-276.

Evans, G.B.A. and Savin, N.E., (1981). Testing for unit roots: 1. *Econometrica*, 49, 753-780.

Evans, G.B.A. and Savin, N.E., (1984). Testing for unit roots: 2. *Econometrica*, 52, 1241-1270.

Fama, E.F., (1975). Short-term interest rates as predictors of inflation. *American Economic Review*, 65, 269-282.

Fama, E.F. and Gibbons, M.R., (1982). Inflation, real returns and capital investment. *Journal of Monetary Economics*, 9, 297-323.

Fuller, W.A., (1976). *Introduction to Statistical Time Series.* New York: John Wiley.

Geweke, J., (1984). Inference and causality in economic time series models, in *Handbook of Econometrics, Vol. 2* (Z. Griliches and M.D. Intriligator, Eds.). Amsterdam: North-Holland.

Godfrey, L.G., (1978a). Testing against general autoregressive and moving average error models when the regressors include lagged dependent variables. *Econometrica*, 46, 1293-1302.

Godfrey, L.G., (1978b). Testing for higher order serial correlation in regression equations when the regressors include lagged dependent variables. *Econometrica*, 46, 1303-1310.

Godfrey, L.G., (1978c). A note on the use of Durbin's h-test when the equation is estimated by instrumental variables. *Econometrica*, 46, 225-228.

Godfrey, L.G., (1979). Testing the adequacy of a time series model. *Biometrika*, 66, 67-72.

Godfrey, L.G., (1981). On the invariance of the Lagrange multiplier test with respect to certain changes in the alternative hypothesis. *Econometrica*, 49, 1443-1456.

Godfrey, L.G., (1983). Comment on "Diagnostic tests as residual analysis". *Econometric Reviews*, 2, 229-233.

Godfrey, L.G., (1987). Discriminating between autocorrelation and misspecification in regression analysis: an alternative test strategy. *Review of Economics and Statistics*, 69, 128-134.

Godfrey, L.G. and Pesaran, M.H., (1983). Tests of nonnested regression models: small sample adjustments and Monte Carlo evidence. *Journal of Econometrics*, 21, 133-154.

Granger, C.W.J. and Newbold, P., (1977). The time series approach to econometric model building, in *New Methods in Business Cycle Research.* (C.A. Sims, Ed.). Minneapolis: Federal Reserve Bank of Minneapolis.

Granger, C.W.J. and Watson, M.W., (1984). Time series and spectral methods in econometrics, in *Handbook of Econometrics, Vol. 2.* (Z. Griliches and M.D. Intriligator, Eds.). Amsterdam: North-Holland.

Hannan, E.J., (1970). *Multiple Time Series.* New York: Wiley.

Hannan, E.J., (1980). The estimation of the order of an ARMA process. *Annals of Statistics*, 8, 1071-1081.

Hendry, D.F., (1977). Comments on Granger-Newbold's "Time series approach to econometric model building", in *New Methods in Business Cycle Research.* (C.A. Sims, Ed.). Minneapolis: Federal Reserve Bank of Minneapolis.

Hendry, D.F., (1986). *Economic Modelling with Cointegrated Variables*, special issue of *Oxford Bulletin of Economics and Statistics.*

Hendry, D.F., Pagan, A.R. and Sargan, J.D., (1984). Dynamic specification, in *Handbook of Econometrics, Vol.2.* (Z. Griliches and M.D. Intriligator, Eds.). Amsterdam: North-Holland.

Hendry, D.F. and Richard, J.-F., (1982). On the formulation of empirical models in dynamic econometrics. *Journal of Econometrics*, 20, 3-34.

Hendry, D.F. and Richard, J.-F., (1983). The econometric analysis of economic time series. *International Statistical Review*, 51, 111-163.

Hosking, J.R.M., (1981). Lagrange multiplier tests of multivariate time series models. *Journal of the Royal Statistical Society, Series B*, 43, 219-230.

Johansen, S. and Juselius, K., (1990). Maximum likelihood estimation and inference on cointegration - with applications to the demand for money. *Oxford Bulletin of Economics and Statistics*, 52, 169-210.

Jones, J.D. and Uri, N., (1987). Money, inflation and causality (another look at the empirical evidence for the USA, 1953-84). *Applied Economics*, 19, 619-634.

King, M.L., (1983). Testing autoregressive against moving average errors in the linear regression model. *Journal of*

Econometrics, 21, 35-51.

King, M.L. and McAleer, M., (1987). Further results on testing AR(1) against MA(1) disturbances in the linear regression model. *Review of Economic Studies*, 54, 649-663.

Liu, L.M. and Hanssens, D.M., (1982). Identification of multiple-input transfer function models. *Communications in Statistics, Part A - Theory and Methods*, 11, 297-314.

Ljung, G.M., (1986). Diagnostic testing of univariate time series models. *Biometrika*, 73, 725-730.

Ljung, G.M. and Box, G.E.P., (1978). On a measure of lack of fit in time series models. *Biometrika*, 65, 297-303.

McAleer, M., McKenzie, C.R. and Hall, A.D., (1986). Testing separate time series models. Paper presented to the European Meeting of the Econometric Society, Budapest, September, 1986.

McLeod, A.I., (1978). On the distribution of residual autocorrelations in Box-Jenkins models. *Journal of the Royal Statistical Society, Series B*, 40, 296-302.

Maddala, G.S. and Rao, A.S., (1973). Testing for serial correlation in regression models with lagged dependent variables and serially correlated errors. *Econometrica*, 41, 761-774.

Mills, T.C. and Stephenson, M.J., (1987). The behaviour of expected short-term real interest rates in the UK. *Applied Economics*, 19, 331-346.

Nankervis, J.C. and Savin, N.E., (1987). Finite sample distributions of t and F statistics in an AR(1) model with an exogenous variable. *Econometric Theory*, 3, 387-408.

Nelson, C.R. and Plosser, C.I., (1982). Trends and random walks in macroeconomic time series: some evidence and implications, *Journal of Monetary Economics*, 10, 139-162.

Newbold, P., (1980). The equivalence of two tests of time series model adequacy. *Biometrika*, 67, 463-465.

Newbold, P., (1983). Model checking in time series analysis, in *Applied Time Series Analysis of Economic Data* (A. Zellner, Ed.), Washington, D.C.: U.S. Department of Commerce, Bureau of the Census.

Neyman, J., (1959). Optimal asymptotic tests of composite statistical hypotheses, in *Probability and Statistics*, (U. Grenander, Ed.), Stockholm: Almquist and Wiksell.

Pagan, A.R., (1984). Econometric issues in the analysis of regressions with generated regressors. *International Economic Review*, 25, 211-247.

Pagan, A.R., (1985). Time series behaviour and dynamic specification. *Oxford Bulletin of Economics and Statistics*, 47, 199-211.

Pagan, A.R. and Hall, A.D., (1983). Diagnostic tests as residual analysis. *Econometric Reviews*, 2, 159-218.

Pagan, A.R. and Wickens, M.R., (1989). A survey of some recent econometric methods, *Economic Journal*, 99, 962-1025.

Park, J.Y. and Phillips, P.C.B., (1988). Statistical inference in regressions with integrated processes: part 1, *Econometric Theory*, 4, 468-498.

Park, J.Y. and Phillips, P.C.B., (1989). Statistical inference in regressions with integrated processes: part 2, *Econometric Theory*, 5, 95-131.

Perron, P., (1989). The great crash, the oil price shock and the unit root hypothesis, *Econometrica*, 59, 1361-1402.

Perron, P. and Phillips, P.C.B., (1987). Does GNP have a unit root?: A re-evaluation. *Economics Letters*, 23, 139-145.

Pesaran, M.H., (1981). Diagnostic testing and exact maximum likelihood estimation of dynamic models, in *Proceedings of the Econometric Society European Meeting, 1979*. (E.G. Charatsis, Ed.), Amsterdam: North Holland.

Phillips, P.C.B., (1986). Understanding spurious regressions in econometrics, *Journal of Econometrics*, 33, 311-340.

Phillips, P.C.B., (1987). Time series regression with a unit root, *Econometrica*, 55, 277-301.

Phillips, P.C.B and Durlauf, S.N., (1986). Multiple time series regression with integrated processes, *Review of Economic Studies*, 53, 473-495.

Phillips, P.C.B. and Ouliaris, S., (1990). Asymptotic properties of residual based tests for cointegration, *Econometrica*, 58, 165-194.

Phillips, P.C.B. and Perron, P., (1988). Testing for a unit root in time series regression, *Biometrika*, 75, 335-346.

Pierce, D.A., (1971a). Least squares estimation in the regression model with autoregressive-moving average errors. *Biometrika*, 58, 299-312.

Pierce, D.A., (1971b). Distribution of residual autocorrelations in the regression model with autoregressive-moving average errors. *Journal of the Royal Statistical Society, Series B*, 33, 140-146.

Pierce, D.A., (1972a). Least squares estimation in dynamic-disturbance time series models. *Biometrika*, 59, 73-78.

Pierce, D.A., (1972b). Residual correlations and diagnostic checking in dynamic-disturbance time series models. *Journal of the American Statistical Association*, 67, 636-640.

Pierce, D.A., (1983). Comment on "Model checking in time series analysis" by Paul Newbold, in *Applied Time Series Analysis of Economic Data*. (A. Zellner, Ed.). Washington, D.C.: U.S. Department of Commerce, Bureau of the Census.

Poskitt, D.S. and Tremayne, A.R., (1980). Testing the specification of a fitted autoregressive-moving average model. *Biometrika*, 67, 359-363.

Poskitt, D.S. and Tremayne, A.R., (1981a). An approach to testing linear time series models. *Annals of Statistics*, 9, 974-986.

Poskitt, D.S. and Tremayne, A.R., (1981b). A time series application of the use of Monte Carlo methods to compare statistical tests. *Journal of Time Series Analysis*, 2, 263-277.

Poskitt, D.S. and Tremayne, A.R., (1982). Diagnostic tests for multiple time series models. *Annals of Statistics*, 10, 114-120.

Poskitt, D.S. and Tremayne, A.R., (1984). Testing misspecification in vector time series models with exogenous variables. *Journal of the Royal Statistical Society, Series B*, 46,

304-315.

Prothero, D.L. and Wallis, K.F., (1976). Modelling macroeconomic time series (with discussion). *Journal of the Royal Statistical Society, Series A*, 139, 468-500.

Rao, C.R., (1948). Large sample tests of statistical hypotheses concerning several parameters with applications to problems of estimation. *Proceedings of the Cambridge Philosophical Society*, 44, 50-57.

Said, S.E. and Dickey, D.A., (1984). Testing for unit roots in autoregressive-moving average of unknown order, *Biometrika*, 71, 599-607.

Sargan, J.D. and Bhargava, A., (1983). Testing the residuals from least squares regression for being generated by the Gaussian random walk, *Econometrica*, 51, 153-174.

Schwert, G.W., (1987). Effects of model specification on tests for unit roots in macroeconomic data, *Journal of Monetary Economics*, 20, 73-103.

Schwert, G.W., (1989). Tests for unit roots: a Monte Carlo investigation, *Journal of Business and Economic Statistics*, 7, 147-159.

Silvey, S.D., (1959). The Lagrangian multiplier test. *Annals of Mathematical Statistics*, 30, 389-407.

Sims, C.A., Stock, J.H. and Watson, M.W., (1990). Inference in linear time series models with some unit roots, *Econometica*, 58, 113-144.

Tiao, G.C. and Box, G.E.P., (1981). Modeling multiple time series with applications. *Journal of the American Statistical Association*, 75, 802-816.

Trivedi, P.K., (1973). Retail inventory investment behaviour. *Journal of Econometrics*, 1, 61-80.

Tsay, R.S., (1985). Model identification in dynamic regression (distributed lag) models. *Journal of Business and Economic Statistics*, 3, 228-237.

Vinod, H.D., (1973). Generalization of the Durbin-Watson statistic for higher order autoregressive processes. *Communications in Statistics*, 2, 115-144.

Walker, A.M., (1967). Some tests of separate families of hypotheses in time series analysis. *Biometrika*, 54, 39-68.

Wallis, K.F., (1972). Testing for fourth order autocorrelation in quarterly regression models. *Econometrica*, 40, 617-636.

West, K.D., (1987). A note on the power of least squares tests for a unit root, *Economics Letters*, 24, 249-252.

Zellner, A., (1985). Bayesian econometrics. *Econometrica*, 53, 253-267.

S. P. BURKE AND L. G. GODFREY

2. TESTING FOR SERIAL CORRELATION IN DYNAMIC SIMULTANEOUS EQUATION MODELS: ALTERNATIVE ASYMPTOTIC PROCEDURES AND THEIR SMALL SAMPLE PERFORMANCE

1. Introduction

This chapter is concerned with asymptotic tests for serial correlation in dynamic simultaneous equation models. Such tests are important because the combination of autocorrelated disturbances and lagged endogenous variables has serious consequences for the usual instrumental variable/two stage least squares (IV/2SLS) procedures, rendering estimators inconsistent and significance tests invalid. Unfortunately, appropriate tests for serial correlation are rarely carried out after IV estimation. This neglect is in marked contrast to practice in the least squares analysis of non-simultaneous (but possibly dynamic) regression models: such analysis almost invariably includes some asymptotically valid check for serial correlation, e.g. by means of Lagrange multiplier (LM) tests of the type discussed by Breusch (1978) and Godfrey (1978b), or Durbin's (1970) h-test.

There are several possible explanations for this neglect. Some popular estimation packages do not compute suitable tests as a routine part of IV/2SLS calculations. Moreover there is little information about the relative merits of existing tests in finite samples, so that there is no clear incentive to undertake the task of separate calculation of any particular test after estimation. In this chapter, evidence is presented on the small sample performance of a number of tests. We shall also outline relationships between these tests and describe how some of them can be implemented using standard programs.

The main aim is to investigate whether it is possible to find a convenient test that performs reasonably well over a wide range of model characteristics. Robustness with respect to model characteristics is important because there is usually only vague information about the values of the parameters of structural models and so tests that are influenced greatly by these values cannot be recommended for general use.

The plan of the chapter is as follows. Section 2 contains details of notation and some preliminary remarks concerning IV/2SLS estimation. The various asymptotic tests for serial correlation are described in Section 3. The experimental design and Monte Carlo methodology are discussed in Section 4. The results obtained from the simulation experiments are summarised and examined in Section 5. Numerical examples are given in Section 6, and Section 7 contains some concluding remarks. A proof of the asymptotic validity of a test procedure is outlined in the Appendix.

2. Notation and preliminary remarks

Tests will be developed for the model

$$\begin{aligned} y &= Y\beta + Z\gamma + u \qquad (1)\\ &= X\delta + u, \end{aligned}$$

where y is a n-dimensional vector of observations on an endogenous variable, Y is a n by g matrix of observations on endogenous regressors, and Z is a n by k matrix containing observations on exogenous and lagged endogenous variables. The n by m, $m = g + k$, matrix of all regressor values is X, i.e. $X = (Y, Z)$. The unrestricted parameter vectors β and γ are g by 1 and k by 1 respectively with $\delta' = (\beta', \gamma')$. It is assumed that the null hypothesis to be tested is that the disturbances u_t are normally and independently distributed with zero mean and common variance

σ_u^2, i.e. u_t NID(0, σ_u^2). The alternative hypothesis is that the disturbances are generated by a stable first order autoregressive (AR(1)) process

$$u = \rho u_{-1} + \varepsilon, \ |\rho| < 1, \qquad (2)$$

where a subscript -1 on a vector or matrix denotes its value lagged by one time period and the n elements of ε are NID(0, σ_ε^2). The null hypothesis can then be denoted by H_0: $\rho = 0$ with the alternative being H_1: $\rho \neq 0$. Results on the invariance of test statistics imply that the tests discussed below are also appropriate for the alternative of a first order moving average process; see Godfrey (1981) for a general treatment of such invariance, and Breusch (1978) and Godfrey (1978b) for a detailed examination in the context of testing the assumption of serial independence. The assumption that the elements of ε are normally distributed is not essential for the asymptotic validity of the test procedures to be examined. These procedures retain their large sample properties under many forms of nonnormality.

Suppose that a set of $p \geq m$ instruments is available for estimating the unknown structural parameters and let the data matrix for these instruments be denoted by W. It will be assumed that the conventional asymptotic theory of IV/2SLS estimators can be applied to the problem of estimating δ when H_0 is true. Thus the probability limits of $n^{-1}(W'X)$ and $n^{-1}(W'W)$ are finite matrices, having ranks equal to m and p respectively. Also the elements of $n^{-1/2}W'u$ are asymptotically normally distributed with zero mean vector and covariance matrix equal to σ_u^2 plim $n^{-1}(W'W)$ when $\rho = 0$. It is worth noting that these standard assumptions rule out variables with time trends as deterministic components and integrated time series processes. As demonstrated by the work of Kramër (1984), Phillips (1986), and Phillips and Hansen (1990), the usual asymptotic theory cannot be relied upon when the processes are non-ergodic and some results and comments relating to this issue will be provided in Section 5 below.

Some of the tests to be described below require that the

instruments of W should be adequate and valid under both null and alternative hypotheses. The instruments used for such tests have to be restricted as follows. There are (m + 1) unknown parameters in the alternative model, viz. the elements of (δ', ρ), and so the number of instruments must satisfy $p \geq (m + 1)$, rather than $p \geq m$. In addition, the instruments must satisfy Sargan's (1959) conditions for estimating models with AR(1) errors. These conditions include the requirement that

$$J = \text{plim } n^{-1}[W'(X - \rho X_{-1}), W'u_{-1}]$$

should be a finite matrix with rank equal to (m + 1). If W contained only exogenous variables, then plim $n^{-1}W'u_{-1}$ would be a vector with every element equal to zero and so the matrix J could not have rank greater than m. Consequently Sargan's condition implies that at least one lagged endogenous variable should be used as an instrument. In particular, the requirement that J should have full rank can be satisfied by using y_{-1} in W, as would be usual when estimating dynamic models with this variable appearing as a regressor.

The actual choice of instruments may pose a dilemma to applied workers. The spirit of misspecification testing is that (1) should be estimated under H_0 and the results then used to test H_0. It is therefore natural to think of forming an IV matrix by choosing instruments only from variables appearing in the null specification, provided that Sargan's (1959) conditions can be satisfied. Let such an IV matrix be denoted by W_N. It could, however, be argued that this strategy places too much weight on H_0 and pays insufficient attention to H_1, thus risking the possibility of generating tests with low power. The justification for such an argument can be seen as follows. If ρ of (2) is nonzero, then the equation to be estimated can be written as

$$y = X\delta + \rho y_{-1} - \rho X_{-1}\delta + \varepsilon. \qquad (3)$$

The form of (3) suggests that it will be useful to include y_{-1} and

X_{-1} in the instrument set when estimating under H_1. Using instrument sets that do not include y_{-1} and X_{-1} may, therefore, lead to imprecise estimation and hence to significance tests that lack power.

In order to investigate this issue, we shall consider the relative merits of using W_N, the IV matrix based upon consideration of the model under H_0, and the augmented IV matrix

$$W_A \underset{r}{=} (W_N, y_{-1}, X_{-1}), \tag{4}$$

where $\underset{r}{=}$ denotes an equality which holds after deletion of redundant variables from the right-hand matrix. We shall make the plausible assumption that the regressors Z in (1) are always used as instruments.

Whatever the choice of instruments, estimation of (1) with $\rho = 0$ leads to the IV/2SLS estimator

$$\hat{\delta} = [X'Q(W)X]^{-1}X'Q(W)y, \tag{5}$$

where Q(W) is the projection matrix $W(W'W)^{-1}W'$. The variance parameter σ_u^2 can then be estimated by

$$\hat{\sigma}_u^2 = (n - m)^{-1}\ \hat{u}'\hat{u}\ , \tag{6}$$

$\hat{u}$ being the residual vector $(y - X\hat{\delta})$. A degrees of freedom adjustment is used in (6). This adjustment is asymptotically irrelevant, but is likely to be useful in finite sample applications; see Binkley and Nelson (1984). The estimate of the asymptotic covariance matrix of $\hat{\delta}$ will be taken to be

$$V(\hat{\delta}) = \hat{\sigma}_u^2[X'Q(W)X]^{-1}. \tag{7}$$

It will be useful to define an easily computed statistic for testing overidentifying restrictions; see Hausman (1983, pp. 433-4) for further discussion. This test statistic is given by

$$\psi_0(W) = \hat{\sigma}_u^{-2}(\hat{u}'Q(W)\hat{u}) \qquad (8)$$
$$= (n - m)[R^2(\hat{u};\ W)],$$

where $R^2(\hat{u};\ W)$ is the (uncentred) coefficient of determination from the OLS regression of $\hat{u}$ on W. The subscript 0 on the left-hand side of (8) is used to indicate that estimation is under H_0: $\rho = 0$. Under the hypothesis of correct specification and valid instruments, $\psi_0(W)$ is asymptotically distributed as $\chi^2(p - m)$. Statistics like $\psi_0(W)$ will be used to construct one of the serial correlation tests to be described in the next section, and have been discussed by Newey (1985b) in an important article on testing for specification errors.

Newey's (1985a, 1985b) general theorems on moment restriction tests provide some unification of results on checking for serial correlation, and so it will be useful to outline some relevant points that emerge from his work. Proofs and a detailed technical statement of regularity conditions are given by Newey (1985a, 1985b).

First, note that $\psi_0(W)$ of (8) can be rewritten as $\psi_0(W) = \hat{u}'W[\hat{\sigma}_u^2(W'W)]^{-1}W'\hat{u}$, i.e. as a quadratic form in the vector $W'\hat{u}$. Thus $\psi_0(W)$ is a check of the joint significance of the elements of $W'\hat{u}$ and so can be interpreted as a method of moments specification error test derived from the moment restrictions that instruments and errors are uncorrelated with $E(W'u) = 0$. Despite first appearances, the matrix $\hat{\sigma}_u^2(W'W)$ is not a consistent estimator of the asymptotic covariance matrix of $W'\hat{u}$ under correct specification. The first order conditions for $\hat{\delta}$, viz. $X'Q(W)\hat{u} = 0$, imply that $W'\hat{u}$ has a singular asymptotic distribution. More precisely, under the assumption of correct specification and valid instruments, $n^{-1/2}W'\hat{u}$ tends to a multivariate normal distribution with zero mean vector and finite covariance matrix with rank equal to (p - m). Consequently equation (8) simply represents the use of a particularly simple form of estimated g-inverse of this covariance matrix.

For certain sequences of local departures from the null model, a general test based upon (p - m) linearly independent elements of $W'\hat{u}$ will be less powerful than a more specific test involving a suitable subset of these elements. Newey (1985b) discusses in detail how potentially useful tests can be derived by checking whether or not linear combinations of the form $H(W'\hat{u})$ have all elements close to zero, H being an h by p matrix that has a finite probability limit with rank equal to h. Different choices of the matrix H will lead to variations in asymptotic power against a given sequence of local alternatives and some results on the optimal form of a test for a specific form of misspecification have been derived; see Newey (1985b). It will be shown below that Godfrey's (1976) π-test for autocorrelation can be obtained by using $h = 1$ and an appropriate choice of H. The π-test and other checks for serial correlation will now be described.

3. Asymptotic tests for serial correlation

The following tests are to be considered.

3.1 The π-test

Suppose first that the instrument set is of the W_N-type so that it permits estimation under H_1, but does not include the variables of (y_{-1}, X_{-1}). A Wald test of H_0 could be carried out by minimizing $\Lambda(\delta, \rho) = \varepsilon' Q(W_N)\varepsilon$, ε being given by (3), to obtain the autoregressive instrumental variable (AIV) estimators $\dot{\delta}$ and $\dot{\rho}$. Sargan (1959) provides the formulae required for calculating the estimated asymptotic variance of $\dot{\rho}$ and hence the t-ratio for testing H_0. Tests of H_0 requiring estimation under H_1 are, however, unattractive and Godfrey (1976) derives a test that requires estimation only under H_0.

Godfrey (1976) modifies Durbin's (1970) approach to make it suitable for IV estimators and considers the minimizer of

$$\Lambda(\hat{\delta}, \rho) = [\hat{u} - \rho\hat{u}_{-1}]Q(W_N)[\hat{u} - \rho\hat{u}_{-1}],$$

in which the unknown parameter δ of $\Lambda(\delta, \rho) = \varepsilon' Q(W_N)\varepsilon$ is replaced by its IV estimate under H_0, viz. $\hat{\delta}$. This minimizer is

$$r_\pi = \hat{u}'_{-1}Q(W_N)\hat{u}/\hat{u}'_{-1}Q(W_N)\hat{u}_{-1}.$$

Godfrey (1976) shows that, under H_0, $\pi = r_\pi/[\hat{avar}(r_\pi)]^{1/2}$ is asymptotically distributed as N(0,1), where $\hat{avar}(r_\pi)$ is a consistent estimator of the asymptotic variance of r_π when $\rho = 0$. An expression for a suitable estimate of this asymptotic variance is provided by Godfrey and an asymptotically valid form of the π-statistic based upon an instrument set W_N is

$$\pi = (\hat{u}'_{-1}Q(W_N)\hat{u})/\{\hat{\sigma}_u^2\hat{u}'_{-1}Q(W_N)\hat{u}_{-1} - \hat{u}'_{-1}Q(W_N)XV(\hat{\delta})X'Q(W_N)\hat{u}_{-1}\}^{1/2}.$$

A test based upon π is asymptotically equivalent to the Wald test of $\rho = 0$ derived from the AIV estimator of ρ, i.e. the two test statistics have the same large sample distributions under the null hypothesis and a sequence of local alternatives. The π-test requires the estimation of only the null model and so these results on its asymptotic equivalence to the AIV Wald test suggest that π might have an LM interpretation. This interpretation can be obtained by considering the Lagrangian

$$\Lambda^*(\delta, \rho, \mu) = \Lambda(\delta, \rho) + \mu\rho,$$

where μ is a multiplier. The elements of the first order derivatives of the Lagrangian with respect to δ, ρ and μ all equal zero when $\delta = \hat{\delta}$, $\rho = 0$ and $\mu = 2\hat{u}'_{-1}Q(W_N)\hat{u}$. Thus the LM principle leads to a test of the significance of $2\hat{u}'_{-1}Q(W_N)\hat{u}$ which is asymptotically proportional to the criterion r_π with

$$n^{-1/2}[2\hat{u}'_{-1}Q(W_N)\hat{u}] = n^{1/2}r_\pi[2\hat{u}'_{-1}Q(W_N)\hat{u}_{-1}/n],$$

in which the term in square brackets on the right-hand side has a finite nonzero probability limit under H_0.

As mentioned above, the π-test can be derived from Newey's (1985b) general methods of moments approach. Under the null hypothesis, the instruments of W_N and the disturbances of (1) are uncorrelated with $E(W_N'u) = 0$. The sample cross-product vector corresponding to this population moment restriction is $W_N'\hat{u}$ and tests can be constructed from linear combinations of the form $H(W_N'\hat{u})$, with H being an h by p matrix with finite probability limit. The π-test is derived in Newey's framework by using

$$H = \hat{u}_{-1}'W_N(W_N'W_N)^{-1}/[\hat{u}_{-1}'Q(W_N)\hat{u}_{-1}/n],$$

since with this choice

$$n^{-1/2}HW_N'\hat{u} = n^{1/2}r_\pi.$$

The interpretation of the π-test as a Newey moment restriction test again indicates the importance of Sargan's (1959) conditions which imply that plim $n^{-1}W_N'\hat{u}_{-1}$ should not be a null vector. If the elements of this probability limit were all equal to zero, then every element of the H vector that yields the π-test would be of the form a_i/b_i with

$$\text{plim } a_i = \text{plim } b_i = 0,\ i = 1, \ \ldots\ , p,$$

and H would not have a well-defined probability limit.

Whatever the interpretation used to justify the π-procedure, it is clear that it is essentially a check of the significance of the covariance between $\hat{u}$ and $Q(W_N)\hat{u}_{-1}$, the vector of OLS predicted values from the regression of $\hat{u}_{-1}$ on W_N. It could be argued that a more direct check for serial correlation would be obtained by inspecting the covariance between $\hat{u}$ and $\hat{u}_{-1}$. This argument leads to tests based upon autocovariances (or equivalently autocorrelations) calculated from the IV residuals. One test of this type is Godfrey's (1978a) θ-test.

3.2 The θ-test

Perhaps the most natural starting point for developing tests against first order autocorrelation is the population moment restriction that u_t and u_{t-1} have a covariance equal to zero, i.e.

$E(u_t u_{t-1}) = 0$. Under the null hypothesis of serial independence, the corresponding sample autocovariance $\hat{c}(1) = n^{-1}\hat{u}'_{-1}\hat{u}$ tends to zero as $n \rightarrow \infty$ with $n^{1/2}\hat{c}(1)$ being asymptotically normally distributed with zero mean and finite variance. Since the variance estimate $\hat{c}(0) = n^{-1}\hat{u}'\hat{u}$ converges to the finite positive constant σ_u^2 under H_0, a test based upon $\hat{c}(1)$ is asymptotically equivalent to one using the first order autocorrelation estimate $r(1) = \hat{c}(1)/\hat{c}(0)$. Godfrey (1978a) derives the asymptotic distribution of r(1) under the null hypothesis and shows that the significance of sample values of r(1) can be assessed by means of the test statistic

$$\theta = n^{1/2}r(1)/(1 - 2vr_{11} + vr_{12})^{1/2},$$

where

$$vr_{11} = \hat{u}'_{-1}XV(\hat{\delta})X'Q(W_N)\hat{u}_{-1}/(n\hat{\sigma}_u^4)$$

and

$$vr_{12} = \hat{u}'_{-1}XV(\hat{\delta})X'\hat{u}_{-1}/(n\hat{\sigma}_u^4),$$

which is asymptotically distributed as N(0, 1) when ρ of (2) equals zero.

The results reported by Godfrey and Tremayne in Chapter 1 indicate that it may be unwise to check for serial correlation by considering only one autocorrelation estimate. Consequently it is of some interest to derive procedures that allow the joint significance of a number of autocorrelation estimates, say, r(1), ... , r(s) to be investigated. Breusch and Godfrey (1981, Appendix B) provide an appropriate statistic thus permitting the construction of an IV analogue of the modified portmanteau test proposed by Ljung (1986) for testing the adequacy of pure time series models. Unfortunately, generalisations of the θ-test are difficult to implement; see Pagan and Hall (1983). Only the original form of the test with s = 1 is examined in the Monte Carlo experiments reported in this chapter.

Unlike the other tests to be examined, the θ-test can be used when the instruments are invalid or inadequate for estimation under H_1, e.g. a set of m exogenous instruments could be employed. In this chapter, however, only θ-statistics calculated using W_N or W_A will be considered because we wish to compare the θ-test to the other procedures. The comparisons are based upon Monte Carlo studies alone. The standard theoretical approach in which tests are compared by examining their asymptotic local powers under a suitable sequence of local alternatives is not helpful in the case under consideration. It is not possible to derive a generally valid ranking of θ and π from such asymptotic theory when the instruments are of the type W_N. On the other hand, if the instruments are of the type W_A, then, since $\hat{u}_{-1} = y_{-1}-X_{-1}\hat{\delta}$, we have $Q(W_A)\hat{u}_{-1} = \hat{u}_{-1}$ and so

$$r_\pi = \hat{u}'_{-1}\hat{u}/\hat{u}'_{-1}\hat{u}_{-1} = r(1)[\hat{u}'\hat{u}/\hat{u}'_{-1}\hat{u}_{-1}],$$

where the term in square brackets equals $1 + O_p(n^{-1})$. It follows that θ and π are asymptotically equivalent when (y_{-1}, X_{-1}) appears in the instrument set.

3.3 **Variable addition tests**

The θ- and π-tests have not been included in many estimation programs and it may be inconvenient to have additional programs written for their implementation. Breusch and Godfrey (1981) and Pagan (1984) suggest that tests which are more convenient than θ and π can be obtained by augmenting the original structural model (1) with the lagged residual $\hat{u}_{-1}$. The augmented model is then

$$y = X\delta + \hat{u}_{-1}\rho + e, \text{ say,} \tag{9}$$

and the check for serial correlation is carried out as a test of $\rho = 0$ in (9). This restriction can be tested by IV analogues of the LM, LR and Wald tests; see Godfrey (1988, Chp. 5) for a discussion of these analogues. It will be convenient to denote the general misspecification test statistic associated with the estimation of (9) by $\psi_1(W)$.

Although a Wald statistic, i.e. the t-ratio of $\hat{u}_{-1}$, is probably the most convenient test criterion in the special case of an AR(1) alternative, the LR analogue will be discussed because it provides a simpler generalisation when testing against AR(s) disturbances, $s > 1$. The IV analogue of the LR test for testing (1) against (9) requires that both models be estimated using the same set of instruments. As pointed out by Breusch and Godfrey (1981, Section 3.3), the test of $\rho = 0$ in (9) is asymptotically equivalent to the π-test if this instrument matrix is either W_N or W_A. It follows that the variable addition test is also asymptotically equivalent to the θ-test when W_A is used in the estimation of (1) and (9).

The choice of the common set of instruments for estimating (1) and (9) merits some discussion. As noted above, researchers may be reluctant to estimate the model of H_0 using W_A, the instruments of H_1, especially if the number of non-redundant variables in (y_{-1}, X_{-1}) is large. On the other hand, a variable addition test based upon W_N alone may lack power if the added variable $\hat{u}_{-1}$ in (9) is only weakly correlated with the instruments of W_N. It is, however, possible to construct an IV set that can be viewed as a compromise between W_N and W_A. Under H_0, the lagged residual is itself a valid instrument with

$$\text{plim } n^{-1}\hat{u}'_{-1}u = \text{plim } n^{-1}u'_{-1}u = 0$$

and it can be used to construct the artificial expanded IV set

$$W_E = (W_N, \hat{u}_{-1}). \tag{10}$$

If a test of H_0: $\rho = 0$ is to be based upon W_E, then the null model must be estimated twice: once using W_N to generate $\hat{u}_{-1}$; and then again using W_E of (10) for comparison with (9). Since the lagged residual $\hat{u}_{-1}$ is a by-product of estimation using W_N, a test using W_E as the common set of instruments for estimating (1) and (9) is not asymptotically equivalent to a π-test.

An attractive feature of the variable addition approach is that the generalisation to cover higher order AR(s), $s > 1$, alternatives is very simple:

(i) the augmented model (9) is replaced by

$$y = X\delta + \rho_1\hat{u}_{-1} + \dots + \rho_s\hat{u}_{-s} + e; \qquad (11)$$

(ii) if W_E is to be used rather than W_N or W_A, then the expanded instrument set corresponding to the matrix of (10) is defined by

$$W_E = (W_N, \hat{u}_{-1},\dots,\hat{u}_{-s});$$

(iii) the s restrictions of H_0: $\rho_1 = \dots = \rho_s = 0$ are tested by means of an appropriate procedure.

Given the form of the restrictions on the coefficients of (11) to be tested in step (iii), it is natural to look for an IV analogue of the least squares F-test. In order to describe a suitable procedure, it will be convenient to introduce some additional notation. Consider the case in which W_E is to be used in the calculation of the test statistic. Let $\tilde{e}_0$ denote the IV residual vector when this instrument matrix is used to estimate (11) subject to $\rho_1 = \dots = \rho_s = 0$, i.e. to estimate (1), and $\tilde{e}_1$ denote the corresponding vector when (11) is estimated with the coefficients ρ_i unrestricted. These residual vectors can be used to obtain the (uncentred) coefficients of determination

$$R^2(\tilde{e}_0;W_E) = \tilde{e}_0'Q(W_E)\tilde{e}_0/\tilde{e}_0'\tilde{e}_0 \qquad (12)$$

and

$$R^2(\tilde{e}_1;W_E) = \tilde{e}_1'Q(W_E)\tilde{e}_1/\tilde{e}_1'\tilde{e}_1. \qquad (13)$$

Next define $\tilde{\ell}$ by

$$\tilde{\ell} = [R^2(\tilde{e}_0;W_E)-R^2(\tilde{e}_1;W_E)]/[1-R^2(\tilde{e}_0;W_E)]. \qquad (14)$$

The test of $\rho_1 = ... = \rho_s = 0$ in (11) is calculated using $\tilde{\ell}$ and is suggested by the work of Morimune and Tsukuda (1984) who consider a statistic corresponding to

$$F(\tilde{\ell}) = [(n - m - s)/s]\tilde{\ell}. \qquad (15)$$

Morimune and Tsukuda examine the finite sample behaviour of the test obtained by comparing $F(\tilde{\ell})$ to critical values of the $F(s, n - m - s)$ distribution. They find that the F-distribution gives a good approximation in finite samples, but express concern about the possibility that the quality of the approximation may be reduced as the degree of simultaneity in the model increases (a more precise statement will be given below). Morimune and Tsukuda also consider the performance of the test obtained by replacing 2SLS estimates in (12)-(15) by the corresponding asymptotically equivalent limited maximum likelihood (LIML) estimates. They find that, when the test statistic of (15) is calculated using LIML estimates, the empirical distribution is very close to the assumed F-distribution when the null hypothesis is valid, and they recommend the use of LIML in practice. Unfortunately, LIML estimators are rarely employed in empirical work, despite the evidence that there are grounds for preferring them to the much more popular IV/2SLS estimators; see Anderson, Kunitomo and Sawa (1982). Attention in this chapter is restricted to IV/2SLS tests, but, if a LIML-type estimator is available, it may be useful to employ it when testing $\rho_1 = ... = \rho_s = 0$ in (11).

It is worth noting that the expression for $\tilde{\ell}$, and hence for the test statistic $F(\tilde{\ell})$, can be simplified. Under the null hypothesis, $R^2(\tilde{e}_0; W_E)$ is $O_p(n^{-1})$ - with $nR^2(\tilde{e}_0; W_E)$ being asymptotically distributed as $\chi^2(p + 1 - m)$ - so that the denominator of (14) can be replaced by unity without affecting the large sample properties of the significance test. Also $(n - m - s)R^2(\tilde{e}_0; W_E)$ is asymptotically equivalent to $(n - m)R^2(\tilde{e}_0; W_E)$. Combining these results yields a modified version of $F(\tilde{\ell})$ that can be written as

$$F(\phi_1) = [\psi_0(W_E) - \psi_1(W_E)]/s, \qquad (16)$$

in which ϕ_1 is the difference between the general test criteria for the null and augmented models, i.e. $\phi_1 = [\psi_0(W_E) - \psi_1(W_E)]$. This test statistic is easy to calculate and, under the assumption of serially independent errors, $F(\phi_1)$ is approximately distributed as $F(s, n - m - s)$ with significantly large values leading to the rejection of the null model.

Expressions for variable addition tests based upon the instrument matrix W_N (resp. W_A) are obtained by replacing W_E in (14) and (16) by W_N (resp. W_A) and reinterpreting the residual vectors as being derived from IV estimation of (1) and (11) using W_N (resp. W_A).

3.4 A general misspecification test

As well as providing a convenient means for calculating a specific test for serial correlation, the expanded model of (9) can also be used to examine the merit of a general misspecification test. The statistic $\psi_0(W_N)$ can be obtained from estimation of the null model employing only instruments that appear in the null specification, provided that they satisfy Sargan's (1959) conditions, and serves as a general diagnostic check. More precisely, $\psi_0(W_N)$ is asymptotically distributed as $\chi^2(p - m)$ under the null hypothesis and is $O_p(n)$ under many misspecifications.

As noted by Godfrey (1988, p. 172), the statistic $\psi_0(W_N)$ can be regarded as the criterion for testing the null model, i.e. (1) with $\rho = 0$, against any more general structural equation with p regressors. In particular, for $p > (m + 1)$, $\psi_0(W_N)$ provides a test of $\{\rho = 0; \alpha^+ = 0\}$ in the model

$$y = X\delta + \rho\hat{u}_{-1} + A^{+}\alpha^{+} + u,$$

where A^{+} is a n by (p - m - 1) matrix of irrelevant variables. Under a sequence of local alternatives with ρ being $O(n^{-1/2})$ and $\alpha^{+} = 0$, $\psi_0(W_N)$ will have the same noncentrality parameter as the test of (1) against (9) - and hence as the π-test - but has a larger degrees of freedom parameter. It follows from the results of Das Gupta and Perlman (1974), that, in terms of asymptotic local power, the general check $\psi_0(W_N)$ is inferior to π and the corresponding variable addition statistic. (Some calculations that illustrate the loss of asymptotic local power due to the inclusion of irrelevant test variables are given by Eastwood and Godfrey in Chapter 3.) This general test is, therefore, not included in the Monte Carlo study of Section 5. General test criteria like $\psi_0(W_N)$ can, however, be used to obtain a useful indirect specific test for serial correlation. This indirect test is a check of the validity of lagged residuals like $\hat{u}_{-1}$ as instruments for the estimation of (1) when the null hypothesis is true.

3.5 A test of the validity of lagged residuals as instruments

Consider the problem of testing the assumption of serial independence against the alternative of a general autoregressive process of order s. As explained above, the variable addition test of Section 3.3 involves estimating the null model (1) twice: the first time using the instruments of W_N; and the second time using the expanded instrument set $W_E = (W_N, \hat{u}_{-1}, \ldots, \hat{u}_{-s})$, where the $\hat{u}_{-i}$ are lagged values of the residual vector obtained in the first estimation, $i = 1, \ldots, s$. Each estimation yields a general misspecification test statistic of the form (8). Let $\psi_0(W_N)$ and $\psi_0(W_E)$ denote the values of this statistic associated with estimating (1) using W_N and W_E respectively. The difference between these two criteria provides a check of the validity of lagged residuals as instruments, thus yielding an indirect test of H_0. This indirect test is based upon Sargan's (1976) results which imply that when H_0 is true

$$\phi_2 = \psi_0(W_E) - \psi_0(W_N) \tag{17}$$

is asymptotically distributed as $\chi^2(s)$ with large values suggesting misspecification. The statistic ϕ_2 of (17) is simply the change in the general IV misspecification test criterion for (1) resulting from the expansion of the instrument set from W_N to W_E. The ϕ_2 test is, therefore, easy to implement. An asymptotically valid F-form of the ϕ_2 test can be carried out by comparing sample values of $F(\phi_2) = \phi_2/s$ to a prespecified critical value for the $F(s, n - m - s)$ distribution.

Godfrey's (1988, Section 5.5.2) discussion of checks of the validity of instruments indicates that the ϕ_2 test can be implemented by the method of variable addition. If $\hat{u}^{\dagger}_{-i}$ denotes the OLS residual from the regression of $\hat{u}_{-i}$ on W_N, $i = 1, \ldots, s$, then the ϕ_2 test is appropriate for testing H_0: $\rho_1 = \ldots = \rho_s = 0$ in the augmented model

$$y = X\delta + \rho_1\hat{u}^{\dagger}_{-1} + \ldots + \rho_s\hat{u}^{\dagger}_{-s} + e,$$

with both null and alternative models being estimated using the instrument matrix $(W_N, \hat{u}^{\dagger}_{-1}, \ldots, \hat{u}^{\dagger}_{-s})$, or equivalently W_E. The asymptotic equivalence of the ϕ_2 test and the test of the joint significance of the regressors $\hat{u}^{\dagger}_{-i}$, $i = 1, \ldots, s$, in the augmented model above is implied by the results given by Godfrey (1988, Section 5.5.2). The Appendix to this chapter contains an outline of a proof of the validity of the general approach underlying the ϕ_2 test. The relationship between this general approach and the method of variable addition is derived as part of the proof.

As with the two variable addition tests $F(\tilde{\ell})$ and $F(\phi_1)$, small sample fluctuations may lead to negative values of ϕ_2 and $F(\phi_2)$ that are inconsistent with their asymptotic distributions under H_0. Monte Carlo evidence on the magnitude of this problem is reported in Section 5.

4. Data generation for the Monte Carlo experiments

Following Hendry and Harrison (1974) and Hendry and Srba (1977), hereafter HH and HS, all data are generated using the two equation model

$$y_t = bY_t + cz_{1t} + dy_{t-1} + u_t \qquad (18)$$

and

$$Y_t = ay_t + \sum_{i=1}^{4} f_i z_{it} + e_t \qquad (19)$$

with

$$u_t = \rho u_{t-1} + \varepsilon_t, \ |\rho| < 1,$$

and $D = |d/(1 - ab)| < 1$. The predetermined variables of this model follow AR(1) processes of the form

$$z_{it} = \zeta_i z_{it-1} + v_{it}, \ |\zeta_i| < 1,$$

with the perturbations v_{it} being contemporaneously and serially independent normal variates with $v_{it} \sim N(0, \omega_{ii})$, $i = 1,...,4$. The disturbances (ε_t, e_t) are independent of the v_{is} for all i, t and s, and are jointly distributed according to

$$(\varepsilon_t, e_t)' \sim IN(0, \Sigma_{\varepsilon e}), \ t = 1,...,n,$$

where $\Sigma_{\varepsilon e} = \{\sigma_{ij}\}$, $i,j = 1,2$.

In their experiments, HH and HS allow $\Sigma_{\varepsilon e}$ to have its elements taken from $\sigma_{11} = (4.0, 0.25)$, $\sigma_{12} = (0.7, 0.0)$ and $\sigma_{22} = 1.0$. The value $\sigma_{12} = 0$ is not used here because the case of independent errors is thought to be of limited practical interest. In order to vary the correlation between ε_t and e_t, the fixed values ($\sigma_{12} = 0.7$, $\sigma_{22} = 1.0$) are combined with $\sigma_{11} = (4.0, 0.8)$. (The HH combination ($\sigma_{11} = 0.25$, $\sigma_{12} = 0.7$, $\sigma_{22} = 1.0$) is not admissible because it implies a covariance matrix which is not positive semi-definite.)

The treatment of other design parameters follows HH more

closely. The parameters of the autoregressive processes used to generate the predetermined variables are exactly as specified by HH (1974, p.165) and are constant for all experiments. The value of c is fixed at unity. Also, b, d, ρ and n are selected from a population similar to that used by HH, viz. b = (± 0.8, ± 0.2), d = (± 0.7, ± 0.4), ρ = (± 0.8, ± 0.5, 0.0) and n = (20, 40, 60). (HH (1974, p.166, fn. 11) perturb the values of ρ by small (possibly zero) amounts to avoid difficulties in their study of a grid-search estimator.)

The value of a is fixed at 0.4 by HH. Conditional upon a = 0.4, values of (b, d) are selected in the analysis of this paper to yield "high", "medium" and "low" values of the associated dynamics parameter $D = |d/(1 - 0.4b)|$. This strategy is adopted in order to investigate how the small sample behaviour of tests might vary with this index of the dynamic structure of the model. The degree of simultaneity of the model is also considered. A natural measure of simultaneity in the context of the limited information estimation of (18) is the correlation between u and Y; see Morimune and Tsukuda (1984) for a formal analysis of the role of the degree of simultaneity in a static system. There is, however, apparently little variation in the correlation between u and Y when a is constrained to equal 0.4. In order to obtain clearer evidence on the sensitivity of tests to variations in this correlation, we use the value a = 1.0 which is selected by trial and error. The value a = 1.0 gives estimated correlations in the range 0.38 to 0.75, whereas using a = 0.4 gives the narrower range 0.44 to 0.55. Setting a = 1.0 also increases the variability of the dynamics parameter D.

The twelve combinations of (σ_{11}, a, b, d) considered are summarised in Table 1 which also contains information on associated values of D and estimated correlations between u and Y for n = 20, 40, 60. These estimated correlations are calculated from the results of 1000 replications in each case. In view of the importance of evaluating the serial correlation tests under null and alternative hypotheses, and of examining the influence of sample size, all fifteen combinations of (n, ρ) are used with each of the twelve combinations of (σ_{11}, a, b, d) given in Table 1.

Table 1
Cases in Monte Carlo Study

						$\hat{r}(u,Y)^b$		
Case	σ_{11}	a	b	d	D^a	n = 20	40	60
1	4.0	0.4	-0.8	0.4	0.30	0.52	0.52	0.52
2	0.8	0.4	-0.8	0.4	0.30	0.55	0.55	0.55
3	4.0	0.4	-0.8	0.7	0.53	0.50	0.50	0.51
4	0.8	0.4	-0.8	0.7	0.53	0.55	0.54	0.55
5	4.0	0.4	0.2	0.7	0.76	0.44	0.43	0.44
6	0.8	0.4	0.2	0.7	0.76	0.50	0.49	0.48
7	4.0	1.0	-0.8	0.4	0.22	0.75	0.75	0.75
8	0.8	1.0	-0.8	0.4	0.22	0.67	0.66	0.66
9	4.0	1.0	-0.8	0.7	0.39	0.72	0.72	0.72
10	0.8	1.0	-0.8	0.7	0.39	0.65	0.65	0.65
11	4.0	1.0	0.2	0.7	0.88	0.45	0.42	0.41
12	0.8	1.0	0.2	0.7	0.88	0.45	0.40	0.38

Notes (a) Recall $D = |d/(1 - ab)|$.
(b) $\hat{r}(u,Y)$ denotes the estimated correlation between u and Y. These estimates are based upon 1000 replications.

All tests of the null hypothesis H_0: $\rho = 0$ are carried out at a nominal significance level of 5 per cent. Estimates of rejection probabilities are calculated using 1000 replications for each experiment. In each replication, (n + 51) observations are generated drawing variates from subroutines in the NAG library and using zeros for starting values. The first 50 observations are then discarded to reduce the effects of fixed starting values and the 51st observation is used as the initial value for the sample. This device appears to be satisfactory. Rerunning some experiments with 100 observations being discarded only results in very small changes.

The two instrument matrices W_N and W_A used to estimate the parameters of (18) and to calculate the various test statistics are defined to have typical rows

$$W_{Nt} = (z_{1t}, z_{2t}, z_{3t}, z_{4t}, y_{t-1})$$

and

$$W_{At} = (z_{1t}, z_{2t}, z_{3t}, z_{4t}, z_{1t-1}, y_{t-1}, Y_{t-1}, y_{t-2}),$$

respectively. If the residual vector for the null model obtained using W_N is denoted by $\hat{u}$, then W_E is constructed as $(W_N, \hat{u}_{-1})$. Godfrey's θ- and π-tests are calculated using W_N and W_A. The $F(\phi_2)$ test is obtained for results derived using W_N and W_E, but is not available when W_A is employed for estimation because $\hat{u}_{-1}$ is a linear combination of the columns of the latter instrument matrix. The variable addition tests $F(\tilde{\ell})$ and $F(\phi_1)$ are calculated using the augmented instrument matrix W_E and W_A. In preliminary work, variable addition tests were also calculated using W_N, rather than W_E, but the exclusion of the lagged residual from the instrument set caused poor performance for nonzero values of ρ. The results for the asymptotically equivalent π-test based upon W_N also indicate relatively low sensitivity to autocorrelation and are given in Section 5.

5. Monte Carlo results

It will be convenient to consider first the rejection frequencies associated with estimation using W_N and/or W_E, and then to examine results obtained using W_A as the instrument matrix. This method of presentation will clarify the impact of taking account of a fixed alternative, i.e. an alternative with $\rho = O(1)$, when selecting the instrumental variables for estimation and hypothesis testing.

5.1 Results obtained with $\rho = 0$ and W_N and/or W_E as the instrument matrix

With 1000 replications and a nominal significance level of 5 per cent for all tests, an estimate outside the range

$$5 \pm 2[5(95)/1000]^{1/2} = 5 \pm 1.4 = (3.6, 6.4)$$

may be viewed as being inconsistent with the assumption that the corresponding finite sample value equals its asymptotically achieved value. As will be seen below, several estimates fall outside this range. Also, for each test, there is evidence that the actual finite sample probabilities of a type I error depend upon the values of the parameters of the data generation process. In such circumstances, it is conventional to identify the significance level of a test for a given sample size with the maximum probability of rejecting the null hypothesis when it is true: the maximum of this function being defined with respect to variations in the parameters, see Rao (1973, p. 446). Consequently the finite sample significance level of a test procedure is estimated by the maximum rejection frequency. The contents of Table 1 do not suggest that the experimental design includes any extreme cases that would render this approach to estimating significance levels uninformative, e.g. neither D nor $\hat{r}(u,Y)$ is close to unity.

The experiments carried out with $\rho = 0$ produce several interesting results about the ways in which the test procedures behave as n, σ_{11}, a, b and d vary. The most clearly visible features reflect the effects of changing n, the sample size, and a, the coefficient of y_t in equation (19). Table 2 contains summary statistics to describe these features. The summary statistics provided are the maximum, minimum, average and standard deviation, denoted by max., min., avg. and s.d. respectively. As with all other tables in this chapter, the rejection frequencies of Table 2 are reported as percentages: in this case, the figures are rounded to one decimal place.

The contents of Table 2 indicate that increasing the value of the parameter a from 0.4 to 1.0 leads to a rather poorer agreement between actual behaviour and the predictions of asymptotic theory, and to a greater variability of performance for every test. The rejection frequencies for the lower value also show a much clearer pattern of estimates approaching the nominal value as the sample size increases.

The following remarks may be made about the tests under consideration. The θ-test appears to be undersized with only one estimate for n = 20 not being smaller than 3.6 per cent and every estimate for the three sample sizes being less than 5 per cent. The estimated significance levels of θ at n = 20, 40, 60 are 3.9, 4.6 and 4.8 per cent respectively. The π and $F(\tilde{\ell})$ tests are, in contrast, likely to reject a true null hypothesis too often. For n = 20, all estimates for π exceed 6.4 per cent which equals the minimum estimate for $F(\tilde{\ell})$ and is the largest value consistent with the hypothesis that the true rejection probability is 5 per cent. The tests are still oversized at the two larger sample sizes with both of them having every estimate exceeding the nominal value, except in case 11 with n = 60. The estimated significance levels for π at n = 20, 40, 60 are 13.2, 9.5 and 7.9 per cent respectively, with the corresponding figures for $F(\tilde{\ell})$ being 12.2, 11.9 and 9.6 per cent.

The $F(\phi_1)$ variable addition test, which is derived as a simplification of the $F(\tilde{\ell})$ test, is generally better behaved, but has high rejection frequencies in cases 7 and 9, the two cases with the highest values of the index of simultaneity; see Table 1. These high rejection frequencies lead to estimated significance levels of 7.0, 9.0 and 8.2 per cent for n = 20, 40, 60. Finally, like $F(\phi_1)$, the indirect test $F(\phi_2)$ appears to be sensitive to the relatively high degrees of simultaneity of cases 7 and 9. Its estimated significance levels are 7.8, 9.0 and 8.6 per cent for n = 20, 40, 60. The performance of the $F(\phi_2)$ test in cases other than 7 and 9 is usually quite reasonable. If the results for these other cases are considered, only 2 of the 20 estimates for n = 40, 60 are outside the range (3.6, 6.4). However, the results for cases 7 and 9 should not be ignored because, in general, there will usually be little idea about the degree of simultaneity and the dynamic characteristics of the model.

Table 2
Rejection frequencies for $\rho = 0$ and $IV = W_N$ and/or W_E: summary statistics

Test		a = 0.4			a = 1.0		
	n =	20	40	60	20	40	60
θ	max	3.9	4.6	4.8	2.7	3.3	4.2
	min.	2.8	3.4	3.5	2.0	1.8	2.7
	avg.	3.2	4.0	4.4	2.4	2.7	3.5
	s.d.	0.4	0.4	0.5	0.3	0.5	0.6
π	max.	7.9	6.8	6.4	13.2	9.5	7.9
	min.	6.8	5.6	4.8	8.9	7.0	5.9
	avg.	7.5	6.2	5.6	11.1	8.3	7.1
	s.d.	0.4	0.4	0.6	2.0	1.2	0.7
$F(\tilde{\ell})$	max.	7.3	6.9	6.2	12.2	11.9	9.6
	min.	6.4	5.2	4.9	8.6	6.7	6.6
	avg.	6.8	5.9	5.7	10.3	8.9	8.1
	s.d.	0.3	0.7	0.5	1.6	2.3	1.2
$F(\phi_1)$	max.	3.6	5.2	5.4	7.0	9.0	8.2
	min.	2.9	3.8	3.8	4.2	5.1	5.7
	avg.	3.2	4.4	4.9	5.5	6.8	6.8
	s.d.	0.3	0.6	0.6	1.1	1.8	1.1
$F(\phi_2)$	max.	4.4	5.6	5.6	7.8	9.0	8.6
	min.	3.0	4.3	4.3	5.1	5.2	6.1
	avg.	3.8	4.8	5.0	6.4	7.0	7.2
	s.d.	0.6	0.5	0.6	1.0	1.6	1.2

The results of Table 2 clearly justify the use of a = 1.0, as well as the original HH value of 0.4. If only a = 0.4 had been considered, an overoptimistic view of the small sample behaviour of the tests would have been obtained. We do not attempt to specify and estimate a response surface that might be used to investigate the dependence of a test's rejection frequency on the value of a (and on the values of other parameters). There is little *a priori* information to guide the specification of a suitable function here, and the dangers of using misspecified response surfaces have been clearly explained by Maasoumi and Phillips (1982). The problems associated with formulating useful response surfaces are exacerbated by the fact that while diagnostic checks of their adequacy are reported in some studies, such checks are typically only valid as the number of combinations of parameters and sample size used in the simulation analysis tends to infinity. Little is known about the value of these tests in practical situations.

To sum up, in terms of finite sample significance levels as estimated by the maximum rejection frequency, the θ-test is the only procedure that is not oversized and its finite sample significance level appears to approach the nominal value from below. However, the significance level is not the only characteristic of practical interest. Another feature that merits some attention is the frequency with which tests are unavailable as a result of sampling fluctuations leading to negative values of statistics that are inconsistent with their asymptotic distributions under H_0. Negative values can occur with the two variable addition tests $F(\tilde{\ell})$ and $F(\phi_1)$, and the $F(\phi_2)$ procedure. Evidence on this problem is presented in Table 3: the estimated frequencies of this table are in percentages and are rounded to one decimal place.

As with the results of Table 2, increasing the value of the parameter a tends to reduce the quality of the approximation provided by asymptotic theory which predicts that there should be no negative values. If the worst outcomes, i.e. the maxima of Table 3, are considered, then the following features are apparent: the problem of negative values is only serious for $F(\tilde{\ell})$ when $n = 20$; this problem is never serious for $F(\phi_2)$ whatever the sample size; and $F(\phi_1)$ is even better behaved in this respect.

Table 3
Estimated frequencies of negative values: summary statistics

Test		a = 0.4			a = 1.0		
		n = 20	40	60	20	40	60
$F(\tilde{\ell})$	max.	7.6	4.8	2.2	13.1	7.4	5.1
	min.	5.9	2.7	1.7	9.2	4.8	2.5
	avg.	6.7	3.6	2.0	11.1	6.0	3.7
	s.d.	0.7	0.8	0.2	1.8	1.1	1.2
$F(\phi_1)$	max.	0.3	0.1	0.1	2.4	0.9	0.4
	min.	0.1	0.0	0.0	0.5	0.1	0.0
	avg.	0.2	0.1	0.0	1.6	0.5	0.2
	s.d.	0.1	0.1	0.1	0.8	0.4	0.2
$F(\phi_2)$	max.	3.9	1.8	0.8	5.8	3.2	2.1
	min.	1.5	1.2	0.4	2.5	1.2	1.2
	avg.	2.7	1.4	0.6	4.1	2.3	1.5
	s.d.	1.0	0.2	0.2	1.2	0.9	0.3

5.2 Results obtained with $\rho \neq 0$ and W_N and/or W_E as the instrument matrix

While it is clearly important that tests be convenient to implement and have finite significance levels that are reasonably close to the nominal value, it is also important that they should be sensitive to the presence of autocorrelation. The results discussed in this section are for cases with $\rho \neq 0$ and throw some light on the usefulness of the checks described above. However, given the differences in estimated significance levels, there are obviously difficulties in comparing estimates of power derived using asymptotically valid critical values. In particular, it could be argued that the π and $F(\tilde{\ell})$ tests should be excluded because their estimated significance levels are sometimes rather greater than those of other tests (and the nominal value). It is however, important to note that, despite being oversized, the π-test rejects H_0 far less frequently than the other procedures when $\rho \neq 0$. As discussed above, this weakness may reflect the fact that π is designed to be asymptotically equivalent to a test based upon an AIV estimate of ρ which is inefficient for $\rho \neq 0$ because it employs W_N, rather than W_A, and so does not use all the instruments appropriate for the alternative model.

In contrast to the poor performance of the π-test, the θ-test, which also only uses W_N, often has estimated rejection frequencies for $\rho \neq 0$ which are similar to those of the $F(\phi_1)$ and $F(\phi_2)$ procedures. The performance of θ relative to $F(\phi_1)$ and $F(\phi_2)$ under $\rho \neq 0$ is noteworthy because the latter tests seem to have two advantages: first, they use W_E, an instrument matrix that takes account of local autocorrelation alternatives; second, θ is relatively (and absolutely) undersized. The differences between the rejection frequencies of $F(\phi_1)$ and $F(\phi_2)$ are neither marked nor systematic. Thus there is little to choose between these two tests on grounds of their ability to detect autocorrelation. In order to provide some examples, rejection frequencies for case 12 are reported in Table 4; these estimates are in percentage terms and have been rounded to the nearest integer.

Table 4 includes estimates of significance levels to assist the reader to evaluate the usefulness of comparisons, especially those involving $F(\tilde{\ell})$ and π with $n = 20$. The rejection frequencies for $\rho \neq 0$ exhibit the general features described above and, except for the π procedure, indicate the expected tendency for tests to become more sensitive as the sample size n increases. The estimates also reveal the dependence on the sign of ρ which is to be expected in dynamic models; see Maddala and Rao (1973). The finding that is most clearly illustrated by the results of Table 4 is the very poor performance of the π-test which rejects a true null hypothesis too often and rejects a false one relatively infrequently. It is difficult to compare $F(\tilde{\ell})$, $F(\phi_1)$ and $F(\phi_2)$ when $n = 20$ because there are substantial differences in size estimates. The apparent superiority of the $F(\tilde{\ell})$ procedure may well reflect the effects of a larger probability of a type I error, and the differences between $F(\phi_1)$, $F(\phi_2)$ and θ are not substantial. For n = 40 and n = 60, there is little to choose between $F(\tilde{\ell})$, $F(\phi_1)$, $F(\phi_2)$ and θ in terms of success in detecting nonzero values of ρ.

Table 4

Rejection frequencies for case 12: $\rho = 0$ and $\rho \neq 0$ using W_N and W_E

(n,ρ)	θ	π	$F(\tilde{\ell})$	$F(\phi_1)$	$F(\phi_2)$
(20, 0.8)	53	19	64	53	49
(20, 0.5)	21	14	36	25	23
(20, 0.0)	2	10	9	4	6
(20,-0.5)	41	10	60	48	50
(20,-0.8)	71	9	86	80	81
(40, 0.8)	96	19	96	95	94
(40, 0.5)	68	13	72	69	66
(40, 0.0)	3	7	7	5	6
(40,-0.5)	85	6	88	86	86
(40,-0.8)	99	8	100	100	100
(60, 0.8)	100	18	100	99	99
(60, 0.5)	89	11	90	88	88
(60, 0.0)	4	6	7	6	6
(60,-0.5)	98	6	98	97	98
(60,-0.8)	100	7	100	100	100

5.3 Results obtained with W_A as the instrument matrix

It was suggested in Section 3 that the effectiveness of test procedures might be improved by using W_A, an instrument matrix for fixed alternatives, rather than W_N, an instrument matrix for the null model. The purpose of this section is to present some evidence on this issue. Only the $F(\phi_1)$, θ- and π-tests are examined. The $F(\tilde{\ell})$ test is not considered because of its poor performance relative to its simplified form $F(\phi_1)$. The $F(\phi_2)$ test is not available when W_A is used as the instrument set because lagged residuals are linear combinations of the variables of W_A, so that expanding W_A to obtain an augmented matrix $(W_A, \hat{u}_{-1})$ would lead to the problem of perfect multicollinearity in the instruments. The π-test is only considered because its method of construction suggests that there may be important increases in power. This improvement in power is expected because, when calculated using W_A, π is asymptotically equivalent to a Wald test based upon an efficient autoregressive 2SLS estimator of ρ.

For the $F(\phi_1)$ and θ-tests, there is no evidence to suggest that performance is systematically and substantially enhanced by adopting the more general instrument set W_A, rather than W_N or W_E. On the other hand, the π-test performs much better when W_A is employed for estimation. The estimated significance levels of this test are in closer agreement with the nominal value, being 5.7, 6.2 and 6.5 per cent for $n = 20, 40, 60$ respectively. As well as being better behaved under the null hypothesis, the π-test is a more effective tool for detecting autocorrelation. This increase in the sensitivity of π is not unexpected for the reason set out in the previous paragraph. Also the performance of the π-test is similar to that of the θ-test. This similarity is to be anticipated because, in the cases under consideration, the difference between the π and θ statistics is only $O_p(n^{-1})$; see Section 3.2 above for a discussion of the order of magnitude of this difference.

It should, however, be noted that the π-test based upon W_A by no means dominates the θ, $F(\phi_1)$ and $F(\phi_2)$ tests calculated using W_N and/or W_E. Even when the rejection frequencies of this variant of the π-test are greater than, for example, those of the $F(\phi_1)$ procedure calculated using W_E estimates, the differences are not large. Further the rejection frequencies of θ and $F(\phi_1)$ for nonzero values of ρ are sometimes reduced by using W_A. Thus the incentive for adding instruments relevant under the alternative hypothesis to those of W_N is not great. Indeed, in the context of testing against general autoregressive (or moving average) errors, the use of W_A may cause difficulties unless the sample size is very large. For example, if the alternative hypothesis is an unrestricted fourth order autoregressive process, then the nonredundant variables of $(y_{-1}, X_{-1}, ..., y_{-4}, X_{-4})$ must be added to those of W_N to form W_A and the number of instruments in the latter matrix may be large relative to the number of observations.

Table 5
Rejection frequencies for cases 1-12 with $n = 40$ and $\rho = 0.5$: some estimates of the effects of using W_A, rather than W_N or W_E

Test	π		θ		$F(\phi_1)$	
IV set	W_N	W_A	W_N	W_A	W_E	W_A
Case 1	10	26	28	24	29	23
2	12	74	72	72	68	68
3	10	42	42	40	43	38
4	12	74	71	72	67	68
5	11	58	56	57	55	53
6	10	65	62	64	57	58
7	12	13	15[a]	11	33	16
8	13	73	71	70	73	70
9	12	20	22[a]	18	39	23
10	14	72	71	70	71	69
11	17	53	53[a]	52	71	56
12	13	72	68[a]	71	69	68

Note: (a) The corresponding estimated probability of a type I error is less than 3 per cent.

The results on the effects of replacing W_N or W_E by W_A are illustrated by the estimates contained in Table 5. These estimates are all twelve cases given in Table 1 with n and ρ fixed at 40 and

0.5 respectively. As can be seen, using W_A, rather than W_N, sometimes results in dramatic increases in the rejection frequencies of the π-test. The effects on other tests are usually much smaller. Overall there is generally little to be gained by taking account of the alternative hypothesis when selecting the instruments; compare the second column of estimates, i.e. those for π with IV matrix W_A, with the third and fifth columns which relate to tests that do not use the IV set for a fixed autocorrelation alternative.

5.4 Nonstationary models

Kiviet (1985) has provided estimates of significance levels for θ and various Lagrange multiplier tests based upon different sets of instruments. His rather mixed results are, however, derived from a model in which the data are constructed to have linear trends and, as noted in Section 2, conventional asymptotic theory uses stationarity assumptions. In order to explore the consequences of nonstationarity in the context of the model of this paper, we use a number of experiments with ($a = 0.4$, $b = 0.8$, $d = 0.7$) implying a dynamics parameter $D = 1.03$. In these experiments, standard asymptotic theory for cases with $D < 1$ provides a very poor approximation to actual behaviour. For example, with $n = 60$, the estimated significance levels of θ, $F(\phi_2)$ and the variable addition tests are about 10 per cent, while those for π are about 20 per cent, i.e. about four times the nominal value.

6. Numerical examples

The results of the previous section suggest that the θ statistic should be used when its calculation does not involve any great difficulty. However, if the estimation program being employed does not include the θ-test in its menu of checks for misspecification and the separate implementation of this test is inconvenient, then the $F(\phi_2)$ test is a flexible alternative procedure which is easy to carry out, requiring only that the program being used allows the residuals to be saved and lagged. The flexibility and simplicity of the $F(\phi_2)$ test are illustrated by the following numerical examples.

The data used in the numerical examples are taken from the tutorial data set of Hendry's (1989) program PC-GIVE. This set consists of 159 simulated values for each variable of a small dynamic model. These values represent quarterly observations for consumption, income, inflation and output over the period 1953(i) to 1992(ii). The numerical examples use the subset of figures for 1975(i) to 1989(iv), so that there are 60 observations used in the initial estimation. The model which is estimated is the simple (and underspecified) equation

$$\Delta c_t = \delta_1 + \delta_2 \Delta y_t + u_t,$$

where Δc_t and Δy_t are the first differences of the logs of consumption and income respectively; see Hendry (1989, pp. 93-94) for details of the actual artificial data generation process. The variable Δy_t is endogenous and the model is estimated by IV methods. Five sets of instruments are employed.

The set of instruments used for estimation under the null hypothesis of serial independence is denoted set A and consists of a constant term, Δy_{t-1} and Δy_{t-4}. Thus, in this example, lagged endogenous variables appear in the instrument set, rather than in the structural equation, and the IV estimator will be inconsistent when the null hypothesis is false. The residuals derived from set A are denoted by $\hat{u}_t$. The other four instrument sets, denoted sets B, C, D and E, are determined by the choice of alternative hypothesis and are obtained by adding lagged values of $\hat{u}_t$ to the variables of set A.

For IV set B, the first four lagged values of $\hat{u}_t$ serve as the additional instruments. This expanded IV set is appropriate for either unrestricted AR(4) or unrestricted MA(4) alternatives, and can also be regarded as a relatively parsimonious portmanteau-type test of the sort recommended by Godfrey and Tremayne in Chapter 1. The remaining three sets are appropriate for the autocorrelation cases examined in Section 5 of Chapter 1. Set C is designed for first order alternatives, with $\hat{u}_{t-1}$ alone being added to the three variables of set A. Set D uses $\hat{u}_{t-4}$ as the only extra instrumental variable, thus providing a check against simple fourth order autocorrelation. Finally, in order to obtain a test relevant to the multiplicative model

$$u_t = \rho_1 u_{t-1} + \rho_4 u_{t-4} - \rho_1 \rho_4 u_{t-5} + \varepsilon_t, \qquad (20)$$

an expanded group of instruments, viz. set E, is constructed by adding both $\hat{u}_{t-1}$ and $\hat{u}_{t-4}$ to set A.

The results for the numerical examples are calculated using the Micro-Fit program written by Pesaran and Pesaran (1987). This program adjusts the starting point for the estimation period when lagged residuals are generated and employed as instruments, rather than leaving the start unaltered and setting pre-sample values of $\hat{u}_{t-i}$ equal to zero. The estimation results relevant to the implementation of the $F(\phi_2)$ test are summarised in Table 6. This table contains the sample values of the general misspecification test statistics $\psi_0(.)$ and the implied values of the serial correlation test statistics ϕ_2; the degrees of freedom (d.f.) parameters for both procedures are given in parentheses. The values of ϕ_2 statistics are found by subtracting the value of the $\psi_0(.)$ function for set A from each of the other four $\psi_0(.)$ functions, d.f. parameters are obtained by the corresponding subtraction, and the values of $F(\phi_2)$ are derived by dividing each ϕ_2 by its d.f. parameter.

The Micro-Fit package provides values of the θ statistic for unrestricted alternatives. Hence, while the θ-test is not available for the alternatives corresponding to IV sets D and E, it can be applied in the other two cases. The sample value of the θ statistic for the unrestricted fourth order autocorrelation alternative is 3.7505 (d.f. = 4) and 0.2760 (d.f. = 1) for the first order case. These outcomes are clearly consistent with those of the ϕ_2 test given in Table 6; see values of ϕ_2 for instrument sets B and C.

Table 6
Some numerical examples

Instrument Set	ψ_0 Value	(d.f.)	ϕ_2 Value	(d.f.)	$F(\phi_2)$ Value
A	0.0349	(1)	–	–	
B	5.5786	(5)	5.5437	(4)	1.3859
C	0.2432	(2)	0.2083	(1)	0.2083
D	0.9376	(2)	0.9027	(1)	0.9027
E	1.7541	(3)	1.7192	(2)	0.8596

It may be worth noting that the $F(\phi_2)$ and θ-tests used in this example never indicate significant residual serial correlation in these examples, despite the fact they are calculated from a null model which is known to be misspecified because it omits two relevant variables from the equation for Δc_t; see Hendry (1989, p. 93) for the correct model. Thus the numerical examples do not just illustrate the ease of calculation and flexibility of the ϕ_2 and $F(\phi_2)$ procedure; they also indicate that tests for serial correlation cannot always be relied upon to detect misspecifications for which they were not designed. The value of information about the nature of specification errors when carrying out tests is discussed in Chapter 4.

7. Conclusions

Several asymptotic tests for serial correlation in the disturbances of dynamic simultaneous equation models have been described and their small sample behaviour has been examined using Monte Carlo experiments. The data generating processes employed in the main set of experiments are specified to produce stationary variables and so the results that are obtained are not relevant to nonstationary processes.

Perhaps the clearest result to emerge is that the π-test of

Godfrey (1976) is not to be recommended unless the researcher wishes to estimate the null model using an instrument set designed for the alternative hypothesis. The θ-test proposed by Godfrey (1978a) tends to be undersized, but is still quite sensitive to serial correlation. Variable addition procedures involving tests of the significance of a lagged residual added to the original regressor set have also been studied. A likelihood ratio-type test based upon 2SLS/IV estimates, denoted by $F(\tilde{\ell})$, has been found to be vulnerable to changes in the degree of simultaneity in the model. A modified test, $F(\phi_1)$, appears to be more robust with estimated probabilities of a type I error being closer to the nominal value, and has also been found to be a useful tool for detecting autocorrelation of the disturbances. The differences between the rejection frequencies of the $F(\phi_1)$ test and the corresponding estimates for an indirect test, denoted by $F(\phi_2)$, are often small, so that similar remarks apply to the latter procedure.

The results suggest that if estimation programs can be easily altered, then it is worth adding routines to calculate the θ statistic. If, however, standard packages are being used and it is difficult to incorporate specially written routines, then the $F(\phi_2)$ test of Section 3.5 is attractive because it is very simple to compute and performs quite well, comparing favourably with the $F(\phi_1)$ procedure. Interesting topics for future research include the finite sample behaviour of the generalisations of the θ and $F(\phi_2)$ tests for AR(s) and MA(s), $s > 1$, alternatives.

References

Anderson, T.W., N. Kunitomo and T. Sawa, 1982, Evaluation of the distribution function of the limited information maximum likelihood estimator, *Econometrica* 50, 1009-27.

Binkley, J.K. and G. Nelson, 1984, Impacts of alternative degrees of freedom corrections in two and three stage least squares, *Journal of Econometrics* 24, 223-33.

Breusch, T.S., 1978, Testing for autocorrelation in dynamic linear models, *Australian Economic Papers* 17, 334-55.

Breusch, T.S. and L.G. Godfrey, 1981, A review of recent work on testing for autocorrelation in dynamic simultaneous models, in: D. Currie, A.R. Nobay and D. Peel, eds., *Macro-economic Analysis* (Croom Helm, London) 63-110.

Das Gupta, S. and M.D. Perlman, 1974, Power of the non-central F-test: effect of additional variates on Hotelling's T^2-test, *Journal of the American Statistical Association* 69, 174-80.

Durbin, J., 1970, Testing for serial correlation in least squares regression when some of the regressors are lagged dependent variables, *Econometrica* 38, 410-21.

Frisch, R. and F. V. Waugh, 1933, Partial time regressions as compared with individual trends, *Econometrica* 1, 387-401.

Gallant, A.R. and D.W. Jorgenson, 1979, Statistical inference for a system of simultaneous, non-linear, implicit equations in the context of instrumental variable estimation, *Journal of Econometrics* 11, 275-302.

Godfrey, L.G., 1976, Testing for serial correlation in dynamic simultaneous equation models, *Econometrica* 44, 1077-84.

Godfrey, L.G., 1978a, A note on the use of Durbin's h-test when the equation is estimated by instrumental variables, *Econometrica* 46, 225-8.

Godfrey, L.G., 1978b, Testing against general autoregressive and moving average error models when the regressors include lagged dependent variables, *Econometrica* 46, 1303-10.

Godfrey, L.G., 1981, On the invariance of the Lagrange multiplier test with respect to certain changes in the alternative hypothesis, *Econometrica* 49, 1443-55.

Godfrey, L.G., 1988, *Misspecification Tests in Econometrics: the Lagrange Multiplier Principle and other Approaches* (Cambridge University Press, Cambridge).

Hausman, J., 1983, Specification and estimation of simultaneous equation models, in: Z. Griliches and M.D. Intriligator, eds., *Handbook of Econometrics*, Volume 1 (North-Holland, Amsterdam), 391-448.

Hendry, D.F., 1989, *PC-GIVE* (Institute of Economics and Statistics, University of Oxford: Oxford).

Hendry, D.F. and R.W. Harrison, 1974, Monte Carlo methodology and the small sample behaviour of ordinary and two-stage least squares, *Journal of Econometrics* 2, 151-74.

Hendry, D.F. and F. Srba, 1977, The properties of autoregressive instrumental variables estimators in dynamic systems, *Econometrica* 45, 969-90.

Kiviet, J.F., 1985, Model selection test procedures in a single linear equation of a dynamic simultaneous system and their defects in small samples, *Journal of Econometrics* 28, 327-62.

Kramër, W., 1984, On the consequences of trend for simultaneous equation estimation, *Economics Letters* 14, 23-30.

Ljung, G.M., 1986, Diagnostic testing of univariate time series models, *Biometrika* 73, 725-30.

Maasoumi, E. and P.C.B. Phillips, 1982, On the behaviour of inconsistent instrumental variable estimators, *Journal of Econometrics* 19, 183-201.

Maddala, G.S. and A.S. Rao, 1973, Testing for serial correlation in regression models with lagged dependent variables and serially correlated errors, *Econometrica* 41,761-74.

Morimune, K. and Y. Tsukuda, 1984, Testing a subset of coefficients in a structural equation, *Econometrica* 52, 427-48.

Newey, W.K., 1985a, Maximum likelihood specification testing and conditional moment tests, *Econometrica* 53, 1047-70.

Newey, W.K., 1985b, Generalized method of moments specification testing, *Journal of Econometrics* 29, 229-56.

Pagan, A.R., 1984, Model evaluation by variable addition, in: D.F. Hendry and K.F. Wallis, eds., *Econometrics and Quantitative Economics* (Blackwell, Oxford) 103-33.

Pagan, A.R. and A.D. Hall, 1983, Diagnostic testing as residual analysis, *Econometric Reviews* 2, 159-218.

Pesaran, M.H. and B. Pesaran, 1987, *Micro-Fit* (Oxford University Press, Oxford).

Phillips, P.C.B., 1986, Understanding spurious regressions in econometrics, *Journal of Econometrics* 33, 311-40.

Phillips, P.C.B. and B.E. Hansen, 1990, Statistical inference in instrumental variables regression with I(1) processes, *Review of Economic Studies*, 57, 99-126.

Rao, C.R., 1973, *Linear Statistical Inference and Its Applications* (Wiley, New York).

Sargan, J.D., 1959, The estimation of relationships with autocorrelated residuals by the use of instrumental variables, *Journal of the Royal Statistical Society, Series B*, 21, 91-105.

Sargan, J.D., 1976, Testing for misspecification after estimating using instrumental variables, unpublished paper, London School of Economics.

Appendix

The purpose of this Appendix is to provide an outline of the proof of the validity of the ϕ_2 test proposed in Section 3.5. The proof will be constructed by examining a general family of procedures which includes the ϕ_2 test for serial correlation as a special case. This family of tests is derived by comparing the two test criteria $\psi_0(.)$ derived from estimations of the null model using an IV matrix W and an expanded IV matrix (W, W_*). The notation adopted in this Appendix will, as far as possible, follow that used in Sections 2 and 3.

Consider the estimation of the structural equation (1) of Section 2, i.e.

$$y = X\delta + u, \; u \sim N(0, \sigma_u^2 I_n), \tag{A.1}$$

using the instrument matrix W. Let $\hat{y}$ and $\hat{X}$ denote Q(W)y and Q(W)X respectively, so that, for example, $\hat{y}$ is the predicted value from the OLS regression of y on W. The IV estimate of δ is then

$$\hat{\delta} = (X'Q(W)X)^{-1}X'Q(W)y \;=\; (\hat{X}'\hat{X})^{-1}\hat{X}'\hat{y}, \tag{A.2}$$

using the symmetry and idempotency of Q(W). As noted by Hendry and Srba (1977), $\hat{\delta}$ is the minimizer of the estimation criterion

$$g(\delta;\ W) = (y - X\delta)'Q(W)(y - X\delta), \tag{A.3}$$

with the associated minimum of g(.; W) being $g(\hat{\delta};\ W) = \hat{u}'Q(W)\hat{u}$, where $\hat{u} = y - X\hat{\delta}$. It follows that $g(\hat{\delta};\ W)$ is the numerator of the general misspecification test statistic $\psi_0(W)$ of (8) in Section 2. Consequently the variate $g(\hat{\delta};\ W)$ is asymptotically distributed as $\sigma_u^2\chi^2(p - m)$ under the assumptions of correct specification and valid instruments.

It will be useful to provide an alternative regression-based interpretation of $g(\hat{\delta};\ W)$. Combining (A.2) and (A.3) yields the result that

$$g(\hat{\delta};\ W) = [y - X(\hat{X}'\hat{X})^{-1}\hat{X}'\hat{y}]'Q(W)[y - X(\hat{X}'\hat{X})^{-1}\hat{X}'\hat{y}]$$

$$= \hat{y}'[I_n - Q(\hat{X})]\hat{y}, \tag{A.4}$$

i.e. $g(\hat{\delta};\ W)$ is the residual sum of squares (RSS) from the OLS regression of $\hat{y}$ on $\hat{X}$.

The ϕ_2 test involves adding lagged residuals to a set of instruments. Consider the case in which the null model (A.1) is to be estimated using the expanded IV matrix $(W,\ W_*)$, with W_* being a n by s matrix of instruments. In order to simplify the development of the proof, it is assumed that the additional instruments of W_* and the original variables of W are orthogonal, i.e. $W_*'W = 0$. This assumption implies that the projection matrix $Q(W,\ W_*)$ can be decomposed and written as

$$Q(W,\ W_*) = Q(W) + Q(W_*).$$

The restriction that $W_*'W = 0$ will be relaxed at a later stage of the argument.

The predicted values corresponding to $\hat{y}$ and $\hat{X}$ that are derived from OLS regressions on (W, W_*) will be written as

$$\bar{y} = Q(W, W_*)y = Q(W)y + Q(W_*)y = \hat{y} + Q(W_*)y,$$

and

$$\bar{X} = Q(W, W_*)X = Q(W)X + Q(W_*)X = \hat{X} + Q(W_*)X,$$

respectively. If $\bar{\delta}$ is the IV estimate of δ based upon (W, W_*) with associated residual vector $\bar{u} = y - X\bar{\delta}$, then $g(\bar{\delta}; W, W_*)$ is asymptotically distributed as $\sigma_u^2\chi^2(p + s - m)$ in the absence of misspecification errors. Using arguments similar to those applied to $g(\hat{\delta}; W)$, it can be shown that $g(\bar{\delta}; W, W_*)$ equals the RSS from the OLS regression of $\bar{y}$ on $\bar{X}$.

In order to prove that the difference between the two g(.;.) functions provides an appropriate large sample test of the validity of the variables of W_* as instruments and that this test can be applied using the method of variable addition, it is useful to consider the residuals from the OLS regression of $\bar{y}$ on W_*. These residuals are the elements of

$$[I_n - Q(W_*)][Q(W) + Q(W_*)]y = Q(W)y = \hat{y},$$

using $W_*'W = 0$. Similarly it can be shown that the residuals from the OLS regression of $\bar{X}$ on W_* are the elements of the matrix $\hat{X}$. Hence $g(\hat{\delta}; W)$, the RSS from the OLS regression of $\hat{y}$ on $\hat{X}$, equals the RSS from the regression of $[I_n - Q(W_*)]\bar{y}$ on $[I_n - Q(W_*)]\bar{X}$. The usual generalisation of the results of Frisch and Waugh (1933) implies that $g(\hat{\delta}; W)$ equals the RSS for the regression of $\bar{y}$ on both $\bar{X}$ and W_*, which is the minimum of the IV estimation criterion when the expanded model

$$y = X\delta + W_*\alpha + u \tag{A.5}$$

is estimated with the variables of $(W,\ W_*)$ employed as instruments.

Thus, instead of thinking of $\{g(\bar{\delta};\ W,\ W_*) - g(\hat{\delta};\ W)\}$ as being calculated by estimating the null model using two sets of instruments, this difference can be regarded as being obtained by estimating two models, viz. (A.1) and (A.5), with a common set of instruments $(W,\ W_*)$. The latter interpretation allows appeal to standard results on IV analogues of the likelihood ratio test; see Gallant and Jorgenson (1979) and Sargan (1976). More precisely, the quantity $\{g(\bar{\delta};\ W,\ W_*) - g(\hat{\delta};\ W)\}$ is the basis of a test of the s restrictions $\alpha = 0$ in (A.5), being asymptotically distributed as $\sigma_u^2\chi^2(s)$ when the null model and the IV set are valid. Dividing the two $g(.;\ .)$ functions by consistent estimators of σ_u^2 yields a feasible test statistic with an asymptotic null $\chi^2(s)$ distribution. If the variance estimators have the general form of (6) in Section 2, then this test statistic is simply the change in the general misspecification test statistic $\psi_0(.)$ when the IV matrix is expanded from W to $(W,\ W_*)$.

The assumption that $W_*'W = 0$ can now be relaxed. If it is desired to test whether or not the variables of W_+ are valid instruments and $W_+'W \neq 0$, then all that is required to modify the results above is to interpret W_* in (A.5) as being the residual matrix from the OLS regression of W_+ on W, i.e. $W_* = [I_n - Q(W)]W_+$. The IV estimation of the null model (A.1) and the alternative model (A.5) can be based upon either $(W,\ W_*)$ or $(W,\ W_+)$; these two IV matrices yield exactly the same estimates and test of $\alpha = 0$.

In order to derive the ϕ_2 test of Section 3.5, the variables of W_+ should be put equal to lagged values of the residual vector obtained when the null model (A.1) is estimated using the IV matrix W. The choice of these lagged values should reflect the nature of the AR or MA model being used as the alternative hypothesis. The corresponding variables of W_* in (A.5) are the artificial regressors $\hat{u}^{\dagger}_{-i}$ that are used to provide the variable addition form of the ϕ_2 test; see Section 3.5.

ALISON EASTWOOD AND L. G. GODFREY

3. THE PROPERTIES AND CONSTRUCTIVE USE OF MISSPECIFICATION TESTS FOR MULTIPLE REGRESSION MODELS

1. Introduction

The need to check the adequacy of econometric models is now widely accepted and tests for misspecification have become commonplace in empirical studies. Applied workers employing up-to-date programs to estimate linear regression models are likely to find that the usual estimation results (such as point estimates of coefficients, the associated standard errors and measures of goodness of fit) are accompanied by a battery of tests of the various assumptions that underpin the textbook model. These tests are usually, but not always, derived from specific alternative hypotheses and so the researcher is faced with a group of separate tests, each being designed for a particular breakdown of the classical assumptions.

The range of tests provided by modern programs is quite comprehensive, e.g. see Microfit (Pesaran and Pesaran, 1987), PC-GIVE (Hendry, 1989) and SHAZAM (White, 1978), and usually covers the following: omitted variables; incorrect functional form; lack of regression parameter constancy/predictive failure; autocorrelation; heteroskedasticity; and nonnormality of the disturbances. Every test procedure is derived and valid under its particular set of assumptions. Consequently the interpretation of the outcomes of a collection of separate tests requires some care. For example, the fact that the sample values of an autocorrelation check and Chow's (1960) statistic for testing parameter constancy are both significant certainly need not imply that the regression parameters vary and the disturbances are autocorrelated. Parameter changes may lead to significant residual correlation (see Carr (1972) for a numerical example that illustrates this possibility) and autocorrelation can lead to a true null hypothesis of parameter constancy being frequently rejected by Chow's procedure (see Consigliere, 1981).

Problems concerning the interpretation of separate tests and the difficulty of determining the overall significance level associated with a battery of such tests have caused some authors to suggest that, wherever possible, a joint test should be used, thus preventing the misinterpretation of individual test outcomes and providing control over the significance level (at least asymptotically); see Godfrey (1988, p. 162). Other authors have, however, argued that separate tests can be useful in isolating and identifying separate causes of model inadequacy. If individual tests had high power against the specific alternative for which they were designed and low power against all other alternatives, then the calculation of separate test statistics could provide a very useful guide when reformulating a model found to be inconsistent with the sample data.

The purpose of this chapter is, in part, to consider these opposing arguments and to investigate the potential constructive value of misspecification tests for linear regression models in the context of ordinary least squares (OLS) estimation. In order to communicate the results that are likely to be important in applied research, the treatment is fairly informal (references to more technical discussions are provided when appropriate).

The first step is to summarise, extend and evaluate the theoretical results that are relevant to the problem. Section 2 contains a discussion of some large sample theory concerning the behaviour of tests under the null hypothesis, local alternatives that move towards the null as the sample size increases, and fixed alternatives that do not. If an individual test is to be a useful source of information for improving specification, then, as noted above, it should be sensitive to the misspecification for which it was designed and insensitive to others. An important feature of

Section 2 is, therefore, that the behaviour of a test under an alternative for which it was not designed receives considerable attention. In such a situation, the test under study will be inappropriate relative to the combination of the assumed (null) model and the true data process. Since there will often be considerable uncertainty about the ways in which the initial specification may be inadequate, the study of the properties of inappropriate tests is important, even if there is no intention to use misspecification checks constructively. The results of Section 2 allow examination of the effects of various types of incorrect choice of test procedure and some numerical examples are provided to illustrate the theoretical findings.

The general theoretical results of Section 2 are used in Section 3 to assess the constructive value of misspecification tests, and, as in Section 2, asymptotic properties under fixed and local alternatives are considered. Section 3 also includes a discussion of a selection procedure, described by Newey (1985), that is designed to isolate errors made in the original specification.

A rather different type of constructive use of misspecification test procedures is examined in Section 4. Most of the test procedures that have become popular in applied econometrics can be carried out by adding specified test variables to the original regressor set and then checking the joint significance of these variables. If a conventional test of the original specification against the expanded version produces a significant outcome, then the former is regarded as inadequate. This 'variable addition' (relative to the original equation) or 'omitted variable' (relative to the expanded equation) approach has been recommended and discussed by several researchers, e.g. see Breusch and Godfrey (1986), Godfrey (1981) and Pagan (1984). It plays a major role in the analyses of Sections 2 and 3, and, in Section 4, the variable addition algorithm is used to obtain estimates of parameters of interest that are corrected for misspecification. Section 4 includes Monte Carlo evidence on the finite sample properties of these corrected estimates. The corrected estimators are compared with unrestricted estimators derived by estimating alternative models, and the work reported in Section 4 can be thought of as an extension of Godfrey's (1981) discussion of locally equivalent alternatives (LEAs) and the relative power of tests developed for such alternative hypotheses.

The results of Sections 2, 3 and 4 are relevant to the tests that are now used routinely in applied econometrics. As an alternative to using these conventional statistics, it is sometimes possible to derive 'robust tests', i.e. procedures that are relatively insensitive to some unconsidered misspecifications. Robust procedures of this type offer the opportunity to increase the constructive value of misspecification tests. An important example is provided by the construction of a check of the adequacy of the regression function that is not sensitive to either autocorrelation or heteroskedasticity. Section 5 contains a discussion of this sort of test in the context of the linear regression model. Wooldridge (1990) provides a more detailed discussion for a wide class of models. The construction and the use of some robust tests are outlined in Section 5.

The findings of the analyses concerning the implementation and constructive use of misspecification tests in regression analysis are summarised in Section 6 and some concluding remarks are made.

2. Asymptotic properties of misspecification tests

The model to be examined for misspecification (hereafter referred to as the null model) is assumed to be a conventional linear regression equation

$$y = X\beta + u, \qquad (1)$$

in which: y is the n-dimensional vector of observations on the regressand; X is the n by k, $n > k$, matrix of observations on the regressors (taken to be at least weakly exogenous) and $\text{rank}(X) = k$; β is the k-dimensional vector of unknown parameters; and u is the n-dimensional vector of disturbances. The elements of u, denoted by u_t, t = 1, ... ,n, are assumed to be independent random variables with zero mean and common variance σ_u^2. These assumptions concerning the disturbances can be summarised using an obvious notation as $u \sim D(0, \sigma_u^2 I_n)$, where the family of distributions D(. , .) is unspecified. More restrictively, it can be assumed that the disturbances are independently and normally distributed (NID), in which case $u \sim N(0, \sigma_u^2 I_n)$. The assumption of normality is usually required to obtain exact finite sample results about distributions of estimators and test statistics and is also needed to justify some asymptotic results, but much of the standard large sample theory for regression models can be derived under weaker distributional assumptions.

The ordinary least squares (OLS) estimator of β in (1) will be written as

$$\hat{\beta} = (X'X)^{-1}X'y, \tag{2}$$

with associated residual vector

$$\hat{u} = y - X\hat{\beta} = (\hat{u}_1, \dots , \hat{u}_n)'.$$

The OLS residuals can be used to obtain estimators of the variance parameter σ_u^2, e.g. $s_u^2 = (n - k)^{-1}\hat{u}'\hat{u}$ or $\hat{\sigma}_u^2 = n^{-1}\hat{u}'\hat{u}$. Under the hypothesis of correct specification, the OLS estimators will enjoy good asymptotic properties, but $\hat{\beta}$ may provide very misleading information if the null model is misspecified. Hence it is important to test that the assumptions underpinning (1) are consistent with the data.

Suppose first that the adequacy of (1) is to be checked by comparing it to expanded models such as

$$y = X\beta + W\gamma + u, \tag{3}$$

and

$$y = X\beta + Z\delta + u, \tag{4}$$

in which W and Z are n by p and n by q, respectively. The

regressors appearing in W and Z may be genuine economic variables or artificial terms constructed, for example, from the OLS results obtained by estimating (1). As explained in detail by Godfrey (1988, Ch. 4) and Pagan (1984), models like (3) and (4) are capable of generating tests designed to be relevant to many specific violations of the assumptions that make up (1) - for example, omitted variables, incorrect functional form, autocorrelation, lack of parameter constancy, endogenous regressors - and can also be used to construct general checks such as Plosser, Schwert and White's (1982) differencing test and Ramsey's (1969) RESET test. Whatever the interpretation of the variables of W and Z, rejection of the parametric restrictions $\gamma = 0$ (resp. $\delta = 0$) in (3) (resp. (4)) on the basis of a suitable test statistic denoted by T_W (resp. T_Z) is taken to indicate that the null model (1) is misspecified.

Augmented models like (3) and (4) do not, however, permit the convenient calculation of all of the popular diagnostic checks used in applied work. In particular, the Lagrange multiplier (LM) tests for heteroskedasticity and nonnormality cannot be easily implemented using the variable addition approach.

The LM test for heteroskedasticity as derived by Breusch and Pagan (1979) is based upon the alternative hypothesis that individual variances are determined by

$$\mathrm{Var}(u_t) = h(\alpha_0 + \Sigma_1^r m_{tj}\alpha_j), \quad t = 1, \ldots ,n,$$

in which h(.) is a continuous function, having first and second order derivatives, and the variables m_{tj} are exogenous. The special case of homoskedasticity is obtained by imposing the r constraints of H_α: $\alpha_1 = \ldots = \alpha_r = 0$ which is equivalent to $\mathrm{Var}(u_t) = h(\alpha_0)$ for all t. In order to satisfy regularity conditions for testing H_α, the function h(.), which is otherwise irrelevant to the form of the test statistic, must be specified so that $h'(\alpha_0) \neq 0$, where $h'(.)$ denotes the first order derivative.

The test statistic proposed by Breusch and Pagan (1979) is obtained from the regression of the scaled squared OLS residuals

$\hat{u}_t^2/\hat{\sigma}_u^2$ on the exogenous variables $m_{t1}, \dots, m_{tr}$ and an intercept term. Under H_α and the assumption that the uncorrelated errors u_t are normally distributed, the explained sum of squares (about the average) from this OLS regression divided by 2 is asymptotically distributed as $\chi^2(r)$, with significantly large values indicating that H_α is not consistent with the data; this procedure is also proposed by Godfrey (1978a).

Although the Breusch-Godfrey-Pagan form of the LM test is provided in the routine calculations of some estimation programs, it suffers from an important drawback. Koenker (1981) has pointed out that this variant is not robust to nonnormality, so that significant values of the test statistic may reflect nonnormality, rather than heteroskedasticity. This property is undesirable, especially if the researcher wishes to use tests constructively. The problem lies in dividing the explained sum of squares from the regression of $\hat{u}_t^2/\hat{\sigma}_u^2$ on the exogenous variables $m_{t1}, \dots, m_{tr}$ and an intercept term by 2. The factor of 2 is used because, under normality, $Var(u_t^2/\sigma_u^2) = 2$ when the disturbances are uncorrelated and homoskedastic. However, if the assumption of normality is relaxed, then $Var(u_t^2/\sigma_u^2)$ may be greater than or less than 2 under H_α. Koenker suggests a simple modification to overcome this problem: the factor of 2 is replaced by the sample variance of the scaled terms $\hat{u}_t^2/\hat{\sigma}_u^2$. The resulting Studentised form is asymptotically distributed as $\chi^2(r)$ when H_α is true, even when the errors u_t are not normal. Koenker's test statistic can be calculated as the product of the sample size, n, and the coefficient of determination, R_α^2, from the OLS regression of $\hat{u}_t^2/\hat{\sigma}_u^2$ (or equivalently and more conveniently, of $\hat{u}_t^2$) on $m_{t1}, \dots, m_{tr}$ and an intercept term. Koenker's test statistic nR_α^2 will be denoted by T_H.

Tests for nonnormality are used quite widely in applied regression analysis and are based upon estimates of measures of skewness and excess kurtosis (relative to the normal distribution). More precisely, if

$$\sqrt{b_1} = n^{-1}\Sigma_1^n \hat{u}_t^3/\hat{\sigma}_u^3,$$

and

$$b_2 = n^{-1}\Sigma_1^n \hat{u}_t^4/\hat{\sigma}_u^4,$$

then

$$T_N = n[(\sqrt{b_1})^2/6 + (b_2 - 3)^2/24]$$

is asymptotically distributed as $\chi^2(2)$ when (1) is correctly specified and its disturbances are NID(0, σ_u^2) variates; see Jarque and Bera (1980, 1987). It could, however, be argued that there is little incentive to calculate the statistic T_N because, if appeal is made to large sample theory to justify the use of the Jarque-Bera procedure, then there is no need to assume normality to derive most of the standard results concerning estimators and tests for linear regression models. Given the uncertainty that will usually be associated with specifying the precise form of the joint distribution of the errors, it is probably best to avoid the relatively small number of procedures that require normality in large samples. For example, Koenker's (1981) Studentised test T_H should be used to check for heteroskedasticity, rather than the Breusch-Pagan (1979) criterion. Several prediction error tests, e.g. Hendry's (1979, p. 222) z_4-test, require normality for their asymptotic validity and may be quite unreliable under weaker distributional assumptions; see Chapter 5 for a discussion of such a test and various more robust modifications.

The rest of this section is devoted to a discussion of the properties of test statistics under: (i) the null hypothesis that (1) is adequate; (ii) a fixed alternative process; and (iii) a sequence of alternative models that approach the null model as the sample size increases. This discussion will be organized to provide a set of results that assist in evaluating the constructive use of misspecification tests; this type of evaluation forms the subject matter of Section 3.

2.1 Asymptotic behaviour under the null hypothesis

Since most diagnostic checks can be computed by the method of variable addition, it seems reasonable to concentrate on tests of the null model against augmented relationships such as (3) and (4). It will be convenient to assume that the test statistics are calculated in χ^2 form, rather than as F-statistics. Thus the tests of (1) against (3) and (4) can be based upon any of the classical tests, viz. the LM, likelihood ratio (LR) and Wald tests. In order to examine the behaviour of these tests under various assumptions about the true data process, it will be useful to note the test statistics for checking the adequacy of (1) against, for example, (3) can be written as:

$$LM_W = n[y'P(X)y - y'P(X, W)y]/[y'P(X)y]; \tag{5}$$

$$LR_W = n\,ln[y'P(X)y/y'P(X, W)y]$$

$$= n\,ln[1 + (y'P(X)y - y'P(X, W)y)/(y'P(X, W)y)]; \tag{6}$$

and

$$Wald_W = n[y'P(X)y - y'P(X, W)y]/[y'P(X, W)y], \tag{7}$$

in which P(.) is the projection matrix yielding residuals from the OLS regression on the regressor matrix that appears as its argument, e.g. $P(X) = I_n - X(X'X)^{-1}X'$, and *ln* denotes the natural logarithm. These statistics test $\gamma = 0$ in (3) and, under the null hypothesis that (1) is correctly specified, all three criteria are asymptotically distributed as $\chi^2(p)$ and are valid candidates to serve as T_W, the check of (1) against (3).

The expressions for statistics designed to test (1) against (4), denoted by LM_Z, LR_Z and $Wald_Z$, are obtained by replacing the test variable matrix W in (5), (6) and (7) by the alternative set Z. Under the assumption of correct specification, these statistics are asymptotically distributed as $\chi^2(q)$.

It is natural to consider the relationships between the two

sets of tests of (1). For example, what can be said about the overall significance level if the null model is accepted only if the restrictions $\gamma = 0$ and $\delta = 0$ are both found to be data consistent on the basis of appropriate separate tests? This question is important because most programs produce a number of separate tests (that could be calculated by variable addition) and presumably applied workers will wish to avoid a high overall probability of a type I error and to have as much information as possible about the nature of the dependencies between such separate tests.

The tests of (1) against (3) and (4) are tests of the significance of the vector of estimated coefficients for the variables of W and Z, respectively. These vectors of OLS estimates can be written as

$$\tilde{\gamma} = (W'P(X)W)^{-1}W'P(X)y$$

and

$$\bar{\delta} = (Z'P(X)Z)^{-1}Z'P(X)y.$$

Under the null hypothesis of correct specification, y is determined by (1) and substitution yields

$$\tilde{\gamma} = (W'P(X)W)^{-1}W'P(X)u$$

and

$$\bar{\delta} = (Z'P(X)Z)^{-1}Z'P(X)u.$$

In order to examine the joint asymptotic distribution of $\tilde{\gamma}$ and $\bar{\delta}$, it will be convenient to adopt assumptions similar to those used by Sargan (1988). More precisely, sample moment matrices, e.g. $n^{-1}X'X$, are taken to have finite probability limits, and a central limit theorem can be applied to relevant linear combinations of disturbances so that, under correct specification,

$$n^{1/2}\begin{bmatrix}\tilde{\gamma}\\ \bar{\delta}\end{bmatrix} \rightarrow N(\begin{bmatrix}0\\0\end{bmatrix}, \begin{bmatrix}V_{\gamma\gamma} & V_{\gamma\delta}\\ V'_{\gamma\delta} & V_{\delta\delta}\end{bmatrix}),$$

in which

$$V_{\gamma\gamma} = \sigma_u^2 \text{ plim } n[W'P(X)W]^{-1},$$

$$V_{\gamma\delta} = \sigma_u^2 \text{ plim } n[W'P(X)W]^{-1}[W'P(X)Z][Z'P(X)Z]^{-1},$$

and

$$V_{\delta\delta} = \sigma_u^2 \text{ plim } n[Z'P(X)Z]^{-1}.$$

Consequently $\tilde{\gamma}$ and $\bar{\delta}$ will be asymptotically independent under the null hypothesis when plim $n^{-1}[W'P(X)Z]$ is a p by q matrix with every element equal to zero. It will be useful to consider the implications of asymptotic independence and to provide some econometric examples.

If $\tilde{\gamma}$ and $\bar{\delta}$ are asymptotically independent and are used in tests with asymptotic significance levels of ε_W and ε_Z, respectively, then the probability that neither will reject the true null model tends to $(1 - \varepsilon_W)(1 - \varepsilon_Z)$ as n increases without limit. In other words, if separate tests of $\gamma = 0$ and $\delta = 0$ are used to induce an overall test of (1), the asymptotic significance level is $1 - (1 - \varepsilon_W)(1 - \varepsilon_Z)$. If ε_W and ε_Z are both small, then this significance level is approximately $(\varepsilon_W + \varepsilon_Z)$.

If the researcher does not wish to attempt to use T_W and T_Z constructively, but is instead only interested in using these statistics to evaluate the null model, then asymptotic independence permits the construction of a suitable joint test without the need to estimate a regression model that includes all the nonredundant variables of (X, W, Z). The additive property of independent χ^2 variates implies that, if (X, W, Z) has full column rank, $(T_W + T_Z)$ is asymptotically distributed as $\chi^2(p + q)$ when T_W and T_Z are asymptotically independent. Hence a valid large sample joint test can be obtained by comparing $(T_W + T_Z)$ to an upper critical value of the $\chi^2(p + q)$ distribution; see Bera and McKenzie (1987) for a general discussion of the additivity of likelihood-based test statistics.

The above analysis has been related to the independence of tests based upon two sets of regressors, viz. W and Z. Sometimes it is useful to consider the relationships between subsets of the regressors of one of these sets. For example, if W is partitioned as $W = [W_1, W_2]$, with W_i being n by p_i, i = 1, 2, then tests of the significance of the corresponding subvectors of $\tilde{\gamma}$ will be asymptotically independent if plim $n^{-1}[W_1'P(X)W_2]$ is a p_1 by p_2 matrix with every element equal to zero.

The nature of the condition for asymptotic independence under the null hypothesis implies that two separate variable addition test statistics will have this property if:

(i) the corresponding regressor matrices are asymptotically orthogonal;

and

(ii) one of these regressor matrices is also asymptotically orthogonal to the original regressor set.

Thus, in the case discussed in the previous paragraph, we require plim $n^{-1}W_1'W_2 = 0$, and either plim $n^{-1}W_1'X = 0$ or plim $n^{-1}X'W_2 = 0$, where 0 denotes a matrix with the appropriate dimensions and every element equal to zero.

These conditions can be illustrated by considering some examples. Suppose that y depends upon a single strongly exogenous variable, so that

$$y_t = \beta_1 + \beta_2 x_t + u_t, \tag{1a}$$

and this specification is to be checked using a RESET-type test derived from the model

$$y_t = \beta_1 + \beta_2 x_t + \gamma_1 x_t^2 + \gamma_2 x_t^3 + \gamma_3 x_t^4 + u_t, \tag{3a}$$

and by means of a test for first order autocorrelation obtained from

$$y_t = \beta_1 + \beta_2 x_t + \delta_1 \hat{u}_{t-1} + u_t, \tag{4a}$$

where $\hat{u}_{t-1}$ is the lagged residual associated with the OLS estimation of (1a). Under correct specification,

$$\text{plim } n^{-1}\Sigma_t \hat{u}_{t-1} x_t^j = \text{plim } n^{-1}\Sigma_t u_{t-1} x_t^j = 0$$

for $j = 1, \ldots, 4$, and so the Z-type regressor of (4a) is asymptotically orthogonal to the original regressor of (1a) and the W-type test variables of (3a). Hence the RESET and autocorrelation tests are asymptotically independent, even though the variables for the former procedure are not asymptotically orthogonal to the regressor, e.g. $\text{plim } n^{-1}\Sigma_t x_t(x_t^3) = \text{plim } n^{-1}\Sigma_t x_t^4$ will be strictly positive since standard regularity conditions require that at least one value of x_t be nonzero.

Next, consider a situation in which (3a) is replaced by a partial adjustment scheme written as

$$y_t = \beta_1 + \beta_2 x_t + \gamma_1 y_{t-1} + u_t, \tag{3b}$$

with (4a) being unaltered. If the null model is valid,

$$\text{plim } n^{-1}\Sigma_t x_t \hat{u}_{t-1} = \text{plim } n^{-1}\Sigma_t x_t u_{t-1} = 0,$$

so that, in the notation of the general discussion above, $W'P(X)Z$ can be replaced by $W'Z$ which in this example equals $\Sigma_t y_{t-1}\hat{u}_{t-1}$. It follows that tests based upon (3b) and (4a) are not asymptotically independent because

$$\text{plim } n^{-1}\Sigma_t y_{t-1}\hat{u}_{t-1} = \text{plim } n^{-1}\Sigma_t y_{t-1}u_{t-1} = \sigma_u^2 > 0.$$

Asymptotic independence of tests involving subsets of regressors can be illustrated by replacing (4a) by the expanded model

$$y_t = \beta_1 + \beta_2 x_t + \delta_1 \hat{u}_{t-1} + \ldots + \delta_4 \hat{u}_{t-4} + u_t, \qquad (4b)$$

which is the basis for a test against unrestricted fourth order autoregressive errors. There may be interest in special cases of the unrestricted alternative, e.g. simple first or fourth order processes, so that the relationship between tests for individual terms $\hat{u}_{t-j}$ could be of interest. It is easy to verify that these tests are asymptotically independent since, under correct specification,

$$\text{plim } n^{-1}\Sigma_t x_t \hat{u}_{t-j} = \text{plim } n^{-1}\Sigma_t x_t u_{t-j} = 0, \ j = 1, \ldots, 4,$$

and

$$\text{plim } n^{-1}\Sigma_t \hat{u}_{t-i}\hat{u}_{t-j} = \text{plim } n^{-1}\Sigma_t u_{t-i} u_{t-j} = 0 \text{ for } i \neq j.$$

Variable addition and heteroskedasticity tests

Clearly checking for asymptotic independence is quite straightforward when the tests under consideration can all be derived by variable addition. However, the LM test for heteroskedasticity requires separate treatment because it cannot be carried out conveniently using this approach. Pagan and Hall (1983) argue that statistics like the criterion T_H will be asymptotically independent of standard variable addition checks if the disturbances are symmetric, i.e. $E(u_t^3) = 0$. A test for skewness is, of course, one of the components of the Jarque-Bera statistic. More specifically, if the u_t are normally distributed, then

$$n(\sqrt{b_1})^2/6 = n^{-1}(\Sigma_t \hat{u}_t^3)^2/6\hat{\sigma}_u^6$$

is asymptotically distributed as $\chi^2(1)$.

If the errors of (1) are symmetric, then an asymptotically valid joint test can be obtained using the sum of T_H and a variable addition check. For example, if T_W denotes a statistic for testing $\gamma = 0$ in (3) and $E(u_t^3) = 0$ for all t, then

$$(T_H + T_W) \rightarrow \chi^2(p + r)$$

under the null hypothesis. A joint test is then derived by comparing the sum of the test statistics to an upper critical value of the $\chi^2(p + r)$ distribution. If T_W, T_Z and T_H are all asymptotically independent, then the probability that at least one of these statistics will be significant tends to

$$1 - (1 - \varepsilon_W)(1 - \varepsilon_Z)(1 - \varepsilon_H)$$

when (1) is correct, and a non-constructive joint test could be based upon the result that the limiting null distribution of $(T_W + T_Z + T_H)$ is $\chi^2(p + q + r)$.

If test statistics are not asymptotically independent, then valid large sample joint tests cannot be obtained by simply adding up the separate statistics, and difficulties arise in determining the overall asymptotic probability of a type I error. When, for example, T_W, T_Z and T_H are asymptotically dependent, the large sample probability that the correct model (1) will be indicated as data inconsistent by at least one of these tests is between $\max(\varepsilon_W, \varepsilon_Z, \varepsilon_H)$ and $\min(1, \varepsilon_W + \varepsilon_Z + \varepsilon_H)$.

2.2 Asymptotic behaviour under a fixed alternative hypothesis

If the usefulness of diagnostic checks in reformulating data inconsistent models is to be examined, it is necessary to examine the behaviour of tests like T_W, T_Z and T_H under data processes other than (1). In this subsection, it is assumed that the data are generated by a fixed alternative model. A quite general form of a fixed alternative model can be written as

$$y = f + v, \quad v \sim D(0, \sigma_v^2 I_n), \tag{8}$$

in which f is a possibly nonlinear function of weakly or strongly exogenous variables and an unknown parameter vector. For the purposes of the asymptotic analysis of this subsection, a model such as (8) is said to be a fixed alternative if the regression function f is not asymptotically equivalent to a linear combination of the regressors of (1), i.e. plim $n^{-1}f'P(X)f > 0$.

Consider first the behaviour of variable addition tests such as T_W and T_Z derived by testing (1) against (3) and (4), respectively. Only one test need be considered since results for the other can be obtained by interchanging W and Z. Suppose then that (1) is tested against (3) and that T_W is one of the classical likelihood-based statistics defined by (5)-(7). The statistics LM_W, LR_W and $Wald_W$ are functions of the residual sums of squares $y'P(X)y$ and $y'P(X, W)y$. If these test statistics are $O_p(n)$ under the true data generation process, then each of them will exceed a finite critical value of the $\chi^2(p)$ distribution with probability tending to 1 as $n \to \infty$, i.e. they will all be consistent.

If T_W is $O_p(n)$, then $n^{-1}T_W$ is $O_p(1)$. Consequently consistency requires that the probability limit of $n^{-1}T_W$ should be nonzero. Examination of (5)-(7) reveals that this requirement is equivalent to the condition

$$\text{plim } n^{-1}[y'Q(X, W)y - y'Q(X)y] > 0,$$

in which $Q(.) = I_n - P(.)$. Under the fixed alternative (8),

$$\text{plim } n^{-1}[y'Q(X, W)y - y'Q(X)y] = \text{plim } n^{-1}[f'Q(X, W)f - f'Q(X)f],$$

the terms involving the disturbance vector v being asymptotically negligible. Standard results concerning projection matrices imply that

$$n^{-1}[f'Q(X, W)f - f'Q(X)f] = n^{-1}[f'Q(P(X)W)f],$$

and this term will only have a nonzero probability limit if, in the pseudo-regression model,

$$f = Xb + Wc + \text{residual}, \tag{9}$$

the OLS estimator of c has a nonzero probability limit. Loosely speaking, the test variables of W must pick up some of the variation of f that X alone cannot capture.

If (3) is actually the true data process, the above condition will certainly be satisfied since in this case

$$\text{plim } n^{-1}[f'Q(X, W)f - f'Q(X)f] = \text{plim } n^{-1}\gamma'W'P(X)W\gamma > 0,$$

using $f = X\beta + W\gamma$, and so T_W is consistent against the alternative for which it is designed. It is interesting to consider the behaviour of T_Z when the data are generated by (3). Replacing W by Z and f by $X\beta + W\gamma$ in the above analysis yields the result that

$$\text{plim } n^{-1}[f'Q(P(X)Z)f] = \text{plim } n^{-1}\gamma'W'Q(P(X)Z)W\gamma$$

and this probability limit must be nonzero for T_Z to be consistent. The consistency of T_Z under (3) cannot be guaranteed and there is an important case in which it is possible to demonstrate that it is inconsistent.

Suppose that T_W and T_Z are asymptotically independent under (1). As shown in the previous subsection, asymptotic independence implies that $\text{plim } n^{-1}W'P(X)Z = 0$ under (1) and, if this condition also holds under (3), then $\text{plim } n^{-1}\gamma'W'Q(P(X)Z)W\gamma = 0$. Thus, T_Z, based upon (4), will be inconsistent under (3). This close connection between asymptotic independence under the null and inconsistency under a fixed alternative for which the test was not designed does not, however, carry over to the heteroskedasticity check T_H.

Heteroskedasticity and variable addition tests

If the regression function of (1) is correctly specified,

i.e. $E(y) = X\beta$, but the disturbances are heteroskedastic with variances given by, e.g.

$$Var(u_t) = \exp(\alpha_0 + \Sigma_1^r m_{tj}\alpha_j),$$

with α_j being $O(1)$, $j = 1, \ldots, r$, then variable addition tests like T_W and T_Z will be inconsistent, even if the u_t are skewed. This result is easily established. Recall that for T_W to be a consistent test

$$\text{plim } n^{-1}[y'Q(X, W)y - y'Q(X)y] = \text{plim } n^{-1}[y'P(X)y - y'P(X, W)y]$$

should be nonzero. When $y = X\beta + u$, this probability limit equals

$$\text{plim } n^{-1}[u'P(X)u - u'P(X, W)u] = \text{plim } n^{-1}(u'u - u'u) = 0,$$

using the fact that u is asymptotically uncorrelated with both X and W. Hence T_W is inconsistent when the only assumption that is violated is that the disturbances are heteroskedastic. Similarly T_Z is also inconsistent in this situation. Note that the inconsistency of the two variable addition tests does not depend upon T_H being asymptotically independent of these tests under the null hypothesis.

Consider next a situation in which data are generated by (3) - a relationship with homoskedastic disturbances - and the adequacy of (1) is tested using T_H. It can be shown that the inappropriate test T_H may have the property of consistency, even if it is asymptotically independent of T_W when $\gamma = 0$. The possibility that T_H will be consistent in such a case will be illustrated by a simple example.

Suppose that y_t is determined by

$$y_t = \beta x_t + \gamma w_t + u_t, \ \gamma \neq 0, \ u_t \text{ iid}(0, \sigma_u^2), \tag{3c}$$

but the null model

$$y_t = \beta x_t + u_t, \; u_t \text{ iid}(0, \sigma_u^2), \tag{1c}$$

is tested by means of a T_H-type statistic with the heteroskedasticity scheme being specified as

$$\text{Var}(u_t) = \alpha_0 + \alpha_1 m_t, \; t = 1, \dots, n.$$

The scalar variables x_t, w_t and m_t are nonstochastic, uniformly bounded and satisfy standard regularity conditions. The T_H statistic is derived from the regression of $\hat{u}_t^2$ on m_t and an intercept, and is essentially a test of the significance of the sample covariance $n^{-1}\Sigma_t\hat{u}_t^2(m_t - \bar{m})$, where $\bar{m}$ denotes the sample average of the m_t, $t = 1, \dots, n$. Suppose that the distribution of the u_t is symmetric, so that, under 1(c), T_H is asymptotically independent of T_W, the test of $\gamma = 0$ in (3c).

Let $\tilde{\beta}$ and $\tilde{\gamma}$ denote the OLS estimates for the true model (3c) with associated residuals $\tilde{u}_t$, then

$$\hat{u}_t = \tilde{u}_t + \tilde{\gamma}w_t^*,$$

where w_t^* is a typical residual from the OLS regression of w_t on x_t, $t = 1, \dots, n$. Hence

$$\begin{aligned} \text{plim } n^{-1}\Sigma_t\hat{u}_t^2(m_t - \bar{m}) = {} & \text{plim } n^{-1}\Sigma_t\tilde{u}_t^2(m_t - \bar{m}) + \\ & 2\gamma \text{ plim } n^{-1}\Sigma_t\tilde{u}_t[w_t^*(m_t - \bar{m})] + \\ & \gamma^2 \text{ plim } n^{-1}\Sigma_t(w_t^*)^2(m_t - \bar{m}), \end{aligned} \tag{10}$$

using plim $\tilde{\gamma} = \gamma$. The first term on the right-hand side of (10) equals zero because $\tilde{u}_t$ is the OLS residual for the true model which has homoskedastic errors. The second term also equals zero since

$$\text{plim } n^{-1}\Sigma_t\tilde{u}_t[w_t^*(m_t - \bar{m})] = \text{plim } n^{-1}\Sigma_t u_t[w_t^*(m_t - \bar{m})] = 0,$$

using plim $\tilde{\beta} = \beta$ and plim $\tilde{\gamma} = \gamma$. However, the third term need not equal zero since there is no reason for $(w_t^*)^2$ to be uncorrelated with m_t as $n \rightarrow \infty$. When the sample covariance between $(w_t^*)^2$ and m_t does not vanish asymptotically, the heteroskedasticity test will be consistent under the unconsidered alternative (3c), despite the fact that T_H is asymptotically independent of T_W under (1c).

Artificial linear alternative models

The link between asymptotic independence and inconsistency can also be broken when two variable addition tests are being considered, if one or other of these tests uses artificial variables constructed from the results of estimating the null model. In such cases, it is possible for the probability limit of $n^{-1}W'P(X)Z$ to be a matrix of zeros under the null, but to have nonzero elements under the alternative. An important example occurs when, say, Z contains lagged OLS residuals so that a test of $\delta = 0$ in (4) provides a check for serial correlation; see Breusch and Godfrey (1981). Suppose again that the null model is (1c) and the true model is (3c). Let T_Z be calculated as a test of $\delta = 0$ in

$$y_t = \beta x_t + \delta\hat{u}_{t-1} + u_t, \tag{4c}$$

where $\hat{u}_{t-1}$ is the lagged value of the OLS residual $\hat{u}_t = y_t - \hat{\beta}x_t$. Under (1c),

$$\text{plim } n^{-1}x_t\hat{u}_{t-1} = \text{plim } n^{-1}w_t\hat{u}_{t-1} = 0,$$

when x_t and w_t are nonstochastic variables, so that T_W and T_Z are asymptotically independent under the null hypothesis of correct specification.

In order to examine behaviour under (3c), it will be

convenient to let x, w and $\hat{u}$ denote the n-dimensional vectors with typical elements x_t, w_t and $\hat{u}_t$, respectively, and to let L be the n by n matrix approximation to the lag operator, i.e. elements of L on the diagonal under the leading diagonal equal one and all other elements equal zero. It is then necessary to consider

$$\text{plim } n^{-1}w'P(x)L\hat{u} = \text{plim } n^{-1}w'P(x)LP(x)y$$

$$= \lim n^{-1}w'P(x)LP(x)w\gamma$$

(ignoring asymptotically negligible terms involving u)

$$= (\lim n^{-1}\Sigma_t w_t^* w_{t-1}^*)\gamma,$$

which, in general, will be nonzero, implying the consistency of the test T_Z which is based upon the false alternative model. (Inconsistency will occur if the first order sample autocovariance of the residuals w_t^* tends to zero.)

The possibility that the test based upon the wrong alternative may be consistent is also present when the roles of (3c) and (4c) are reversed. Thus, in contrast to the case in which homoskedasticity is the only basic assumption to be violated, a variable addition test using actual economic variables can be consistent when the disturbances are autocorrelated. Suppose that the data are generated by (1c) with

$$u_t = \rho u_{t-1} + v_t, \ |\rho| < 1, \ v_t \ \text{NID}(0, \sigma_v^2),$$

i.e. the disturbances are first order autoregressive [AR(1)], so that the test based upon (4c) is asymptotically optimal. The function f of (8) has, in this case, typical element

$$f_t = \beta x_t + \rho(y_{t-1} - \beta x_{t-1}),$$

so that the consistency of T_W can be examined by considering the pseudo-regression corresponding to (9), i.e.

$$\beta x_t + \rho(y_{t-1} - \beta x_{t-1}) = x_t b + w_t c + \text{residual}.$$

Given that x_t is nonstochastic, T_W will be consistent if w_t is able to explain a nonvanishing part of the variation of the term $(y_{t-1} - \beta x_{t-1}) = u_{t-1}$ as $n \rightarrow \infty$. If w_t is strictly exogenous, this will not be possible. (More generally, when X and W are strictly exogenous, the variable addition test T_W of (1) will be inconsistent under autocorrelated disturbances.) However, if w_t of the example is y_{t-1}, then the variable addition test T_W will be consistent because plim $n^{-1}\Sigma_t y_{t-1}u_{t-1} \neq 0$.

The variable addition test T_W can also be consistent under AR(1) errors when W is strictly exogenous, provided that the original regressor matrix X contains lagged values of y. The consistency of T_W under AR(1) errors requires that

$$\text{plim } n^{-1}W'P(X)u_{-1} \neq 0,$$

in which the subscript -1 denotes the one-period lagged value. If W includes only strictly exogenous variables, this condition is equivalent to

$$\text{plim } n^{-1}W'X(X'X)^{-1}X'u_{-1} \neq 0,$$

and this requirement may be met when X contains lagged values of the dependent variable because some elements of plim $n^{-1}X'u_{-1}$ will be nonzero.

The results on behaviour under fixed alternatives so far obtained can be summarised as follows:

(i) Consistency cannot be guaranteed when a test is derived for the wrong alternative.

(ii) If W and Z in (3) and (4) include only genuine economic variables and the data are generated by (3) (resp. (4)), then T_Z (resp. T_W) will be inconsistent if T_W and T_Z are asymptotically independent under the null model.

(iii) Variable addition tests of the type referred to in (ii) are inconsistent when only the assumption of homoskedasticity is violated. In contrast, the heteroskedasticity test T_H can be consistent under models such as (3) and (4), whether or not it is asymptotically independent of T_W and T_Z under the null model.

(iv) If X, W and Z are strictly exogenous, then T_W and T_Z are inconsistent when only the assumption of serial independence is violated, but relaxing the assumption of strict exogeneity to allow for lagged dependent variables in the regressors of the null or expanded models introduces the possibility that variable addition tests can be consistent. Also autocorrelation tests can be consistent when relevant regressors have been omitted in specifying the null model.

The practical importance of the property of consistency should not be overemphasised. It is not the case that a consistent test will always have high finite sample power. For example, if the data are generated by (1c) with AR(1) errors

$$u_t = \rho u_{t-1} + v_t, \ |\rho| < 1, \ v_t \ \text{NID}(0, \sigma_v^2),$$

then autocorrelation estimates derived from OLS residuals will be consistent so that if $r_j = \Sigma_{j+1}^n \hat{u}_{t-j}\hat{u}_t / \Sigma_1^n \hat{u}_t^2$, then plim $r_j = \rho^j$, $j = 1, 2, 3, \ldots$. Consequently, r_4 is $O_p(1)$, having probability limit equal to ρ^4. Under the null hypothesis, $\rho = 0$ and nr_4^2 is asymptotically distributed as $\chi^2(1)$. It follows that, under AR(1) errors, a large sample test in which nr_4^2 is compared to a finite critical value of the $\chi^2(1)$ distribution will be consistent. However, Monte Carlo estimates provided by Godfrey and Tremayne indicate that, in this situation, the finite sample power of the consistent test $T_2 = nr_4^2$ can be very low, being approximately equal to the nominal significance level; see Chapter 1, Section 5.

The method of approximate slopes

In examining whether or not tests are consistent, it has been convenient to consider the probability limits of test statistics divided by the sample size. The test statistics discussed in this chapter are all asymptotically distributed as χ^2 under the null hypothesis, so that such probability limits equal the approximate slopes of the corresponding test procedures; see Geweke (1981, Theorem 1). The concept of the approximate slope was proposed by Bahadur (1960, 1967) and has been used to compare the asymptotic power properties of tests; see Geweke (1981) and Magee (1987) for econometric examples of the application of Bahadur's method.

Approximate slope analysis provides an approach to overcoming the problem of comparing the powers of large sample tests when, under a fixed alternative and with a given significance level, these tests are all consistent. Useful comparisons require that powers be bounded away from unity and one way in which this can be achieved is to reduce the significance level as $n \rightarrow \infty$; see Kendall and Stuart (1973, p. 191). Bahadur's approximate slope measure is obtained by considering how the asymptotic significance level must be decreased to hold power constant under a fixed alternative. For a test statistic T that is asymptotically distributed as χ^2 under the null hypothesis, the approximate slope, denoted hereafter by AS(T), equals plim $n^{-1}T$; see Geweke (1981) and Pesaran (1984) for a discussion of tests for non-nested models. If two test procedures test the same parametric restrictions, then, in the absence of other information, the one with the larger approximate slope is presumably to be preferred.

In order to illustrate the calculation of approximate slopes, consider the Wald-variants of the variable addition tests of (1) based upon (3) and (4) when (3) is the actual data generation process. After taking probability limits under (3), a little manipulation yields the following expressions for the approximate slopes of Wald_W and Wald_Z,

$$\begin{aligned} AS(\text{Wald}_W) &= \text{plim } n^{-1}\text{Wald}_W \\ &= \sigma_u^{-2} \text{ plim } n^{-1}[\gamma' W' P(X) W\gamma] \qquad (11) \end{aligned}$$

and

$$\mathrm{AS}(\mathrm{Wald}_Z) = \operatorname{plim} n^{-1}\mathrm{Wald}_Z$$

$$= (\sigma_u^2 + \operatorname{plim} n^{-1}[\gamma' W' P(X, Z) W\gamma])^{-1} \times$$

$$\operatorname{plim} n^{-1}[\gamma' W' P(X) Z (Z' P(X) Z)^{-1} Z' P(X) W\gamma]. \quad (12)$$

Now the first term on the right-hand side of (11), viz. σ_u^{-2}, is greater than the corresponding term in (12) because $\gamma' W' P(X, Z) W\gamma$ is the residual sum of squares from the OLS regression of $W\gamma$ on (X, Z) and so must be positive. Similarly the second term on the right-hand side of (11) is greater than that of (12); the difference between these terms being equal to the probability limit of the residual variance for the OLS regression of $W\gamma$ on (X, Z). It follows that the approximate slope of the variable addition test derived from the correct fixed alternative is greater than that of the test based upon the incorrect alternative, i.e. $\mathrm{AS}(\mathrm{Wald}_W) > \mathrm{AS}(\mathrm{Wald}_Z)$ under (3).

As in the above example, the approximate slope method is often simple to use and Magee (1987) has argued that it can provide tractable analytical expressions that are useful in understanding and designing simulation experiments. Despite these advantages, the use of approximate slopes as predictors of relative finite sample power is not without problems; there being no analytical relation between power and approximate slope. If a set of parametric restrictions is tested by two statistics that share the same asymptotic χ^2 distribution under the null, then, given a common nominal significance level, a direct comparison of approximate slopes may be useful. However, as stressed by Geweke (1981), it should not be concluded that the test with the greater approximate slope is to be preferred without first examining the comparative suitability for the two tests of the asymptotic distribution theory under the null hypothesis.

When two statistics check the same null model, but are designed for different alternatives, then additional problems arise because the approximate slope method takes no account of the number of restrictions being tested. Using extra test variables cannot reduce the approximate slope of a variable addition test (or reduce the number of alternatives against which the test is consistent), but can lead to smaller finite sample power against the process that actually generated the data. For example, in Chapter 1, Godfrey and Tremayne examine the power of tests of serial independence under the AR(1) process

$$u_t = \rho u_{t-1} + v_t, \ |\rho| < 1, \ t = 1, \ldots, n,$$

by means of the following statistics: $T_1 = nr_1^2$; $T_2 = nr_4^2$; $T_3 = n(r_1^2 + r_4^2)$; and $T_4 = n(r_1^2 + \ldots + r_6^2)$, in which r_j denotes the autocorrelation estimate $[\Sigma_{j+1}^n \hat{u}_t \hat{u}_{t-j}]/[\Sigma_1^n \hat{u}_t^2]$, $j = 1, \ldots, 6$. Since the disturbances are AR(1), plim $r_j^2 = \rho^{2j}$, $j = 1, \ldots, 6$; so that the approximate slopes are given by: $AS(T_1) = \rho^2$; $AS(T_2) = \rho^8$; $AS(T_3) = \rho^2 + \rho^8$; and $AS(T_4) = \rho^2 + \ldots + \rho^{12}$. It is obvious that $AS(T_1) < AS(T_3) < AS(T_4)$, but the Monte Carlo results presented for Case (ii) in Table 3 of Chapter 1 give exactly the opposite ranking by finite sample power estimates (estimated significance levels being quite similar). Indeed the rejection frequencies of T_4, the statistic with the largest approximate slope, are only about half the corresponding values obtained for the statistic T_1. One explanation for the conflict in rankings is that the approximate slope measure does not involve a penalty for the inclusion of irrelevant test variables. An approach in which the number of restrictions being tested is taken into account, therefore, merits consideration. One such approach is based upon the idea of local alternatives and this method for comparing large sample tests will now be considered.

2.3 Asymptotic behaviour under local alternatives

Local alternative hypotheses that converge to the null provide a second approach for comparing large sample tests that are consistent. Rather than following Bahadur's method by allowing test statistics to be $O_p(n)$ and then comparing them to critical values that also increase with the sample size, the alternative hypothesis is made to grow closer to the null model in such a way that the test statistics remain $O_p(1)$. Having restricted the test statistics to be $O_p(1)$, the use of a fixed finite critical value will yield asymptotic powers that are less than unity and can be compared usefully.

The sequences of local alternatives that are used to limit the stochastic asymptotic order of magnitude of the test criteria are not intended to represent actual economic processes, but are instead proposed as tools for using asymptotic theory to gain information about power characteristics. Suppose that the alternative model of (8) has a (k + g + 1)-dimensional parameter vector $\theta' = (\beta', \tau', \sigma_v^2)$, and that the regression function f is such that if $\tau = \tau^o$, a g-dimensional vector of constants, then $f(\beta, \tau^o) = X\beta$. Thus imposing the g restrictions of H_τ: $\tau = \tau^o$ on the alternative model yields the null specification.

Given the standard assumptions on the orders of magnitudes of the sampling variances of estimators, conventional χ^2 criteria for testing H_τ (and hence (1) against (8)) will be $O_p(1)$ under the sequence $H_{\tau n}$: $\tau = \tau^o + n^{-1/2}\eta$, $\eta'\eta < \infty$. (Strictly speaking, it would be more appropriate to use the notation τ_n, rather than τ, in $H_{\tau n}$ to reflect the dependence on the sample size.) Clearly, $H_{\tau n}$ represents a sequence of vectors that approach τ^o as $n \longrightarrow \infty$ implying that the associated alternative models grow closer (local) to the null.

The use of local, as opposed to fixed, alternatives allows several useful results to be derived. These results are obtained by considering linearisations of the regression function $f(\beta, \tau)$. For example, a Taylor series expansion about $\tau = \tau^o$ yields

$$f(\beta, \tau) = X\beta + [\partial f(\beta, \tau^o)/\partial\tau][\tau - \tau^o] + f_R,$$

in which the equality $f(\beta, \tau^o) = X\beta$ is used and f_R denotes a vector-valued remainder term. Now, in the local alternative approach, attention is confined to departures from the null that are $O(n^{-1/2})$. The term f_R in the Taylor series expansion can, therefore, be ignored since the elements of this vector involve second and higher order terms in $(\tau - \tau^o) = O(n^{-1/2})$ under $H_{\tau n}$. Thus, for local asymptotic theory, $f(\beta, \tau)$ of (8) may be replaced by

$$X\beta + [\partial f(\beta, \tau^o)/\partial\tau][\tau - \tau^o]. \qquad (13)$$

Godfrey (1981) points out that there may be several different forms of $f(\beta, \tau)$ that share the same linearisation (13) and calls such forms "locally equivalent alternatives" (LEAs) with respect to H_τ; see Godfrey (1981, Chp. 3) for a detailed discussion of LEAs. If the LM principle is used to test (1) against (8), then the same LM statistic is appropriate for all members of the class of LEAs. If a test is based upon either the LR or the Wald approach, so that the alternative model must be estimated, then the test statistic will differ from the common LM statistic by asymptotically negligible terms for all combinations of test approach and LEA. Thus, in asymptotic local power analysis, there is no gain associated with knowing which LEA generated the data; see Bera and McKenzie (1986) and Godfrey (1981) for simulation experiments on the finite sample implications of this result. Perhaps the best known econometric example of this result is the fact that, when testing the assumption of serial independence, the choice between the gth-order autoregression and gth-order moving average alternatives is asymptotically irrelevant; see Breusch (1978) and Godfrey (1978b).

Equation (13) can be modified to obtain a form of LEA that permits the construction of many variable addition tests. More precisely, following linearisation of f about $(\beta = \hat{\beta}, \tau = \tau^o)$, the approximation

$$X\beta + [\partial f(\hat{\beta}, \tau^o)/\partial\tau][\tau - \tau^o], \qquad (14)$$

can be used, which implies an LEA that can be written as

$$y = X\beta + F(n^{-1/2}\eta) + u, \qquad (15)$$

in which F denotes the matrix of estimated first order partial derivatives $[\partial f(\hat{\beta}, \tau^o)/\partial\tau]$. The adequacy of the null model (1) can then be assessed by testing the joint significance of the artificial regressors of F.

Suppose first that the test model (3) represents a correct choice of the family of LEAs so that W = F and p = g. The restrictions $\gamma = 0$ can be tested by any of the criteria given in (5)-(7). Since, in this case, the coefficient vector of W is $\gamma = n^{-1/2}\eta$, which is only $O(n^{-1/2})$,

$$\text{plim } [y'P(X)y/y'P(X, W)y] = 1$$

under $H_{\tau n}$, and the differences between LM_W, LR_W and $Wald_W$ are asymptotically negligible. Further it is well known that their common asymptotic distribution under $H_{\tau n}$ is noncentral $\chi^2(p, \nu_W^2)$, where the noncentrality parameter ν_W^2 is equal to

$$\nu_F^2 = \text{plim } n^{-1}\eta'F'P(X)F\eta/\sigma_u^2, \qquad (16)$$

evaluated with F = W.

Next consider the case mentioned in the discussion of approximate slopes in which irrelevant test variables are used. (The specific example used was one in which the correct alternative was an AR(1) scheme, but tests were derived for more complex AR alternatives.) Let $W = [W_1, W_2]$, where $W_1 = F$ and W_2 is an n by p_2 matrix of irrelevant variables, so that the selected test model is

$$y = X\beta + W_1\gamma_1 + W_2\gamma_2 + u,$$

and, under $H_{\tau n}$, $\gamma_1 = \tau - \tau^o$ is $O(n^{-1/2})$ and γ_2 is a p_2-dimensional vector of zeros. The statistic for testing $\gamma = 0$ is asymptotically distributed as noncentral $\chi^2(p, \nu_W^2)$. The variables of $W_1 = F$ pick up the term $W_1\gamma_1 = F(\tau - \tau^o)$ that has been omitted from (1) (relative to the LEA) and the variables of W_2 are irrelevant. The noncentrality parameter ν^2_W, therefore, equals ν^2_F of (16), but the presence of irrelevant terms means that $p > g$. It follows from the results of Das Gupta and Perlman (1974) that adding irrelevant test variables to the correct set will cause a loss of asymptotic local power.

There are other mistakes that can be made in selecting test variables for variable addition tests of (1). Suppose that T_Z is based upon an incorrect choice of variables and does not include all relevant test variables. The results obtained by Godfrey (1988, Section 4.2.1) imply that when (15) is an LEA for the true process, T_Z is asymptotically distributed as noncentral $\chi^2(q, \nu^2_Z)$ under $H_{\tau n}$ with

$$\nu^2_Z = \nu^2_F - \sigma_u^{-2}\Delta, \qquad (17)$$

where Δ is the probability limit of the residual sum of squares from the OLS regression of $F(\tau - \tau^o)$ on (X, Z). As in a previous example, suppose that (3) is an LEA for the true data process with $F = W$ and $g = p$, so that ν^2_W equals ν^2_F of (16). Equation (17) then implies that ν^2_Z is less than ν^2_W, but it does not follow that T_Z always has smaller asymptotic local power than the correct test, T_W. If q is less than p, it is possible for the incorrect procedure to have the greater asymptotic local power. This sort of result has been discussed in detail by Holly (1982) in the context of tests of the type proposed by Hausman (1978). Table 1 illustrates the loss of asymptotic local power associated with the inclusion of irrelevant test variables, and also the relative magnitudes of ν^2_W and ν^2_Z that lead to equal asymptotic local power when $p > q$.

The correct test statistic T_W is distributed as $\chi^2(p, \nu^2_W)$, $p = 1, \ldots, 6$, and ν^2_W is selected to obtain power of 75 per cent with a significance level of 5 per cent. The incorrect test statistic T_Z is distributed as $\chi^2(q, \nu^2_Z)$ and is also compared to a 5 per cent critical value. All figures in Table 1 are expressed as percentages. Entries with $p > q$, i.e. in the upper triangle of Table 1, show the ratios of noncentrality parameters, ν^2_Z/ν^2_W, that imply equality of powers. For example, if the correct test T_W requires five test variables, but the inappropriate test T_Z uses only one variable, then v_Z^2 must be 60 per cent of v_W^2 to equalise powers. Entries with p < q, i.e. in the lower triangle of Table 1, are for cases with $v_W^2 = v_Z^2$ and show the ratio of powers when irrelevant variables are included. For example, if T_W uses only one variable and T_Z employs six variables, the power of the latter test will only be about 46 per cent of that of the former. This example is consistent with the Monte Carlo results of Table 3 in Chapter 1 concerning the autocorrelation tests T_1 and T_4 mentioned in the discussion of approximate slopes above.

Table 1

Ratios of powers and noncentrality parameters of T_W and T_Z

q \ p	1	2	3	4	5	6
1	*	81	71	65	60	56
2	65	*	88	80	74	70
3	59	69	*	91	84	80
4	54	64	70	*	93	87
5	50	60	66	71	*	94
6	46	57	63	68	72	*

Note: all figures are given as percentages

Davidson and MacKinnon (1985a, 1987) have provided interesting discussions of the impact of the choice of test variables on asymptotic local power, and give two definitions that correspond to the extreme values of v_Z^2, the noncentrality parameter of the variable addition test based upon an incorrect

LEA. A noncentrality parameter must be nonnegative and so (17) implies that $0 \leq \nu_Z^2 \leq \nu_F^2$, where ν_F^2 is given by (16). Consider the largest possible value of ν_Z^2, viz. ν_F^2. Data processes with local models such that $\nu_Z^2 = \nu_F^2$ are defined by Davidson and MacKinnon (1987) to form the implicit alternative hypothesis of T_Z. Godfrey's (1981) discussion makes it clear that, in the absence of irrelevant variables in Z, the implicit alternative hypothesis is just the set of models having (4) as an LEA with respect to (1). If $\nu_Z^2 = \nu_F^2$ and Z includes irrelevant terms, then, as shown in Table 1, T_Z will have lower asymptotic local power than the correct test. (If Z does not include all relevant variables, $\nu_Z^2 < \nu_F^2$ since Δ of (17) will be positive.)

Consider next the smallest possible value of ν_Z^2, viz. zero. If $\nu_Z^2 = 0$, then the distribution of T_Z is central $\chi^2(q)$ under both the null model and the appropriate sequence of local alternatives. Consequently, when $\nu_Z^2 = 0$, the asymptotic local power of T_Z equals its nominal significance level. Davidson and MacKinnon (1987) define the implicit null hypothesis of T_Z to be the set of processes with local models that yield $\nu_Z^2 = 0$. As with the case of inconsistency under fixed alternatives, it is possible to link the characteristics of models in the implicit null hypothesis to the property of asymptotic independence under the null model.

Suppose that the true data process is such that (3) serves as LEA with regressor matrix defined by (14), so that F = W and g = p. Let T_W and T_Z be asymptotically independent under the null. This asymptotic independence also holds under $H_{\tau n}$ and the additive property of independent noncentral χ^2 variates implies that $(T_W + T_Z)$ is asymptotically distributed as $\chi^2(p + q, \nu_W^2 + \nu_Z^2)$. Under the sequence of local alternatives $H_{\tau n}$, no test - including the one based upon the sum of T_W and T_Z - can have a noncentrality parameter greater than ν_F^2 which is equal to ν_W^2 in this case, so that ν_Z^2 must equal zero. Hence if T_W and T_Z are asymptotically independent under the null model, then any data process for which (3) is an LEA is part of the implicit null hypothesis of T_Z.

As noted in Section 2.1 above, asymptotic independence of T_W and T_Z also implies that, provided there are no redundant variables in (X, W, Z), a valid large sample joint test can be obtained by obtaining the sum $(T_W + T_Z)$ which is asymptotically distributed as $\chi^2(p + q)$ under H_τ. Under the general LEA of (15), the sum of the two individual test statistics is asymptotically distributed as $\chi^2(p + q, \nu_W^2 + \nu_Z^2)$, with $\nu_W^2 + \nu_Z^2 \leq \nu_F^2$ of (16). Such joint tests are of interest when no attempt is to be made to employ separate tests constructively. When T_W and T_Z are not independent, a joint test T_{WZ} can be obtained by testing (1) against

$$y = X\beta + W\gamma + Z\delta + u,$$

and the large sample distribution of T_{WZ} under $H_{\tau n}$ is $\chi^2(p + q, \nu_{WZ}^2)$, with

$$\max(\nu_W^2, \nu_Z^2) \leq \nu_{WZ}^2 \leq \nu_W^2 + \nu_Z^2,$$

provided there are no linear dependencies in (X, W, Z).

Comparison of the power of T_{WZ} with that of either T_W or T_Z will be complicated by the fact that, while T_{WZ} cannot have a smaller noncentrality parameter than either separate test, the joint test has more degrees of freedom than the individual tests. The joint test T_{WZ} could also be compared to the induced test based upon T_W and T_Z in which the null model is only accepted if both separate tests are insignificant. In this case, there is the further complication that the asymptotic significance level of the induced procedure will generally be unknown, although it will be between $\max(\varepsilon_W, \varepsilon_Z)$ and $(\varepsilon_W + \varepsilon_Z)$, where, as before, ε_i denotes the nominal size of the test T_i, i = W, Z.

Heteroskedasticity and variable addition tests

Attention has so far been confined to variable addition tests. There, therefore, remains the task of considering the check for heteroskedasticity of the disturbances. (Tests for autocorrelation can be implemented using variable addition and do not require separate treatment.) Analysis of behaviour under fixed alternatives revealed that variable addition tests, e.g. for omitted variables, are inconsistent under heteroskedasticity, but heteroskedasticity tests can be consistent when there are omitted variables. This asymmetry does not occur when asymptotic behaviour under local alternatives is examined.

If the regression function of (1) is correct, but the disturbances of this model are locally heteroskedastic with

$$Var(u_t) = h(\alpha_0 + \Sigma_j \alpha_j m_{tj}),$$

in which α_j is $O(n^{-1/2})$ for $j = 1, \dots, r$, then the asymptotically negligible heteroskedasticity will not affect either the asymptotic properties of the OLS estimators of γ and δ in (3) and (4), respectively, or the large sample validity of the standard OLS expressions for the estimated variance-covariance matrices of those estimators. It follows that variable addition tests such as T_W and T_Z will be asymptotically distributed as central χ^2 variates under local heteroskedasticity and will have asymptotic local power equal to nominal size.

Consider next the behaviour of Koenker's heteroskedasticity check T_H when an LEA to the true data process is of the form (15) with homoskedastic disturbances. In this situation, it can be shown that, under $H_{\tau n}$, T_H is asymptotically distributed as central $\chi^2(r)$ - a result that is not suggested by the results concerning the possible consistency of T_H under fixed, unconsidered alternatives. The lack of sensitivity of T_H under local misspecification of the regression function can be illustrated using the simple example of Section 2.2 in which data are generated by (3c),

$$y_t = \beta x_t + \gamma w_t + u_t, \ u_t \ \text{iid}(0 \ , \ \sigma_u^2),$$

but OLS is used to estimate the null model (1c),

$$y_t = \beta x_t + u_t,$$

and to obtain residuals $\hat{u}_t = y_t - \hat{\beta}x_t$, $t = 1, \ldots, n$. If the assumption that the disturbances of (1c) are homoskedastic is checked by the LM test involving the OLS regression of $\hat{u}_t^2$ on m_t and an intercept, then the slope estimator to be tested for significance is $\Sigma_t \hat{u}_t^2(m_t - \bar{m})/\Sigma_t (m_t - \bar{m})^2$. Equation (10) implies that this estimator can be written as

$$\Sigma_t \tilde{u}_t^2(m_t - \bar{m})/\Sigma_t (m_t - \bar{m})^2 +$$

$$2\tilde{\gamma}\Sigma_t \tilde{u}_t[w_t^*(m_t - \bar{m})]/\Sigma_t (m_t - \bar{m})^2 +$$

$$\tilde{\gamma}^2\Sigma_t [(w_t^*)^2(m_t - \bar{m})]/\Sigma_t (m_t - \bar{m})^2. \tag{10a}$$

The first term of (10a) is the slope estimator for the LM test algorithm using OLS residuals from the correct model (3c). Since the disturbances of (3c) are homoskedastic, this term is $O_p(n^{-1/2})$. Next, in considering the stochastic orders of magnitudes of the other two terms of (10a), it is useful to note that $(\tilde{\gamma} - \gamma)$ is $O_p(n^{-1/2})$ and γ is $O(n^{-1/2})$ under local alternatives, so that $\tilde{\gamma}$ is $O_p(n^{-1/2})$ under such alternatives. The second term of (10a) is then $O_p(n^{-1})$, as is the third term ($\tilde{\gamma}^2$ being $O_p(n^{-1})$). Consequently the LM statistic T_H for the underspecified model is asymptotically equivalent to the corresponding statistic derived for the correct model, under local alternatives. Hence the asymptotic local power of T_H equals its nominal significance level.

3. The constructive use of misspecification tests

The fact that several separate misspecification tests are usually provided by an estimation program presumably reflects a belief that such separate tests can be used to identify the errors made when specifying the original model. Otherwise a joint test covering all the individual specification errors could be reported and would have the advantage of a known asymptotic significance level. The justification for belief in the constructive value of misspecification tests will be examined using the theoretical results given in Section 2 above.

However, before employing theoretical analysis to assess the constructive properties of misspecification tests, it will be useful to make some preliminary remarks about the information available to a researcher faced with the problem of testing a specified empirical model. Some writers, e.g. Ramsey (1983), have argued that such testing will often have to be based upon very little information and that considerable uncertainty will be associated with the selection of precise forms of alternative models. It, therefore, seems reasonable to acknowledge that mistakes may be made in choosing misspecification tests and to consider the impact of such mistakes on the constructive value of test statistics. In order to discuss these mistakes, it will be convenient to introduce some additional notation.

Suppose that m separate test statistics T_i, i = 1, ..., m, are being employed. A typical test statistic T_i is appropriate for testing the null model against any member of a particular family of LEAs and this family of alternative models will be denoted by $\mathcal{F}_i$, i = 1, ..., m. The fact that a statistic T_i belongs to a family of alternatives is most clearly seen when the statistic is calculated using an LM-type approach because the numerical value of T_i need not vary over members of $\mathcal{F}_i$. But, under the null and local alternatives, this lack of uniqueness also applies to Wald and LR criteria in so far as variations of test statistics over members of $\mathcal{F}_i$ are asymptotically negligible.

The families $\mathcal{F}_i$, i = 1, ..., m, can be used to derive a set of super-models. Selecting one member from each family $\mathcal{F}_i$, i = 1, ..., m, yields a collection of m different extensions of the null model, e.g. specified forms of both autocorrelation and heteroskedasticity of the disturbances of (1). A typical super-model $\mathcal{S}_j$ can then be obtained by extending the null model to allow simultaneously for the presence of all of these m extensions. As selections from the families of LEAs $\mathcal{F}_i$ vary, so do the precise characteristics of the derived super-model, but any super-model $\mathscr{S}_j$ will include, as a special case, a model from the implicit alternative hypothesis of every test T_i, i = 1, ..., m. The relationships between the true data process, the families of LEAs, and the collection of super-models define the various situations that are the outcomes associated with the choice of test statistics. These situations can be described as follows.

Case (i)

The true data process is not included in any of the possible super-models $\mathscr{S}_j$ that are associated with the test statistics T_i. A case of this type will arise if either one or more of the specification errors made is not considered, or insufficient terms are used to characterise an alternative.

Case (ii)

The true data process is a special case of a super-model $\mathscr{S}_j$, but this super-model represents an overparameterisation. In other words, the test statistics include checks for types of errors that have not been made and/or too many terms have been used to characterise the errors that have been made. As in conventional regression analysis, "general-to-simple" searches seem preferable to "simple-to-general" approaches, and so simplification searches will be of interest. If irrelevant terms can be detected and removed, the remaining problems are as for the next case.

Case (iii)

The true process is one of the super-models associated with the statistics T_i, i = 1, ..., m. Although there is no overparameterisation, even this case poses considerable difficulties because the correct combination of LEAs must be identified if reformulation and re-estimation is required.

It may be useful to illustrate the ideas set out above by considering a simple example.

An example

Suppose that the null model is

$$y_t = \alpha + \beta ln(x_t) + u_t, \ u_t \ \text{iid}(0, \sigma_u^2),$$

and that the adequacy of this specification is to be tested using m = 2 test statistics. The first of these is a test of $\gamma = 0$ in the expanded model

$$y_t = \alpha + \beta ln(x_t) + \gamma[ln(x_t)]^2 + u_t,$$

and the second is a test of $\delta = 0$ in

$$y_t = \alpha + \beta ln(x_t) + \delta\hat{u}_{t-1} + u_t,$$

i.e. it is a check of the significance of the lagged value of the residual $\hat{u}_t = y_t - \hat{\alpha} - \hat{\beta}ln(x_t)$ from the OLS estimation of the null model.

The corresponding families of LEAs, denoted by $\mathcal{F}_1$ and $\mathcal{F}_2$, can be investigated using the methods of Godfrey (1988, Chp. 3). Godfrey shows that $\mathcal{F}_1$ includes the Box-Cox regression model

$$y_t = \pi + \beta x_t(\lambda) + u_t,$$

in which

$$\begin{aligned} x_t(\lambda) &= (x_t^{\lambda} - 1)/\lambda, \ \lambda \neq 0, \\ &= ln(x_t), \ \lambda = 0, \end{aligned}$$

and, of course, the regression model in which $E(y_t)$ is a quadratic function of $ln(x_t)$. The other family of LEAs, viz. $\mathcal{F}_2$, consists of autocorrelation models for the disturbances u_t and contains the first order autoregressive and moving average (MA) processes,

$$u_t = \rho u_{t-1} + v_t \qquad \text{[AR(1) errors]}$$

and

$$u_t = v_t + \rho v_{t-1}; \qquad \text{[MA(1) errors]}$$

see Godfrey (1981).

Given two members from each of the two families $\mathcal{F}_i$, $i = 1, 2$, it is possible to construct four super-models. These super-models are quite different specifications of the data process. For example, combining the Box-Cox functional form alternative with the MA(1) error model yields a super-model

$$y_t = \pi + \beta x_t(\lambda) + u_t,$$

with

$$u_t = v_t + \rho v_{t-1},$$

which may be contrasted with a second super-model consisting of

$$y_t = \alpha + \beta ln(x_t) + \gamma[ln(x_t)]^2 + u_t,$$

with

$$u_t = \rho u_{t-1} + v_t.$$

If the true data process is the latter super-model, then the use of T_1 and T_2 (but no other statistic) corresponds to case (iii). If, however, the actual process is the special case of this super-model derived by setting $\rho = 0$, then T_2 is irrelevant and there is overparameterisation, i.e. an example of case (ii). Finally, if the data are generated by

$$y_t = \alpha + \beta x_t + \gamma x_t^2 + u_t, \; u_t \text{ iid}(0, \sigma_u^2),$$

then the two tests being employed do not lead to a super-model containing this process and an example of case (i) is obtained.

Having illustrated some of the basic ideas, the constructive use of test statistics in each of the three cases will now be examined.

3.1 Asymptotic results and case (i)

If the true data generation process is not contained in the set of super-models derived from the test statistics T_i, $i = 1, \dots, m$, then consideration of these statistics cannot lead to correct reformulation. The results of Section 2 suggest that if the test statistics T_i that have been used are all asymptotically independent of the correct statistics when the null model is valid, then the poor specification of the null model may only be detected with low probability. If, on the other hand, the statistics T_i and the correct statistics are asymptotically dependent under the null, then false signals about misspecifications may be given. Also the results obtained for fixed alternatives in Section 2 suggest that a test for heteroskedasticity may produce a significant outcome as a result of incorrect functional form or omitted regressors when the disturbances of the correct model are actually homoskedastic.

3.2 Asymptotic results and case (ii)

There are two sources of irrelevant terms in the test statistics T_i, $i = 1, \dots, m$. First, it may be that the m different types of error being entertained are all present, but one (or more) of the test statistics is based upon an overparameterised form of its specific alternative, e.g. the assumption of serial independence is tested against an AR(6) alternative when the u_t are actually generated by some stable AR(4) scheme. Second, it is possible that only m*, m* < m, different types of error have been made, e.g. checks have been carried out for both heteroskedasticity and autocorrelation, but only the former is present in the actual data process.

Now, for any given type of specification error, the test statistic provided by an estimation program is almost always given as a joint test of the validity of the parametric restrictions that yield the null model from the corresponding extension. Thus, for variable addition tests such as T_W and T_Z, only the χ^2 statistics (or the corresponding F-statistics) relevant to testing $\gamma = 0$ and $\delta = 0$ in models such as (3) and (4) are provided, rather than a full set of estimation results for the expanded models, including individual t-statistics. Consequently current practice implies that there is little opportunity to eliminate the first source of overparameterisation.

It follows that the misspecification tests currently available cannot be useful constructive tools unless they can solve some of the problems associated with the second source of overparameterisation. In order to study the situation in which irrelevant types of error are entertained, suppose that the number of misspecifications actually present is m*, m* < m, and, without loss of generality, assume that the data process is a super-model derived from the first m* alternatives. Under the null hypothesis of correct specification, every statistic T_i is asymptotically distributed as χ^2 and so is $O_p(1)$. As a first step towards examining the scope for detecting the (m* - m) irrelevant types of misspecification, it will be useful to consider the asymptotic stochastic orders of magnitude of the test statistics under fixed alternatives.

It seems reasonable to assume that, except for relatively rare combinations of parameter values, test statistics that are $O_p(n)$ under the misspecification for which they were designed will still be $O_p(n)$ when other misspecifications are also present in the null model. If the presence of other misspecifications caused some T_i, $i \leq m^*$, to be $O_p(1)$, then the probability of correctly detecting the presence of the ith type of error would not tend to one as $n \longrightarrow \infty$, thus indicating the possibility of underspecification in a reformulated model, even asymptotically. The behaviour of the (m - m*) statistics corresponding to the irrelevant alternatives must now be considered. If, for some $i > m^*$, T_i is $O_p(n)$ under the data process, then the probability of correctly accepting the assumption that the ith misspecification is absent tends to zero as $n \longrightarrow \infty$, thus indicating the possibility of overspecification of a reformulated model, even asymptotically. The effects of overspecification are not as serious as those of underspecification, since the former leads to inefficient estimation and testing, while the latter leads to inconsistent estimation and invalid testing.

These general remarks about the orders of magnitude of separate test statistics can be illustrated by considering the statistics T_W and T_Z which test (1) against (3) and (4), respectively. The statistic T_W will be $O_p(n)$ when the OLS estimator of γ in (3), viz. $\tilde{\gamma}$, has a vector with some nonzero elements as its probability limit, i.e. $\tilde{\gamma}$ is $O_p(1)$. Under (3), plim $\tilde{\gamma} = \gamma \neq 0$ and so T_W is $O_p(n)$, i.e. T_W is consistent against its "specific" alternative. Suppose, however, that the data are generated by the comprehensive model

$$y = X\beta + W\gamma + Z\delta + u.$$

The probability limit of $\tilde{\gamma}$ from (3) is

$$\gamma_\delta = \gamma + \text{plim } (W'P(X)W)^{-1}(W'P(X)Z)\delta, \tag{18}$$

under the comprehensive model, which is, in general, a vector with nonzero elements. Clearly the test statistic T_W will be $O_p(n)$ in the presence of unconsidered misspecification, provided that γ_δ is not a vector of zeros.

Equation (18) indicates that if W and Z represent genuine economic variables with plim $n^{-1}W'P(X)Z$ being a matrix of zeros, then, not only will T_W and T_Z be asymptotically independent under the null, but also T_W will be consistent when both W and Z have been omitted, since $\gamma_\delta = \gamma \neq 0$.

Suppose next that the actual data process is (4), so that the test variables of W are irrelevant. It is clear that the probability limit of $\tilde{\gamma}$ equals the second term on the right-hand side of (18). In general, this second term contains nonzero elements and so T_W will usually be $O_p(n)$, even though W should not appear in the specified regressor set. In the case in which plim $n^{-1}W'P(X)Z = 0$, there is asymptotic independence of T_W and T_Z, and T_W is $o_p(n)$ since $\gamma_\delta = 0$, i.e. the test derived from T_W is inconsistent.

Asymptotic independence of variable addition tests is, however, a rather special property and matrices such as W and Z are selected because they correspond to interesting alternative models, and not because they have the property that plim $n^{-1}W'P(X)Z = 0$. Consequently asymptotic theory for variable addition tests under fixed alternatives suggests that the opportunities for isolating the errors actually present is quite limited.

It remains to consider the behaviour of the heteroskedasticity test T_H under fixed alternatives. When the specification of (1) is correct, except that

$$Var(u_t) = h(\alpha_0 + \Sigma_j m_{tj}\alpha_j),$$

with $\alpha_j = O(1)$, $j = 1, \ldots, r$, this LM test is consistent with T_H being $O_p(n)$. The statistic T_H can, however, be $O_p(n)$ under other types of heteroskedasticity, or other types of specification error, e.g. omitted variables. (The latter possibility was illustrated by a simple example in Section 2.2.) In contrast, variable addition tests are inconsistent when only the assumption of homoskedasticity is violated. Consequently theoretical considerations suggest that if the only difference between the actual process and the null model is that the former has heteroskedastic disturbances, then this specification error will, in general, be isolated and identified with high probability in large samples. It is difficult to be so optimistic about identifying the precise form of heteroskedasticity. Incorrect choice of the form of the function h(.), or failure to include all relevant variables m_{tj} will lead to inconsistent estimators of asymptotic standard errors. The problem of obtaining asymptotically valid inference can be solved by calculating White (1980)-type standard errors for the elements of $\hat{\beta}$. The results of Chesher and Austin (1987) suggest that it may be useful to examine the leverage values associated with X before obtaining White's heteroskedasticity-consistent covariance matrix estimator.

The value of asymptotic results concerning behaviour under fixed alternatives is, of course, not to be overstated. it is difficult to believe that, for sample sizes of the magnitude typically available in applied econometrics, tests are so powerful against their specific alternatives that rejection occurs on almost every occasion. Asymptotic local theory may, therefore, provide a more useful guide to finite sample performance.

Consider then the usefulness of this theory in identifying overparameterisations of the first type. The remarks made above concerning the limitations imposed by current ways of calculating and reporting test statistics still apply. Without more detailed results, it will not be possible to determine that, for example, the order of an autocorrelation alternative that has been employed is too high, or that its parameterisation should allow for seasonal gaps. One way in which more detail could be provided is to give full estimation results - point estimates, asymptotic standard errors, etc. - for the following LEA to a super-model:

$$y = X\beta + F_1(\tau_1 - \tau_1^o) + \dots + F_m(\tau_m - \tau_m^o) + u, \qquad (19)$$

in which F_i is $[\partial f_i(\hat{\beta}, \tau_i^o)/\partial\tau_i]$, $f_i(\beta, \tau_i)$ is the regression function for the ith alternative, and $f_i(\beta, \tau_i^o) = X\beta$, $i = 1, \dots, m$. The LEA of (19) is a generalisation of (15) to allow for the simultaneous consideration of several types of departure from the null.

Asymptotic local theory suggests that simplification searches could be conducted in the context of LEAs of the general form (19). For example, in a simple case with $m = 1$ and the only misspecification under consideration being autocorrelation, an attempt could be made to discover which restrictions cause the rejection of $\rho_1 = \dots = \rho_6 = 0$ in the expanded model

$$y_t = \beta x_t + \rho_1\hat{u}_{t-1} + \dots + \rho_6\hat{u}_{t-6} + v_t,$$

by examining individual t-statistics or χ^2-tests of the significance of subsets of the lagged OLS residuals. (Although much of his discussion is confined to the classical normal regression model, Savin (1984) provides many interesting comments on the issues associated with this kind of problem.)

The LEA (19) can also serve as a vehicle for checking the joint validity of sets of exclusion restrictions like $(\tau_i - \tau_i^o) = 0$, and can, therefore, be used against the second source of overparameterisation in case (ii), viz. the inclusion of irrelevant types of misspecification. For example, if F_i contains g_i test variables, $i = 1, ..., m$, then testing the $(g_1 + g_2)$ restrictions of $(\tau_1 - \tau_1^o) = 0$ and $(\tau_2 - \tau_2^o) = 0$ is equivalent to testing whether the first two types of alternative are required to characterise the actual data process. Clearly intercorrelations between the test variables in (19) may limit the effectiveness of this approach, while asymptotic independence between tests with $\text{plim}\ n^{-1}F_i'P(X)F_j = 0$, $i \neq j$, will allow sharper inferences, *ceteris paribus*.

If all that is available to the researcher is the set of m sample values of the separate χ^2 statistics, then it may prove difficult to determine which alternatives are irrelevant. As explained in Section 2.3, it is possible that a test based upon an irrelevant alternative will have greater asymptotic local power than the correct test for some local departures from the null, provided that the incorrect procedure uses fewer test variables than the correct one. More generally, tests for irrelevant alternatives will have some asymptotic local power and significant values of test statistics may mislead researchers into developing overelaborate reformulated models. As noted above, asymptotic independence will assist the researcher since, if the incorrect and correct test statistics are asymptotically independent, the asymptotic local power of the test for the irrelevant alternative will only equal its significance level.

It is likely that the heteroskedasticity test statistic T_H will be included in the set of misspecification checks that accompany the results of estimating the null model. It is not convenient to analyse T_H by considering a super-LEA like (19), but asymptotic local theory provides potentially useful information. As illustrated by an example discussed in Sections 2.2 and 2.3, when the assumption of homoskedasticity is satisfied, the statistic T_H will continue to have an asymptotic central χ^2

distribution in the presence of unconsidered local misspecifications such as omitted variables. Similarly standard variable addition tests will not be affected asymptotically when the only local misspecification is the assumption that (1) has homoskedastic disturbances. Thus, like the asymptotic results for fixed alternatives, theory derived from local alternatives suggests that there is scope for isolating heteroskedasticity when it is the only specification error. Local theory, however, also suggests that T_H will not be sensitive to misspecifications of the regression function such as omitted variables.

3.3 Asymptotic results and case (iii)

Suppose now that the actual data process is one of the super-models associated with the set of misspecification tests under consideration. This situation corresponds to the correct selection of the types of misspecification to be entertained and a correct parameterisation of each type of alternative. Even in this case, substantial problems remain under fixed alternatives. As demonstrated by the example above in which the null specification is a simple regression model with regressor $ln(x_t)$ and disturbances that are iid(0, σ^2) variates, the super-models associated with a battery of tests can be very different, e.g. compare the Box-Cox model with MA(1) errors and the model consisting of a regression function that is quadratic in $ln(x_t)$ and an AR(1) error model.

Since the choice of the wrong super-model will, in general, lead to inconsistent estimators under fixed alternatives, standard misspecification tests are not very helpful in themselves because they cannot guide the choice between super-models $\mathcal{S}_j$. One route that could be taken would be to test the super-models against each other using tests of nonnested hypotheses or other procedures; see, for example, the literature on testing AR(1) and MA(1) error models (Burke, Godfrey and Tremayne, 1990; Godfrey and Tremayne, Section 3 in Chapter 1 of this volume; and King and McAleer, 1987.) However, since estimation programs now provide several tests, the number and complexity of the associated super-models may make such testing an onerous task. Once again it is difficult to be optimistic about the constructive value of test statistics, despite the accuracy of the information used in choosing these statistics.

Under local alternatives, the position is quite different. As far as the estimation of the parameters appearing in the null model is concerned, the choice of super-model $\mathscr{S}_j$ is irrelevant because all such super-models are LEAs with respect to the null specification. In the absence of heteroskedasticity, the artificial LEA of (19) provides an extremely convenient algorithm for computing corrected estimators of β that take account of local misspecifications. The usefulness of LEAs like (15) and (19) as sources of corrected estimators of β is examined in Section 4 in the context of linear models with autocorrelated disturbances. A more general discussion of corrected estimators in nonlinear models is provided by Orme in Chapter 6.

If heteroskedasticity is included in the set of misspecifications, one strategy is to estimate an artificial model such as (19) to adjust for other types of misspecification error and then to estimate standard errors from White's (1980) heteroskedasticity-consistent covariance matrix estimator. This strategy should be more convenient than taking account of the specific form of the heteroskedasticity because many regression programs provide White's standard errors as part of routine calculations.

3.4 Newey's selection procedure

In his discussion of specification testing, Newey (1985) mentions the possibility of using individual test statistics as reliable guides to identifying the form of misspecification present in an inadequate model. In order to illustrate his approach, Newey considers a pair of alternative hypotheses H_1 and H_2 with corresponding test statistics T_1 and T_2. It is argued that if

$$\text{plim } n^{-1}T_1 > \text{plim } n^{-1}T_2 \text{ under } H_1$$

and

$$\text{plim } n^{-1}T_2 > \text{plim } n^{-1}T_1 \text{ under } H_2,$$

then inferring the form of misspecification from which test statistic is the larger will lead to a correct decision with probability tending to one as $n \longrightarrow \infty$; see Newey (1985, pp. 1508-9). Clearly the conditions required to justify this selection procedure are that T_i has the larger approximate slope under H_i, $i = 1, 2$.

Newey's conditions for the application of the selection rule are satisfied by variable addition tests such as T_W and T_Z when expanded models such as (3) and (4) are genuine alternative specifications of the data process, rather than artificial relationships introduced to simplify the calculation of test statistics. Examination of the expressions for test statistics given in (5)-(7) indicates that Newey's conditions are equivalent to

$$\text{plim } [y'P(X, W)y/y'P(X, Z)y] < 1 \text{ under } (3)$$

and

$$\text{plim } [y'P(X, Z)y/y'P(X, W)y] < 1 \text{ under } (4).$$

If the data are generated by (3), i.e.

$$y = X\beta + W\gamma + u, \; u \; D(0, \sigma_u^2 I_n),$$

then

$$\text{plim } n^{-1}y'P(X, W)y = \sigma_u^2$$

and

$$\text{plim } n^{-1}y'P(X, Z)y = \sigma_u^2 + \text{plim } n^{-1}\gamma'W'P(X, Z)W\gamma > \sigma_u^2;$$

so that

$$\text{plim } [y'P(X, W)y/y'P(X, Z)y] < 1$$

under (3). By interchanging W and Z and replacing γ by δ, it is simple to show that

$$\text{plim } [y'P(X, Z)y/y'P(X, W)y] < 1$$

under (4). Newey's procedure may, therefore, be useful when two variable addition tests are considered and both sets of test regressors correspond to genuine economic variables.

The importance of this result for variable addition tests should, however, not be overestimated. First, it is clear that for Newey's result to be valid, one of the expanded models used to test the null model must be the correct specification. If all the alternatives being considered are misspecified, then selecting the alternative corresponding to the largest test statistic will lead to an inadequate model. Second, it may be that several misspecifications are present, so that selecting only one of the alternatives will lead to underspecification. Third, the test regressors used in many variable addition procedures are artificial terms derived from an LEA, rather than real variables: in such cases, the alternative model is not obtained by adding regressors to a linear regression equation, but is instead some nonlinear generalisation of the null model.

As with linear alternatives, the comments on the need to include the correct specification in the set of alternatives under consideration and the problems of underspecification in the presence of more than one misspecification are relevant when nonlinear generalisations of (1) that satisfy Newey's conditions are used as alternatives. Newey (1985, p. 1059) points out that his conditions will be satisfied by tests against nonlinear alternatives when departures from the null model are small. This finding is implied by the results that, under local departures, a nonlinear model can be approximated by an LEA derived by adding extra regressors to (1) and that the inequalities on test statistics are satisfied when the null and alternatives are all linear regression equations. However, as emphasised above, if departures from the null model are assumed to be local, a test statistic can only be identified with a family of LEAs and not with a particular member of this family, so that the selection procedure can only point to a class of alternative models.

Also it should be noted that, under local alternatives, the approximate slope comparisons central to Newey's approach reduce the comparisons of noncentrality parameters; see Magee (1987, p. 252) and Newey (1985, p. 1059). Consequently such a comparison does not take the degrees of freedom parameters into account. The critical value of a $\chi^2(\omega)$ distribution for a given significance level increases with the degrees of freedom parameter ω. It would, therefore, be possible for T_1 to be larger than T_2 with the former test statistic being insignificant and the latter being significant, if T_1 tested more restrictions than T_2. In such a situation, it would seem odd to deduce that the misspecification corresponding to T_1 was present and that corresponding to T_2 was absent. The idea of looking for significant outcomes, rather than numerical magnitudes, is used in the multiple comparison procedure (MCP) proposed by Bera and Jarque (1982). The MCP and its usefulness in finite samples are discussed in Chapter 4.

4. Corrected estimators and LEAs

If the data were generated by (3) with W being a matrix of genuine economic variables, then the OLS estimator $\hat{\beta}$ derived from the inadequate specification (1) would, in general, be biased and inconsistent, as would the associated estimators of the sampling variances of the elements of $\hat{\beta}$. The remedy is straightforward in this case. Estimators with good properties are obtained by using (X, W), rather than X, as the regressor matrix to produce OLS estimators denoted by $\tilde{\beta}$ and $\tilde{\gamma}$. The vector of estimator differences $(\tilde{\beta} - \hat{\beta})$ can be regarded as a vector of corrections that treats the problems caused by misspecification.

The problem of deriving corrected procedures under fixed alternatives is more difficult when the true data generation process is some nonlinear generalisation of the null model. Corrected estimates of β can be obtained quite simply, however, if attention is restricted to local misspecifications, even though we can isolate only a family of LEAs. The analysis of this section is undertaken for such local alternatives and covers only the case in which the null model is a linear regression equation; see Orme's work in Chapter 6 of this volume for more general discussions.

If there are only local departures from the null and the test statistics being employed generate a super-model that includes the actual data process (at least locally), then the algorithm used to construct misspecification tests can also be used to produce a corrected estimate of β. The basis for calculating test statistics and corrected estimators is provided by equation (19) and its regressor matrix $(X, F_1, \dots, F_m)$. This artificial relationship is just an expanded version of (1) and so will be simple to estimate because it has the form of a linear regression model. More importantly, (19) is an LEA (with respect to (1)) to the super-models $\mathscr{S}_j$ associated with T_i, $i = 1, \dots, m$, and is, therefore, locally equivalent to the true data process. It follows that OLS estimation of (19) not only permits the testing of hypotheses such as $\tau_i - \tau_i^o = 0$, but also generates a corrected estimate of β. More precisely, if the OLS estimator of β derived from (19) is denoted by β^* and the unrestricted MLE for a super-model $\mathscr{S}_j$ is denoted by $\dot{\beta}(\mathscr{S}_j)$, then β^* and $\dot{\beta}(\mathscr{S}_j)$ are asymptotically equivalent under local alternatives, i.e. $\beta^* - \dot{\beta}(\mathscr{S}_j)$ is $O_p(n^{-1})$ when $\tau_i - \tau_i^o$ is $O(n^{-1/2})$ for $i = 1, \dots, m$.

The vector of corrections $(\beta^* - \hat{\beta})$ implied by this method of estimation adjusts $\hat{\beta}$ to remove "local inconsistency"; see Kiefer and Skoog (1984) and Godfrey (1988, p. 212). For local estimator inconsistency to exist under the ith type of misspecification, the unrestricted MLE of β and τ_i should not be asymptotically independent; see Kiefer and Skoog (1984). For example, there will be no local inconsistency of estimators if the regressors of (1) are strictly exogenous and the only departure from the assumptions of the null model of (1) is that the disturbances are generated by an AR(1) scheme with a parameter that is $O(n^{-1/2})$. On the other hand, if the regressors of (1) include y_{t-1}, then the results of Cooper (1972) imply that local first order autocorrelation will lead to local inconsistency of $\hat{\beta}$. A regression model containing y_{t-1} in the regressors and with locally autocorrelated errors can be used to illustrate the general procedure for deriving corrected estimates.

Consider a case in which the null specification is

$$y_t = \beta_1 + \beta_2 x_t + \beta_3 y_{t-1} + u_t, \; u_t \; N(0, \sigma_u^2), \tag{20}$$

in which $|\beta_3| < 1$, the regressor x is strictly exogenous, and the disturbances u_t are falsely assumed to be independent. Suppose that the only misspecification check that is carried out is to test $\gamma_1 = 0$ in

$$y_t = \beta_1 + \beta_2 x_t + \beta_3 y_{t-1} + \gamma_1 \hat{u}_{t-1} + u_t, \tag{21}$$

where, as usual, $\hat{u}_{t-1}$ denotes the one-period lagged value of the residual from the OLS estimation of the null model. Let the OLS estimators of β_i from the LEA (21) be denoted by β_i^*, i = 1, 2, 3. The artificial model of (21) is locally equivalent to the alternative models consisting of the regression function of (20) and either AR(1) or MA(1) processes for the disturbances u_t. Consequently, under local departures from serial independence, the estimator β_i^* is asymptotically equivalent to the corresponding MLE derived by estimating (20) under either AR(1) or MA(1) errors, whichever of these processes actually provides the true sequence of local alternatives. This sort of result extends the work by Bera and McKenzie (1986) and Godfrey (1981) on the behaviour of test statistics to cover the properties of estimators.

If results on using models such as (19) to generate corrected estimators are to have practical value, local asymptotic theory must provide a reasonable approximation to finite sample behaviour. A pilot Monte Carlo study based upon (20) and (21) is reported below in order to provide some evidence on the quality of this approximation.

4.1 The design of the Monte Carlo experiments

The data in this pilot study are generated by the dynamic regression model (20) with the regressor values x_t and the disturbances u_t being obtained from AR(1) processes that can be written as

$$x_t = \rho_x x_{t-1} + a_t, \ |\rho_x| < 1, \ a_t \ \mathrm{NID}(0, \sigma_a^2),$$

and

$$u_t = \rho u_{t-1} + \varepsilon_t, \ |\rho| < 1, \ \varepsilon_t \ \mathrm{NID}(0, \sigma_\varepsilon^2),$$

respectively. Four sets of estimates of the coefficients β_i are obtained for each set of n observations. Alternative estimators of (20) provide three of these sets: the estimators are ordinary least squares (OLS-H_0), maximum likelihood assuming AR(1) errors (ML-AR) and maximum likelihood assuming MA(1) disturbances (ML-MA). The final set of estimates (OLS-LEA) is derived by applying OLS to the expanded model (21).

The parameter values that we use to obtain the artificial data include the following: $\beta_1 = 0.0$, $\beta_2 = 1.0$, $\beta_3 = (0.6, 0.9)$, $\rho_x = (0.75, 0.95)$ and $\sigma_a^2 = 9.0$. The remaining parameters are σ_ε^2 and ρ. Following Maddala and Rao (1973), we use the signal-noise ratio, denoted by G, to control the value of σ_ε^2. In order to throw some light on the relevance of asymptotic local theory over a range of situations, values of ρ and G are selected by trial and error in order to give estimated probabilities of rejecting the false null hypothesis H_0: $\rho = 0$ that are approximately equal to 25, 50, 75 and 95 per cent when the nominal size is 5 per cent. The four target values can be interpreted as reflecting slight, moderate, substantial and very severe misspecification. The test of H_0 is based upon the t-ratio of $\hat{u}_{t-1}$ obtained after OLS estimation of (21).

Computer routines that are required for generating the data and maximizing the estimation criteria for ML-AR and ML-MA are taken from the NAG library. Initial values u_0, x_0 and y_0 are all set equal to zero and 110 observations are obtained for each set of values of the design parameters. In order to reduce the impact of constraining the initial values, the first 49 observations are discarded, so that n = 60 in all cases. The data generation process is carried out 1000 times for each case under consideration.

4.2 Monte Carlo results

If the asymptotic theory of LEAs and corrected estimators is to be useful in finite samples, then the OLS-H_0 estimator should exhibit greater biases than the other three estimators. Moreover, the ML-AR, ML-MA and OLS-LEA procedures should produce similar results. These predictions of large sample theory are examined in the Monte Carlo study by considering, for each estimation procedure, differences between the average estimate over 1000 replications and the corresponding true value. It is convenient to express these differences as percentages. For example, if $\hat{\beta}_{2j}$ denotes the OLS-H_0 estimate of β_2 that is obtained for the jth replication, j = 1, ... , 1000, then we compute

$$pb(\hat{\beta}_2) = [(\Sigma_j \hat{\beta}_{2j})/1000 - \beta_2]/\beta_2 \times 100 \text{ per cent.}$$

Loosely speaking, pb(.) can be thought of as an estimated percentage bias. Table 2 contains values of the pb(.) measure for cases with $\beta_3 = 0.6$, $\rho_x = 0.95$ and G = 10. The average values of the R^2 statistics from the OLS estimation of the null model are quite similar, being between 0.90 and 0.94, and are given under "avg R^2" in this table. The value of the parameter of the autoregression, denoted by "AR ρ" in the first column of the table, is also provided along with the frequency with which OLS estimation of (21) leads to the rejection of the assumption of serially independent errors. This frequency is given under "rej freq" in the second column of Table 2 and is obtained by calculating the percentage of times that the OLS estimate γ_1^* from (21) is significant.

The results of Table 2 appear to be broadly consistent with what might be expected from large sample theory. Although OLS-H_0 is not markedly inferior to the other estimators for the lowest level of autocorrelation, its performance worsens as ρ increases. The effects of increasing ρ on the other three estimators are mixed. The estimators that are derived for LEAs to the true process, viz. OLS-LEA and ML-MA, are very similar for all values of ρ under consideration, as predicted by asymptotic theory. For the two lower values of ρ, 0.23 and 0.34, however, these two estimators outperform ML-AR, the MLE for the true data process, and this is not expected. When ρ = 0.45 and "rej freq" = 76 per cent, the overall quality of performance of OLS-LEA, ML-AR and ML-MA is similar, but the directions of bias vary. For the largest value of ρ = 0.6 and "rej freq" = 93 per cent, ML-AR outperforms all other estimators. Thus, in the presence of the most severe misspecification, knowledge of the correct form of autocorrelation is useful, but in other cases the LEA (21) appears to provide a useful source of easily computed corrected estimators.

Table 2

Performance of alternative estimators of β_2 and β_3

(β_1 = 0.0, β_2 = 1.0, β_3 = 0.6, ρ_x = 0.95, G = 10)

	Values of		Values of the index pb(.) for							
AR	rej	avg	OLS-H_0		OLS-LEA		ML-AR		ML-MA	
ρ	freq	R^2	β_2	β_3	β_2	β_3	β_2	β_3	β_2	β_3
0.23	25	0.90	5	4	3	-3	4	5	3	-3
0.34	51	0.91	-9	8	1	-1	3	-3	1	-1
0.45	76	0.93	-13	11	-2	1	1	-2	-2	1
0.60	93	0.94	-17	15	-6	5	0	-1	-4	4

Variations in ρ are not the only important influence on the relative performance of the estimators being compared. In particular, the coefficient of the lagged dependent variable in (20) plays a major role. Table 3 contains results for cases with values of "rej freq" and "Avg R^2" that are very similar to those of Table 2, but that have $\beta_3 = 0.9$, rather than $\beta_3 = 0.6$. Comparison of Tables 2 and 3 indicates that there is a remarkable change in the relative position of OLS-H_0 associated with the increase in β_3. On the basis of the measure pb(.), OLS-H_0 is superior to the other three estimators for all values of ρ when $\beta_3 = 0.9$. This finding may reflect the effects of increasing β_3 on the inconsistencies of the OLS estimators of the false null model (20); see the expressions provided by Maddala and Rao (1973, Appendix A). (It should not, however, be interpreted as justifying the use of standard OLS analysis in the presence of autocorrelation because effects on sampling variance estimates and tests of significance may be important.) Overall there is only mixed support for the use of LEA-based corrections in the context of estimating dynamic regression models with autocorrelation.

Table 3

Performance of alternative estimators of β_1, β_2 and β_3
($\beta_1 = 0.0$, $\beta_2 = 1.0$, $\beta_3 = 0.9$, $\rho_x = 0.95$, $G = 3$)

	Values of		Values of the index pb(.) for							
AR	rej	avg	OLS-H_0		OLS-LEA		ML-AR		ML-MA	
ρ	freq	R^2	β_2	β_3	β_2	β_3	β_2	β_3	β_2	β_3
0.21	27	0.90	9	-3	17	-6	19	-7	17	-6
0.30	52	0.91	5	-1	15	-5	17	-6	10	-4
0.41	76	0.93	1	0	12	-4	14	-6	10	-4
0.56	94	0.95	-3	2	7	-3	10	-8	5	-3

5. Robust testing for generalised regression models

In the previous sections, it has been observed that autocorrelation and heteroskedasticity can have important effects on tests of the adequacy of the regression function. Some authors have discussed the possibility of obtaining tests of the regression function that are robust to autocorrelation and/or heteroskedasticity; see, for example, Davidson and MacKinnon (1985b), and Godfrey (1984, 1987). Such tests are useful because they can be employed in sequential procedures to identify and isolate specification errors. For example, Godfrey (1987) considers the problem of discriminating between autocorrelation and misspecification of the regression function when the regressors of this function are strictly exogenous. The first step of his procedure is to conduct a test for omitted variables and incorrect functional form that is not sensitive to autocorrelation. If the robust test of this first stage produces an insignificant outcome, then a standard test of the assumption that the disturbances are independent (or have some prespecified autoregressive structure) can be used; see Godfrey (1987) for details.

The construction of a robust test can be illustrated by considering a generalised regression model

$$y = X\beta + u, \tag{22}$$

in which X is a n by k regressor matrix with rank(X) = k, β is an unknown k-dimensional parameter vector, and u is a n-dimensional random vector with E(u) = 0 and $E(uu') = \Omega \neq \sigma_u^2 I_n$. Suppose that the adequacy of the regression function of (22) is to be assessed by testing the p restrictions of $\gamma = 0$ in the augmented model

$$\begin{aligned} y &= X\beta + W\gamma + u \\ &= S\theta + u, \end{aligned} \tag{23}$$

in which W is a n by p matrix of test variables, S = (X, W) with rank(S) = k + p, and $\theta' = (\beta', \gamma')$. Let the OLS estimator for (23) be denoted by θ^*, with its subvectors being denoted by β^* and γ^*. The OLS residual vector for (23) will be written as $u^* = y - S\theta^*$. It is assumed that, under $\gamma = 0$, the variables of (23) satisfy the appropriate specializations of the regularity conditions of Newey and West (1987).

Let R be the p by (k + p) selection matrix such that $R\theta = \gamma$. If the elements of Ω were known, then the parametric restrictions $\gamma = 0$ could be tested using the criterion

$$(R\theta^*)'[R(S'S)^{-1}(S'\Omega S)(S'S)^{-1}R']^{-1}R\theta^*$$

which is asymptotically distributed as $\chi^2(p)$ under the null hypothesis. Unfortunately the covariance matrix Ω will rarely be known (even up to a constant of proportionality) and it will usually be necessary to estimate it. If an estimator Ω^* has the property that

$$\text{plim } (S'\Omega^*S)^{-1}(S'\Omega S) = I,$$

under $\gamma = 0$, then the feasible test statistic

$$(R\theta^*)'[R(S'S)^{-1}(S'\Omega^*S)(S'S)^{-1}R']^{-1}R\theta^*$$

can be computed and compared to an upper critical value of the $\chi^2(p)$ distribution. The choice of estimator Ω^* has received some attention in the literature.

White and Domowitz (1984) suggest that the OLS residuals u_t^*, $t = 1, \ldots, n$, be used to calculate a covariance matrix estimator with elements defined by

$$\omega_{ts}^* = u_t^* u_s^* \text{ if } |t - s| \leq \tau, \text{ and } \omega_{ts}^* = 0 \text{ if } |t - s| > \tau,$$

where τ is some specified truncation parameter. While this estimator of Ω is simple to compute, it has an undesirable property. Newey and West (1987) draw attention to the fact that the White-Domowitz procedure is not guaranteed to yield positive semi-definite estimates. This property is troublesome in the context of testing $\gamma = 0$ because it implies that the derived estimate of the covariance matrix of $R\theta^* = \gamma^*$ may not be positive definite, so there is the possibility of obtaining negative sample values of a statistic that is asymptotically distributed as χ^2. Fortunately this problem is not difficult to overcome, and Newey and West propose a modification that guarantees a semi-definite estimate of Ω.

The construction of a robust statistic for testing $\gamma = 0$ is simplified and can be based on a regression-based approach if cross-section data are being used and heteroskedasticity is the only potential obstacle to valid large sample inference. The Frisch-Waugh result implies that the OLS estimator γ^* can be obtained from the OLS regression of $P(X)y = \hat{u}$ on $P(X)W = \ddot{W}$, say. A heteroskedasticity-robust test of the significance of γ^* is required and it can be constructed by applying the data transformations suggested by Messer and White (1984) to $\hat{u}$ and $\ddot{W}$ to generate the regressand and regressors for an artificial regression. More precisely, the transformed dependent variable has typical observation equal to $\hat{u}_t/\hat{u}_t = 1$ and the corresponding term for the transformed regressors is $\hat{u}_t\ddot{w}_{ti}$, where $\ddot{w}_{ti}$ is a typical element of $\ddot{W}$, $t = 1, \ldots, n$ and $i = 1, \ldots, p$. The product of the sample size n and the uncentred R^2 from the regression of 1 on $\hat{u}_t\ddot{w}_{ti}$ is a suitable criterion for testing $\gamma = 0$. Under the null hypothesis, it is asymptotically distributed as $\chi^2(p)$ with significantly large values taken to indicate misspecification. This regression-based approach to deriving an asymptotically valid heteroskedasticity-robust test is a special case of the general algorithm proposed by Wooldridge (1990).

As always, it is important to assess the practical value of asymptotically valid procedures and it has been found that the above autocorrelation and heteroskedasticity consistent covariance matrix estimators can perform badly in some finite sample situations; see Andrews (1990). An improved estimator has been proposed by Andrews and Monahan (1990). A comparison of the finite sample effects of alternative estimators of Ω on autocorrelation and hetroskedasticity robust tests of regression functions is an interesting topic for future research. It would also be useful to obtain evidence on the finite sample behaviour of the robust regression-based tests proposed in recent work by Wooldridge (1990).

6. Conclusions

In the past ten years, the reporting of regression results has undergone a significant change. As well as the usual collection of point estimates, standard errors, etc., estimation programs now provide the sample values of quite a large number of separate tests for different types of misspecification. These tests were derived by different authors at different times, but this does not imply that they should be presented as a collection of separate tests, rather than be combined to form an appropriate joint test. If misspecification tests are assumed to have only destructive value, i.e. they can be used only for detecting inadequate specifications, it is difficult to see the justification for the current practice. There is the obvious drawback that the overall significance level of the induced test of the hypothesis of correct specification based upon the set of individual tests is generally unknown, even asymptotically. Also there is no reason to suppose that this induced test will usually be more powerful than the appropriate joint test.

The use of a collection of individual tests would, of course, be justified if their outcomes provided reliable guidance in isolating and identifying misspecification errors. The purpose of this chapter has been to provide an analysis of the potential constructive value of individual test procedures in discovering the causes of model inadequacy and in respecifying models that have been found to be inconsistent with the data.

It has been argued that, in order to have constructive value, misspecification tests should be sensitive to the error for which they were derived and insensitive to other violations of the assumptions underpinning the null model. Any test is obtained under a set of assumptions and it seems sensible to make tests robust to "unconsidered alternatives" whenever this is convenient; see the recent work by Wooldridge (1990). In particular, researchers frequently have to rely upon large sample theory and many forms of nonnormality are asymptotically irrelevant, provided appeal can be made to a suitable Central Limit theorem. Misspecification tests that require normality even for their asymptotic validity should, therefore, be avoided in order to make inference more robust. This requirement is not very restrictive. The heteroskedasticity check proposed by Breusch and Pagan (1979) can be replaced by Koenker's (1981) Studentised form, and, as discussed by Burke, Godfrey and McAleer in Chapter 5, some prediction error tests can be modified. It is also possible to extend the range of robust procedures by constructing tests of the adequacy of the regression function that are robust to autocorrelation and heteroskedasticity; see Section 5.

The examination of the constructive value of test statistics has been carried out in two stages. In the first stage, the large sample properties of tests under the null specification, fixed alternatives, and sequences of local alternatives have been considered in order to derive a body of results upon which the second part of the analysis can be based. This second stage uses these results in the context of a three-fold division of the situations that can be encountered in empirical work.

In the first type of situation, either the range of alternatives being considered does not include all the errors actually made, or some alternatives are underparameterised. Inspection of the sample values of the test statistics cannot lead to an adequate specification. Moreover, in such a situation, the test statistics may produce misleading signals either by indicating the presence of misspecifications that are absent, or by failing to detect some of the errors that are present in the original model. The properties of tests that are likely to produce one or other of these misleading signals have been discussed.

The second type of situation involves overparameterisation of relevant alternatives and/or consideration of irrelevant alternatives. As illustrated in Section 2, overparameterisation of relevant alternatives reduces power. Also the separate tests for irrelevant alternatives may be sensitive to the misspecifications that are present unless conditions that ensure asymptotic independence under the false null are also satisfied under the actual data process. Overall the results of asymptotic theory suggest that the opportunity for isolating those errors that have been made is quite limited, except in the special case in which heteroskedasticity is the only specification error.

In the third and final group of situations, it is assumed that the test statistics that are used reflect a correct choice of the types of alternatives to be entertained and a correct parameterisation of each of these alternatives. Since, in practice, there will often be considerable uncertainty about the nature of specification errors, this type of situation is unlikely to be common and is in any case not without problems. It has been emphasised that many test statistics can be regarded as being appropriate for a family of (locally equivalent) alternative models. Hence a collection of individual test statistics will point not to a single super-model incorporating various different extensions of the null model, but instead to a group of such super-models. If asymptotic local theory gives a good approximation to finite sample behaviour, then this problem will not be serious from the point of view of estimating the parameters characterising the null model because all these super-models are locally equivalent. (This equivalence cannot, however, be expected to hold when the departures from the null are not small.) The results of a pilot Monte Carlo study involving the estimation of linear dynamic regression models with autocorrelated errors suggest, however, that the usefulness of asymptotic local theory is sometimes limited. A more general set of Monte Carlo experiments is discussed in the next chapter. These experiments provide evidence on the finite sample relevance of several of the theoretical results contained in this chapter and are used in a reappraisal of the "multiple comparison procedure" described by Bera and Jarque (1982).

References

Andrews, D.W.K. (1990). Heteroskedasticity and autocorrelation consistent covariance matrix estimation, *Econometrica*, 59, 817-58.

Andrews, D.W.K. and J.C. Monahan (1990). An improved heteroskedasticity and autocorrelation consistent covariance matrix estimator, Cowles Foundation Discussion Paper No. 942, Yale University.

Bahadur, R.R. (1960). Stochastic comparison of tests, *Annals of Mathematical Statistics*, 31, 276-95.

Bahadur, R.R. (1967). Rates of convergence of estimates and test statistics, *Annals of Mathematical Statistics*, 38, 303-24.

Bera, A.K. and C.M. Jarque (1982). Model specification tests: a simultaneous approach, *Journal of Econometrics*, 20, 59-82.

Bera, A.K. and C.R. McKenzie (1986). Alternative forms and properties of the Score test, *Journal of Applied Statistics*, 13, 13-25.

Bera, A.K. and C.R. McKenzie (1987). Additivity and separability of the Lagrange multiplier, likelihood ratio and Wald tests, *Journal of Quantitative Economics*, 3, 53-63.

Breusch, T.S. (1978). Testing for autocorrelation in dynamic linear models, *Australian Economic Papers*, 17, 334-55.

Breusch, T.S. and L.G. Godfrey (1981). A review of recent work on testing for autocorrelation in dynamic simultaneous equation models, in *Macroeconomic Analysis*, ed. D. Currie, A.R. Nobay and D. Peel, 63-105. London: Croom Helm.

Breusch, T.S. and L.G. Godfrey (1986). Data transformation tests, *Economic Journal*, 96, 47-58.

Breusch, T.S. and A.R. Pagan (1979). A simple test for heteroskedasticity and random coefficient variation, *Econometrica*, 47, 1287-94.

Burke, S.P., L.G. Godfrey and A.R. Tremayne (1990). Testing AR(1) against MA(1) disturbances in the linear regression model: an alternative procedure, *Review of Economic Studies*, 57, 135-45.

Carr, J. (1972). A suggestion for the treatment of serial correlation: a case in point, *Canadian Journal of Economics*,

5, 301-6.

Chesher, A.D. and G. Austin (1987). Finite sample behaviour of heteroskedasticity robust Wald statistics, unpublished paper, University of Bristol (DP87/187).

Chow, G.C. (1960). Tests of equality between sets of coefficients in two linear regressions, *Econometrica*, 28, 591-605.

Consigliere, I. (1981). The Chow test with serially correlated errors, *Rivista Internatazionale di Scienze Sociali*, 89, 125-37.

Cooper, J.P. (1972). Asymptotic covariance matrix of procedures for linear regression in the presence of first order serially correlated disturbances, *Econometrica*, 40, 305-10.

Das Gupta, S. and M.D. Perlman (1974). Power of the non-central F-test: effect of additional variates on Hotelling's T-test, *Journal of the American Statistical Association*, 69, 174-80.

Davidson, R. and J.G. MacKinnon (1985a). The interpretation of test statistics, *Canadian Journal of Economics*, 18, 38-57.

Davidson, R. and J.G. MacKinnon (1985b). Heteroskedasticity-robust tests in regression directions, *Annales de l'INSEE*, 59, 183-218.

Davidson, R. and J.G. MacKinnon (1987). Implicit alternatives and the local power of test statistics, *Econometrica*, 55, 1305-29.

Geweke, J. (1981). The approximate slopes of econometric tests, *Econometrica*, 49, 1427-42.

Godfrey, L.G. (1978a). Testing for multiplicative heteroskedasticity, *Journal of Econometrics*, 8, 227-36.

Godfrey, L.G. (1978b). Testing against general autoregressive and moving average error models when the regressors include lagged dependent variables, *Econometrica*, 46, 1293-1302.

Godfrey, L.G. (1981). On the invariance of the Lagrange multiplier test with respect to certain changes in the alternative hypothesis, *Econometrica*, 49, 1443-55.

Godfrey, L.G. (1984). On the use of misspecification checks and tests of nonnested hypotheses in empirical econometrics, *Economic Journal*, Supplement, 96, 69-81.

Godfrey, L.G. (1987). Discriminating between autocorrelation and misspecification in regression analysis: an alternative test strategy, *Review of Economics and Statistics*, 69, 128-134.

Godfrey, L.G. (1988). *Misspecification tests in econometrics: the Lagrange multiplier principle and other approaches*. Cambridge: Cambridge University Press.

Hausman, J. (1978). Specification tests in econometrics, *Econometrica*, 46, 1251-71.

Hendry, D.F. (1979). Predictive failure and econometric modelling in macroeconomics: the transactions demand for money, in *Economic Modelling*, ed. P. Omerod, 217-42, London: Heinemann Educational Books.

Hendry, D.F. (1989). *PC-GIVE: an interactive econometric modelling system*. Oxford: Institute of Economics and Statistics.

Holly, A. (1982). A remark on Hausman's specification test, *Econometrica*, 50, 749-59.

Jarque, C.M. and A.K. Bera (1980). Efficient tests for normality, homoscedasticity and serial independence of regression residuals, *Economics Letters*, 6, 255-9.

Jarque, C.M. and A.K. Bera (1987). An efficient large sample test for normality of observations and regression residuals, *International Statistical Review*, 55, 163-72

Kendall, M.G. and A. Stuart (1973). *The Advanced Theory of Statistics, Vol.II*. London: Griffin.

Kiefer, N.M. and G.R. Skoog (1984). Local asymptotic specification error analysis, *Econometrica*, 52, 873-85.

King, M.L. and M. McAleer (1987). Further results on testing AR(1) against MA(1) disturbances in the linear regression model, *Review of Economic Studies*, 54, 649-63.

Koenker, R. (1981). A note on studentizing a test for heteroskedasticity, *Journal of Econometrics*, 17, 107-12.

Maddala, G.S. and A.S. Rao (1973). Tests for serial correlation in regression models with lagged dependent variables and serially correlated errors, *Econometrica*, 41, 761-74.

Magee, L. (1987). Approximating the approximate slopes of LR, W, and LM test statistics, *Econometric Theory*, 3, 247-71.

Messer, K. and H. White (1984). A note on computing the

heteroscedasticity consistent covariance matrix using IV techniques, *Oxford Bulletin of Economics and Statistics*, 46, 181-4.

Newey, W.K. (1985). Maximum likelihood specification testing and conditional moment tests, *Econometrica*, 53, 1047-70.

Newey, W.K. and K.D. West (1987). A simple, positive semi-definite, heteroskedasticity and autocorrelation consistent covariance matrix, *Econometrica*, 55, 703-8.

Pagan, A.R. (1984). Model evaluation by variable addition, in *Econometrics and Quantitative Economics*, ed. D.F. Hendry and K.F. Wallis, 103-33. Oxford: Blackwell.

Pagan, A.R. and A.D. Hall (1983). Diagnostic tests as residual analysis, *Econometric Reviews*, 2, 159-218.

Pesaran, M.H. (1984). Asymptotic power comparisons of tests of separate parametric families by Bahadur's approach, *Biometrika*, 71, 245-52.

Pesaran, M.H. and B. Pesaran (1987). *Data-fit: an interactive econometrics software package*. Oxford: Oxford University Press.

Plosser, C.I., G.W. Schwert and H. White (1982). Differencing as a test of specification, *International Economic Review*, 23, 535-52.

Ramsey, J.B. (1969). Tests for specification errors in classical least squares regression analysis, *Journal of the Royal Statistical Society*, Series B, 31, 350-71.

Ramsey, J.B. (1983). Comments on diagnostic checks as residual analysis, *Econometric Reviews*, 2, 241-8.

Sargan, J.D. (1988). *Lectures on Advanced Econometric Theory,* (edited by M. Desai). Oxford: Basil Blackwell.

Savin, N.E. (1984). Multiple hypothesis testing, in *Handbook of Econometrics*, Vol.II, ed. by Z. Griliches and M.D. Intriligator, 827-79. Amsterdam: North Holland.

White, H. (1980). A heteroskedasticity-consistent covariance matrix and a direct test for heteroskedasticity, *Econometrica*, 48, 421-48.

White, H. and I. Domowitz (1984). Nonlinear regression with

dependent observations, *Econometrica*, 52, 143-61.

White, K. (1978). A general computer program for econometric methods - SHAZAM, *Econometrica*, 46, 239-40.

Wooldridge, J.M. (1990). A unified approach to robust, regression-based specification tests, *Econometric Theory*, 6, 17-43.

ALISON EASTWOOD AND L. G. GODFREY

4. FINITE SAMPLE PROPERTIES OF MISSPECIFICATION TESTS AND THE MULTIPLE COMPARISON PROCEDURE

1. Introduction

In the previous chapter, a general discussion of the asymptotic properties of tests was supplemented by several simple analytical examples. The purpose of this chapter is, in part, to complement this work by presenting simulation evidence on the usefulness of results derived from asymptotic theory. Monte Carlo methods are used to explore the finite sample relevance of asymptotic theory in the context of the identification and isolation of specification errors. As in the previous chapter, the model being tested is assumed to be a multiple regression equation. The Monte Carlo results discussed below are also used in a reappraisal of the Multiple Comparison Procedure (MCP) recommended by Bera and Jarque (1982). This MCP consists of inspecting the sample values of a group of separate tests and then extending the original specification in directions that correspond to the significant outcomes. The important study by Bera and Jarque (1982) is discussed in Section 2 and features of its Monte Carlo design are linked to the predictions of asymptotic theory given in Chapter 3.

Section 3 contains a description of the various Monte Carlo experiments that are employed in our own work. These experiments are suggested by the analysis of Chapter 3 and are used in a reappraisal of the practical value of the MCP, and also to illustrate some problems that may be encountered in applied econometrics. The results obtained in the experiments are summarised and discussed in Section 4. These results suggest that it would be unwise to be optimistic about the ability of the conventional tests to reveal the precise nature of the inadequacies of a model. Finally Section 5 contains some concluding remarks.

2. The Multiple Comparison Procedure

Bera and Jarque (1982), hereafter BJ, provide an interesting analysis of the use of an MCP. In order to simplify comparisons with their work, we shall adopt the notation used by BJ with only a few minor changes. The null model is written as

$$y_t = \sum_{j=1}^{k} x_{tj}\beta_j + \sum_{j=1}^{m} d_{tj}\mu_j + u_t, \; u_t \; \text{NID}(0, \sigma_u^2), \qquad (1)$$

$t = 1, \dots, n$, where: y_t denotes an observation on the dependent variable; the regressors with typical observations x_{tj} and d_{tj} are strictly exogenous, with one of the d_{tj}, say d_{t1}, being equal to unity for all t; and NID denotes "normally and independently distributed". BJ consider a case in which the adequacy of (1) is assessed by means of individual tests against the following misspecifications: a combination of incorrect functional form and omitted variables; autocorrelation; heteroskedasticity; and nonnormality.

In order to develop a general framework to encompass these individual misspecifications, BJ set up the super-model

$$y_t = \sum_{j=1}^{k} x_{tj}(\lambda_j)\beta_j + \sum_{j=1}^{m} d_{tj}\mu_j + \sum_{j=1}^{\ell} w_{tj}\delta_j + u_t, \qquad (2)$$

in which $x_{tj}(\lambda_j)$ is the Box-Cox transformation of the strictly positive variable x_{tj}, i.e.

$$x_{tj}(\lambda_j) = (x_{tj}^{\lambda_j} - 1)/\lambda_j, \; \lambda_j \neq 0,$$

$$= ln(x_{tj}), \; \lambda_j = 0,$$

$j = 1, \ldots, k$; and w_{tj} denotes an observation on a strictly exogenous variable, $j = 1, \ldots, \ell$. The disturbances u_t of (2) are generated by a stable autoregressive process

$$u_t = \sum_{j=1}^{p} \gamma_j u_{t-j} + v_t, \tag{3}$$

where the errors v_t are independent variates with $E(v_t) = 0$. BJ specify that the density of v_t is a member of the Pearson family of distributions and allow for additive heteroskedasticity of the form

$$\mathrm{Var}(v_t) = \alpha_0 + \sum_{j=1}^{q} z_{tj}\alpha_j, \tag{4}$$

$t = 1, \ldots, n$, where the variables z_{tj} are nonstochastic and do not include an intercept dummy.

Equations (2)-(4) and the distributional assumption define a data generation process that is included in the set of super-models associated with the test statistics that are considered by BJ. These individual statistics are as follows.

(i) A Lagrange multiplier (LM) test of the $(k + \ell)$ restrictions of the hypothesis

$$H_{\lambda\delta}: (\lambda_1 - 1) = \ldots = (\lambda_k - 1) = \delta_1 = \ldots = \delta_\ell = 0.$$

Thus the hypothesis $H_{\lambda\delta}$ represents the assumptions that the effects on y_t of variations in the x_{tj} can be modelled using a simple linear combination, as in (1), and also that the variables w_{tj} are irrelevant.

The LM statistic relevant to testing $H_{\lambda\delta}$ will be denoted by T_{FOV} and it can be calculated by the method of variable addition using the expanded model

$$y_t = \sum_{j=1}^{k} x_{tj}\beta_j + \sum_{j=1}^{m} d_{tj}\mu_j + \sum_{j=1}^{k} x^*_{tj}\pi_j + \sum_{j=1}^{\ell} w_{tj}\delta_j + u_t, \qquad (5)$$

$t = 1, \ldots, n$, in which the disturbances are assumed to be iid$(0, \sigma^2_u)$ variates, $x^*_{tj} = x_{tj}ln(x_{tj}) - x_{tj} + 1$, $j = 1, \ldots, k$, and the intercept term has been redefined by subtracting the sum of the slopes β_j. (This unimportant redefinition of the intercept is required because $x_{tj}(\lambda_j)$ reduces to $(x_{tj} - 1)$, not x_{tj}, when $\lambda_j = 1$ is imposed.)

The standard F-test of (1) against (5) provides a check against incorrect functional form and omitted variables. If the F-statistic derived from the OLS estimation of (1) and (5) is denoted by F_{FOV}, then

$$F_{FOV} \sim F(k + \ell, n - 2k - \ell - m)$$

under $H_{\lambda\delta}$, when all regressors are strictly exogenous and the disturbances are NID$(0, \sigma^2_u)$ variates. An asymptotically valid test of $H_{\lambda\delta}$ in the presence of predetermined regressors, such as y_{t-1}, and/or nonnormal errors can be based upon comparing the LM statistic

$$T_{FOV} = (k + \ell)F_{FOV} + o_p(1),$$

to critical values of the $\chi^2(k + \ell)$ distribution. This LM statistic can be calculated as n times the R^2 statistic from the regression of the OLS residual vector from the null model (1) on the regressors of the alternative model (5). The relationship between T_{FOV} and F_{FOV} contains a term that is $o_p(1)$ under $H_{\lambda\delta}$ because these statistics use restricted and unrestricted estimators of σ^2_u, respectively. The two estimators of σ^2_u differ by terms that are asymptotically negligible when (1) is valid.

(ii) The assumption of serial independence is checked using the statistic

$$T_I = n(r_1^2 + \ldots + r_p^2),$$

where r_j denotes an estimate of the autocorrelation coefficient of order j derived from the OLS residuals from (1). Results on the finite sample behaviour of tests of this form are provided by Godfrey and Tremayne in Chapter 1.

(iii) In order to test against the single misspecification of heteroskedasticity, BJ use the Breusch-Pagan (1979) variant of the LM procedure with the scaled squared OLS residuals $\hat{u}_t^2/\hat{\sigma}_u^2$, $\hat{\sigma}_u^2 = n^{-1}\Sigma\,\hat{u}_t^2$, being regressed on an intercept and the fixed variables z_{tj}, $t = 1, \ldots, n$ and $j = 1, \ldots, q$. Provided that the regression function of (1) is specified correctly and the disturbances u_t are independent normal variates, the Breusch-Pagan criterion, denoted by T_{BP}, is asymptotically distributed as $\chi^2(q)$ under homoskedasticity.

(iv) Finally, BJ use their test statistic T_N as a check of the assumption that the disturbances are normally distributed; see Jarque and Bera (1987) for details of this statistic. The asymptotic null distribution of T_N is $\chi^2(2)$.

Since all the regressor variables that appear in the super-model (2) are strictly exogenous and the disturbances are symmetric under the null hypothesis, the test statistics T_{FOV}, T_I, T_{BP} and T_N are all asymptotically independent when (1) is the correct specification. Consequently, as explained in Chapter 3, a joint test against all the misspecifications being entertained can be obtained by adding the four separate statistics together, with

$$T_J = T_{FOV} + T_I + T_{BP} + T_N, \tag{6}$$

being asymptotically distributed as $\chi^2(k + \ell + p + q + 2)$ under the null hypothesis.

This asymptotic independence leads BJ (1982, p. 73) to favour an MCP based upon T_{FOV}, T_I, T_{BP} and T_N. The nominal significance level of this MCP is given by

$$\varepsilon_{MCP} = 1 - (1 - \varepsilon_{FOV})(1 - \varepsilon_I)(1 - \varepsilon_{BP})(1 - \varepsilon_N),$$

where, for example, ε_I is the nominal marginal significance level of the test using T_I. Reformulation of a data inconsistent null model is determined by which individual test statistics are significant. It is important to note that only one test, viz. T_{FOV}, is used to check for both incorrect functional form and omitted variables. The single statistic T_{FOV} is not capable of distinguishing between these two causes of model adequacy and this limits the usefulness of the MCP.

For example, if $\lambda_1 = ... = \lambda_k = 1$, but $\delta_j \neq 0$, some j, then there are two quite different potential difficulties. First, if the influence of the omitted variables is large, significant values of T_{FOV} may lead to the use of the Box-Cox transformation when it is unnecessary. Second, if k is large relative to ℓ, then the presence of irrelevant test variables may lead to low power and model (1) may be accepted when it is misspecified; see Table 1 of Chapter 3.

It would be possible to obtain separate tests of H_δ: $\delta_1 = ... = \delta_\ell = 0$ and H_λ: $\lambda_1 = ... = \lambda_k = 1$ by testing (1) against

$$y_t = \sum_{j=1}^{k} x_{tj}\beta_j + \sum_{j=1}^{m} d_{tj}\mu_j + \sum_{j=1}^{\ell} w_{tj}\delta_j + u_t,$$

and

$$y_t = \sum_{j=1}^{k} x_{tj}\beta_j + \sum_{j=1}^{m} d_{tj}\mu_j + \sum_{j=1}^{k} x^*_{tj}\pi_j + u_t,$$

respectively, to obtain statistics denoted by T_{OV} and T_F. In general, however, these separate statistics will be asymptotically dependent under the null hypothesis (conditions for asymptotic independence can be deduced from the results of Section 2.1 of Chapter 3). As argued in Section 3 of Chapter 3, such asymptotic dependence is likely to make it difficult to unscramble the causes of model rejection when there are problems of either incorrect functional form or omitted variables.

While the MCP advocated by BJ can only give a general signal about the adequacy of the regression function of (1), it appears, at first sight, to offer greater scope for identifying and isolating errors made in specifying the joint distribution of the disturbances u_t, $t = 1, \ldots, n$, because three separate test statistics are employed. The predictions of asymptotic theory concerning the properties of an MCP involving these statistics do not, however, fully support this initial impression.

Under fixed alternatives to $H_{\lambda\delta}$, the omission of relevant variables and the use of incorrect functional form can both lead to situations in which the true assumptions of normality, independence and homoskedasticity are rejected with probability tending to 1 as $n \to \infty$; see Chapter 3 for examples relevant to serial correlation and heteroskedasticity tests. Consequently effects arising from the misspecification of the regression function may spill over to produce false signals of inadequacy of the error model. Under local alternatives, however, with $\lambda_j = 1 + O(n^{-1/2})$, $j = 1, \ldots, k$, and $\delta_j = O(n^{-1/2})$, $j = 1, \ldots, \ell$, asymptotic theory suggests that incorrect functional form and omitted variables will not affect the asymptotic distributions of statistics such as T_{BP} and T_I. Note that the assumption of strictly exogenous variables is used in establishing the robustness of T_I. If the regressors w_{tj} included y_{t-1}, then T_I would not be asymptotically distributed as a central $\chi^2(p)$ variate under local misspecification of the regression function. It would instead have a positive noncentrality parameter and so its asymptotic power would be greater than the nominal significance level. It is not clear on *a priori* grounds which type of asymptotic theory gives the better guidance in any given situation. Some evidence can be obtained by carrying out Monte Carlo experiments with different degrees of misspecification; see Sections 3 and 4 below.

It should also be noted that problems may arise in using T_{BP},

T_I and T_N in an MCP, even when the regression function is correctly specified. Koenker (1981) points out that T_{BP} is sensitive to nonnormality. Thus the presence of nonnormality alone might produce significant values of both T_{BP} and T_N. The serial correlation check T_I, and asymptotically equivalent variable addition procedures, are, however, robust to nonnormality in large samples.

Another issue relevant to the usefulness of an MCP involving T_{BP} and T_I is the robustness of the former test to autocorrelation and of the latter test to heteroskedasticity. The asymptotic theory outlined in Chapter 3 implies that, under fixed alternatives, the test based upon T_{BP} (resp. T_I) will not be asymptotically valid in the presence of autocorrelation (resp. heteroskedasticity). In contrast, the results of asymptotic local theory suggest that the tests will be robust. Loosely speaking, these two sets of predictions suggest that the impact on a test of the error for which it was not designed will depend on the seriousness of this error. Some evidence on the magnitude of the problems associated with lack of robustness is provided by Epps and Epps (1977) who conclude that the autocorrelation tests are quite robust to moderate heteroskedasticity, but heteroskedasticity checks are sensitive to autocorrelation. It is inevitable that the generality of these conclusions is limited by the specification of the models from which they were derived; see Epps and Epps (1977, Sections 2 and 3) for descriptions of their models. A fairly large set of Monte Carlo results on this and related issues is provided by BJ. These results will be considered after the Monte Carlo design employed by BJ has been discussed and described.

2.1 Monte Carlo design

The null model of the Monte Carlo experiments carried out by BJ is

$$y_t = \sum_{j=1}^{4} d_{tj}\mu_j + u_t,\ u_t\ \text{NID}(0, \sigma_u^2), \tag{7}$$

$t = 1, \dots, n$, in which $d_{t1} = 1$, and the regressors d_{t2}, d_{t3}, and d_{t4} are generated from normal, uniform and $\chi^2(10)$ distributions, respectively. This model is a special case of (1) and the absence of variables x_{tj} from (7) implies that, instead of a joint test against incorrect functional form and omitted variables, a single "specific" test for omitted variables, T_{OV}, is employed.

Extensions of the null model are derived as follows.

(i) There is a single omitted variable that equals the square of the second regressor of (7), i.e. $\ell = 1$ and $w_{t1} = d_{t2}^2$, $t = 1, \dots, n$. The extended regression model is, therefore, of the form

$$y_t = \sum_{j=1}^{4} d_{tj}\mu_j + w_{t1}\delta_1 + u_t. \tag{8}$$

(ii) Autocorrelated disturbances are introduced using the first order autoregressive process

$$u_t = \rho u_{t-1} + v_t, \quad |\rho| < 1, \tag{9}$$

with the v_t being independently distributed with common mean equal to zero.

(iii) Heteroskedasticity is specified to be of the form

$$\mathrm{Var}(v_t) = 25 + \alpha_1 z_t, \tag{10}$$

in which z_t is generated as the square of an N(10, 25) variable.

(iv) Finally nonnormality is characterised by either the t(5) or log-normal distribution, denoted by t and Λ, respectively, with linear transformations being applied to achieve the required means and variances.

The one-directional alternatives of (8)-(10) require appropriate special cases of the general forms of the test statistics T_{OV}, T_I, and T_{BP}. For example, T_{BP} is calculated as one half of the explained sum of squares from the OLS regression of $\hat{u}_t^2/\hat{\sigma}_u^2$ on z_t and an intercept, where $\hat{u}_t = y_t - \Sigma d_{tj}\hat{\mu}_j$, $t = 1, \dots, n$, and, under correct specification, this statistic is asymptotically distributed as $\chi^2(1)$. The form of the test criterion T_N is not affected by the choice of alternative in (iv).

For comparison with the LM procedures, BJ consider the Durbin-Watson test for serial correlation, the Goldfeld-Quandt test for heteroskedasticity, White's (1980) test for heteroskedasticity, and a modification of the Shapiro-Francia (1972) test for nonnormality discussed by White and MacDonald (1980) - these statistics will be denoted by DW_I, GQ_H, W_H and SF_N, respectively. The DW_I and T_I statistics are closely related since

$$T_I = nr_1^2 \simeq n(1 - 0.5DW_I)^2.$$

The GQ_H test is calculated after reordering the data by the values of the variable z_t in (10) and omitting the central set of n/5 observations. White's (1980) procedure W_H can be regarded as a variant of Koenker's (1981) Studentised test for heteroskedasticity that uses the wrong set of test variables, namely the nonredundant terms of $d_{ti}d_{tj}$, $i, j = 1, \dots, 4$. Given this interpretation, it is to be anticipated that the power of W_H will be inferior to that of T_{BP} when the underlying error distribution is normal. The third heteroskedasticity test under consideration, viz. GQ_H, is a small sample test that, like T_{BP}, requires normality for its validity. (The asymptotic distribution of GQ_H under correct specification is degenerate since plim $GQ_H = 1$ when (7) is valid.)

Clearly it is important to examine the behaviour of test statistics under the null hypothesis before attempting either to compare "power" estimates for the LM procedures and other tests, or to assess the usefulness of the MCP. (A test may appear to be relatively powerful just because it is relatively oversized.) BJ, therefore, conduct experiments with data generated by (7). They fix the sample size at a value of n = 50 and set $\delta_j = 0$, $j = 1, \ldots, 4$, and $\sigma_u^2 = 1$. The results of Breusch (1980) imply that imposing these restrictions on the coefficients of (7) involves no loss of generality. Marginal critical values associated with an empirical significance level of 10 per cent are then estimated for the individual test statistics and for the joint test T_J with general form given by (6). These critical values are estimated using 100 replications. BJ also obtain estimates of the marginal critical values that correspond to an empirical significance level of 3 per cent for each of the individual tests T_{BP}, T_I, T_N and T_{OV}. This smaller value is of interest because the empirical significance level of the MCP is close to 10 per cent when each of its four component tests is carried out at the 3 per cent level. Consequently it is possible to make comparisons of the power estimates of the MCP and the joint test T_J that are not invalidated by substantial differences in significance levels. Such comparisons are an important part of the evaluation of the MCP because it will be unattractive if it only has low power relative to the joint test.

BJ regard the estimation of marginal critical values as an important part of their MCP and do not rely upon the use of asymptotically valid critical values in finite samples; see BJ (1982, p. 75). Their procedure can, therefore, be described as follows:

(i) generate R, e.g. 1000, replications of n observations from an N(0, 1) distribution, n being the sample size available to the researcher;

(ii) compute the values of T_{FOV}, T_I, T_{BP}, and T_N for each of the R replications;

(iii) if ε_{MCP} denotes the desired significance level for the MCP, determine the int[R(100 - ε_{MCP}/4)] largest value of each of the four individual tests and use these values as marginal critical values, where int[e] denotes the integer part of the number e;

(iv) calculate the sample values of T_{FOV}, T_I, T_{BP}, and T_N, using the actual data, and reject the assumption that a particular type of misspecification is absent if its corresponding test statistic is not less than the appropriate marginal critical value that is estimated in (iii).

The results obtained by BJ in their Monte Carlo experiments will now be considered.

2.2 Monte Carlo results

Using the estimated empirical values derived by simulating data under the null model (7), BJ obtain various sets of power estimates by specifying nonnormal distributions and/or nonzero values of δ_1, ρ and α_1 in (8),(9) and (10), respectively. Each of these estimates is calculated using 500 replications, and is reported below as a percentage.

Although the main aim here is to consider the performance of the tests used in the MCP, the power estimates for the other diagnostic checks that are reported by BJ merit some brief comments. The estimates for the modified Shapiro-Francia test for nonnormality, SF_N, are very similar to those of the LM procedure T_N discussed by Jarque and Bera (1987), whatever the data process. There are, however, greater differences between the results for the three tests for heteroskedasticity. As expected, White's (1980) general check W_H is markedly less powerful than the Breusch-Pagan test T_{BP} when heteroskedasticity, as specified in (10), is the only extension of the null model that is present in the true data generation process. In contrast, the Goldfeld-Quandt test GQ_H produces very good results. This test is even superior to the asymptotically optimal procedure T_{BP}, but both tests require fairly precise information about the nature of heteroskedasticity to achieve a good performance. BJ (1982, p. 71, fn. 9) report that the powers of GQ_H and T_{BP} are greatly reduced when an incorrect specification of the alternative model is used. This feature of the results is one indication of the value of *a priori* information when constructing misspecification tests, and others will be mentioned later.

Turning now to the four separate LM tests, the joint test T_J and the MCP, some results from Tables 1 and 2 of the article by BJ are summarised in Table 1. Table 1 includes estimates of power, in terms of percentages, for T_N, T_{BP}, T_I, T_{OV}, T_J and MCP. The four tests for separate misspecifications are all used with an empirical significance level of 3 per cent, so that the implied empirical significance level for the MCP is about 10 per cent, which is the value used for T_J. In order to illustrate the usefulness of precise and correct information, Table 1 also contains, for each case, results for the asymptotically optimal test T_* using an empirical significance level of 10 per cent.

The rows of Table 1 correspond to different data generation processes that are defined by a choice of the error distribution and the specification of the parameters of (8)-(10). The normal, t(5) and log-normal distributions are denoted by N, t and Λ, respectively, in the first column. The values of parameters that determine heteroskedasticity (α_1), autocorrelation (ρ) and the contribution of the omitted variable (δ_1) are given in the next three columns. The fifth column contains estimates for the asymptotically optimal test T_*, with the next two columns providing estimates for the joint test T_J and MCP, respectively. These three sets of power estimates are based upon an empirical significance level of 10 per cent. The last four columns of Table 1 give estimates of the powers of the separate tests that are used in the MCP, the empirical significance level for these estimates being 3 per cent. Finally estimates for separate tests are underlined if they correspond to an extension of the null model that is present in the actual data process.

Consider first the estimates for T_*, T_J and MCP. It is clear that T_* dominates T_J and MCP. The gap between the power estimates of T_* and those of T_J and MCP reflects the fact that the latter two procedures use irrelevant alternatives. In the first four rows of results, there is only one type of misspecification present in (7) and the gap is large - as much as 35 per cent. The next six

Table 1

Power estimates for test procedures of Bera-Jarque study

Design				Test procedure						
D	α_1	ρ	δ_1	T^a_*	T^a_J	MCP^a	T^b_N	T^b_{BP}	T^b_I	T^b_{OV}
t	0.00	0.0	0.00	51	43	38	$\underline{33}$	9	2	4
N	0.25	0.0	0.00	73	55	48	20	$\underline{42}$	3	1
N	0.00	0.3	0.00	58	32	43	4	2	$\underline{39}$	1
N	0.00	0.0	0.05	63	28	27	3	3	2	$\underline{22}$
t	0.25	0.0	0.00	74	68	70	$\underline{46}$	$\underline{46}$	3	3
Λ	0.00	0.3	0.00	100	100	100	$\underline{100}$	45	$\underline{32}$	4
t	0.00	0.0	0.05	65	61	61	$\underline{28}$	5	5	$\underline{42}$
N	0.85	0.3	0.00	89	83	77	29	$\underline{59}$	$\underline{34}$	1
N	0.85	0.0	0.05	76	72	72	31	$\underline{64}$	3	$\underline{2}$
N	0.00	0.7	0.10	100	99	99	6	0	$\underline{91}$	$\underline{71}$
t	0.25	0.3	0.00	82	78	80	$\underline{43}$	$\underline{41}$	$\underline{37}$	3
t	0.85	0.0	0.05	91	90	89	$\underline{61}$	$\underline{65}$	6	$\underline{26}$
Λ	0.00	0.7	0.10	100	100	100	$\underline{100}$	8	$\underline{42}$	$\underline{83}$
N	0.85	0.7	0.10	98	97	100	8	$\underline{12}$	$\underline{100}$	$\underline{5}$
t	0.25	0.3	0.10	90	90	88	$\underline{33}$	$\underline{31}$	$\underline{23}$	$\underline{58}$

Notes:
(i) [a] and [b] denote tests with empirical significance levels of 10 per cent and three per cent, respectively;
(ii) all estimates are given as percentages; and
(iii) estimates corresponding to alternatives that form part of the actual data process are underlined.

rows correspond to data processes possessing two extensions of the null model, so that, in a sense, MCP and T_J have greater relevance, and the gap does not exceed 6 per cent. The next four rows represent situations in which only one irrelevant alternative is entertained in T_J and MCP, and T_* is only slightly more powerful than these two tests. The last row has all four assumptions violated and the asymptotically optimal test T_* is the joint test T_J. These comparisons illustrate how the effectiveness of testing for misspecification can be improved if *a priori* information can be used to eliminate irrelevant extensions of the null model.

It was noted above that, for the MCP to be attractive, it should not be markedly less powerful than the joint test T_J. If the use of four separate tests greatly reduced the probability of rejecting an inadequate specification, then the price of obtaining individual signals would be high. The figures in Table 1 indicate that there is no evidence that the MCP is usually much less powerful than T_J. The results for the two procedures are often quite similar, and MCP sometimes outperforms T_J. These findings, however, do not in themselves lead to the recommendation of the MCP because they relate only to its destructive value. The MCP is proposed by BJ on the grounds that, by identifying errors, it can lead to the better specification of econometric models. It is, therefore, necessary to examine the constructive value of the MCP by considering the outcomes of the four components, viz. T_N, T_{BP}, T_I and T_{OV}.

Recall that underlined estimates correspond to tests that are relevant to the actual data process, and that, for the MCP to be useful, tests should not be sensitive to errors for which they are not designed. Consequently the use of the MCP would receive some support if nonunderlined estimates were all small. Inspection of the results in Table 1 reveals that T_I and T_{OV} both have low power when they are irrelevant. The results for T_{BP} and T_N are, however, not so encouraging. The sensitivity of T_{BP} to nonnormality has been mentioned above, and figures in the sixth row of results reveal that, in the presence of log-normal errors, the MCP leads to the rejection of the true hypothesis of homoskedasticity in 45 per cent of the replications. Also T_N, the Jarque-Bera (1987) test for nonnormality, is not always robust to heteroskedasticity (see the second row of the power estimates).

Given the simple form of Koenker's (1981) Studentised test for heteroskedasticity and the robustness of many asymptotic results concerning estimators and tests to nonnormality of the disturbances, it may be worthwhile to consider a modification of the MCP. If procedures that require normality for asymptotic validity are avoided, e.g. by using Koenker's (1981) statistic rather than T_{BP}, then T_N need not be included in the set of separate tests. However, even if this strategy is adopted, some issues still require consideration.

As BJ are careful to point out, the generality of their results may be limited by the design of their Monte Carlo experiments. For example, their conclusion that the MCP is capable of identifying the sources of model inadequacy is based upon the analysis of models with strictly exogenous regressors. The methods used by BJ are valid in the context of such models, but require modification when the regressors include lagged values of the dependent variable. The joint test T_J, for example, can no longer be calculated using (6); see Bera and McKenzie (1986, p. 59).

Even in the class of regression models with strictly exogenous regressors, there are special features of the designs used by BJ that may influence the performance of T_I and T_{OV} under autocorrelation and/or omitted variables. As noted by BJ (1982, p. 71), it is important to note that the regressors d_{tj} of (7), and hence the omitted variable $w_{t1} = d_{t2}^2$, are generated as serially independent variables. This serial independence implies that OLS inference is asymptotically robust to autocorrelation and, in particular, that the test T_{OV} will not be sensitive to nonzero values of ρ in (9). Similarly the check for autocorrelation T_I will not be sensitive to the omission of w_{t1} when this variable is relevant; see the example provided in Section 2.2 of Chapter 3.

Since the Monte Carlo design used by BJ implies the robustness of the omitted variables test to autocorrelation, the poor performance of T_{OV} in the presence of omitted variables, autocorrelation and heteroskedasticity - as indicated by the penultimate row of results in Table 1 - suggests that the power of this test can be very adversely affected by the presence of heteroskedasticity. The effects of heteroskedasticity on T_{OV} can also be observed in the results for the case in which heteroskedasticity and omitted variables are the only two violations of the assumptions of (7). The power estimate of the test T_{OV} for the case with (D = N, $\alpha_1 = 0.85$, $\rho = 0.0$, $\delta_1 = 0.05$) is only about one-tenth of its value for the case with (D = N, $\alpha_1 = 0.0$, $\rho = 0.0$, $\delta_1 = 0.05$) and is even smaller than its empirical significance level.

While the results provided by BJ serve as a valuable step towards an understanding of the usefulness of the MCP, there are several reasons for extending their analysis.

(i) As argued in the previous chapter, it is often possible to restrict attention to procedures that are asymptotically robust to nonnormality. In particular, Koenker's (1981) statistic T_H (equal to n times the R^2 statistic from the regression of the squared OLS residual on an intercept term and the variables assumed to determine variances) can be used in place of the less robust Breusch-Pagan criterion T_{BP}. If this approach is adopted, there is no need to include a check for nonnormality in the battery of tests for misspecification.

(ii) For time series applications, it is important to recognise that regressors are often serially correlated and to investigate the consequences of such serial correlation for the sensitivity of T_{OV} to autocorrelated disturbances and of T_I to omitted variables.

(iii) It is also useful to employ a Monte Carlo design that has a lagged value of the dependent variable in the regressors of the alternative specifications because the relationships between tests are sensitive to the inclusion of this variable.

(iv) Ramsey's (1983) comments on the paucity of information that is available when constructing tests for misspecification merit attention. It is, therefore, worthwhile to explore the effects on power of having "vague", rather than "precise", information.

The structure of a set of Monte Carlo experiments designed to cover these points is described in the next section, and the results obtained in these experiments are discussed in Section 4.

3. Monte Carlo experiments

The null model used in these experiments is of the form

$$y_t = \beta_1 + \beta_2 x_t + u_t, \; u_t \text{ iid}(0, \sigma_u^2), \tag{11}$$

and the adequacy of this relationship is investigated using separate checks for autocorrelation, heteroskedasticity and omitted variables. The actual data processes are special cases of the super-model

$$y_t = \beta_1 + \beta_2 x_t + \beta_3 x_t^2 + u_t, \tag{12}$$

$$u_t = \rho u_{t-1} + v_t, \; |\rho| < 1, \tag{13}$$

the variates v_t being independent with $E(v_t) = 0$ and

$$\text{Var}(v_t) = \sigma_v^2(1 + \lambda_1 x_t^2) = \alpha_0 + \alpha_1 x_t^2, \tag{14}$$

in which $\alpha_0 \equiv \sigma_v^2$ and $\alpha_1 = \lambda_1 \sigma_v^2$. The distributions used to generate the disturbances are normal, t(5), log-normal, $\chi^2(2)$ and uniform: artificial drawings from these distributions are transformed to have the required population mean and variances.

The regressor variable x_t follows the AR(1) process

$$x_t = \rho_x x_{t-1} + a_t, \; |\rho_x| < 1, \; a_t \text{ NID}(0, \sigma_a^2). \tag{15}$$

Some Monte Carlo experiments involve a variable w_{ti}, i = 1 or 2, where

$$w_{ti} = x_t^i + z_{ti}, \tag{16}$$

in which the z_{ti} are serially uncorrelated variates with $E(z_{ti}) = 0$, $E(x_t^i z_{ti}) = 0$ and $Var(z_{ti}) = Var(x_t^i)$, i = 1, 2. This specification of the terms w_{ti} implies that the squared correlation coefficient between w_{ti} and x_t^i is 0.5, i = 1, 2. The variable w_{t1} can be interpreted as the observed value of the regressor when the true value x_t is contaminated by a measurement error z_{t1}. The variable w_{t2} can be thought of as an incorrect test regressor that is correlated with the true omitted variable x_t^2 when the regression equation is given by (12).

All tests of the null model are assumed to be calculated from OLS results. Consequently, while standard procedures retain their asymptotic validity under a variety of nonnormal distributions such as those used in this study, it cannot be assumed that asymptotic optimality is achieved. Although several tests are only asymptotically valid, all are implemented in F-form by dividing the χ^2 variant by the number of restrictions under test, and then selecting a critical value from the (approximately valid) F distribution. For example, when testing the null hypothesis that $\rho = 0$, conditional upon $\beta_3 = \alpha_1 = 0$, a check of model adequacy is obtained by investigating whether or not the F-statistic for testing $\delta_1 = 0$ in

$$y_t = \beta_1 + \beta_2 x_t + \delta_1 \hat{u}_{t-1} + u_t, \tag{17}$$

exceeds the prespecified critical value of the F(1, n - 3) distribution, $\hat{u}_{t-1}$ denoting the one-period lagged value of the residual obtained by estimating the null model (11) by OLS. These F-forms are used partly because applied workers often employ the F-test when calculating variable addition tests and also because these variants sometimes have better small sample properties than the corresponding χ^2 procedures.

The tests for autocorrelation, heteroskedasticity and omitted variables that are examined in this and the next section are classified as being either "precise" or "vague". Precise tests correspond to one of the alternative models of (12)-(14). Vague tests correspond to overparameterised versions of these alternative models. Consideration of vague tests permits investigation of the consequences of overparameterisation; see the discussion of Case (ii) in Section 3 of Chapter 3 for a summary of theoretical predictions concerning these consequences. The details of the two groups of tests are as follows.

3.1 Precise tests

There are precise tests for omitted variables (OV), autocorrelation (A), and heteroskedasticity (H). Brief descriptions of the statistics for the three separate tests and various joint tests are provided in this subsection. The approximate distribution of each statistic under the assumption of correct specification is given in square brackets.

Omitted variables

The test statistic, denoted by POV, is the F-statistic for testing $\beta_3 = 0$ in

$$y_t = \beta_1 + \beta_2 x_t + \beta_3 x_t^2 + u_t, \; u_t \text{ iid}(0, \sigma_u^2).$$

[F(1, n - 3)]

Autocorrelation

The test statistic, denoted by PA, is the F-statistic for testing $\delta_1 = 0$ in

$$y_t = \beta_1 + \beta_2 x_t + \delta_1 \hat{u}_{t-1} + u_t, \; u_t \text{ iid}(0, \sigma_u^2),$$

where $\hat{u}_{t-1}$ is defined in (17).

[F(1, n - 3)]

Heteroskedasticity

The test statistic, denoted by PH, is (n - 3)/(n - 2) times the F-statistic for testing $\alpha_1 = 0$ in

$$\hat{u}_t^2 = \alpha_0 + \alpha_1 x_t^2 + \text{residual},$$

$\hat{u}_t^2$ being the OLS residual from (11). This criterion is asymptotically equivalent to Koenker's (1981) Studentised LM statistic nR_α^2, where R_α^2 is the R^2 statistic from the OLS regression of $\hat{u}_t^2$ on x_t^2 and an intercept term. This equivalence holds because

$$PH = \{(n - 3)/n\}\{nR_\alpha^2/(1 - R_\alpha^2)\},$$

and, under correct specification, R_α^2 is $O_p(n^{-1})$ so that plim $(1 - R_\alpha^2) = 1$ (and, of course, lim $\{(n - 3)/n\} = 1$). The artificial regression employed to compute R_α^2 suggests the use of the degrees of freedom correction (n - 2), rather than (n - 3). We use the latter quantity because the alternative model has three parameters to be estimated, viz. β_1, β_2 and α_1.
[F(1, n - 3)]

In addition to these individual tests, various joint tests are also employed to allow the analysis of overtesting (overspecified alternatives) and undertesting (underspecified alteratives) - discussion of these topics is also provided by BJ (1982).

Autocorrelation and omitted variables

The test statistic, denoted by PAOV, is the F-statistic for testing $\beta_3 = \delta_1 = 0$ in the expanded model

$$y_t = \beta_1 + \beta_2 x_t + \beta_3 x_t^2 + \delta_1 \hat{u}_{t-1} + u_t.$$

Since the regressor x_t is strictly exogenous, it would be asymptotically valid to employ (POV + PA)/2 as the test statistic.
[F(2, n - 4)]

Autocorrelation and heteroskedasticity

Assuming asymptotic independence of PA and PH when (11) is valid, a suitable test statistic is

PAH = (PA + PH)/2.

[F(2, n - 4)]

Heteroskedasticity and omitted variables

Assuming asymptotic independence of PH and POV when (11) is valid, a suitable test statistic is

PHOV = (PH + POV)/2.

[F(2, n - 4)]

Autocorrelation, heteroskedasticity and omitted variables

Assuming asymptotic independence of PH and PAOV when (11) is valid, a suitable statistic for testing against all three misspecifications is

PAHOV = (2PAOV + PH)/3.

An asymptotically equivalent form is (PA + POV + PH)/3.

[F(3, n - 5)]

It should be noted that all the joint tests, except PAOV, require the heteroskedasticity test statistic PH to be asymptotically independent of other test statistics under correct specification. Pagan and Hall (1983) argue that this condition is satisfied only if $E(u_t^3) = 0$, i.e. the error distribution is symmetric. As the Monte Carlo experiments include the use of the $\chi^2(2)$ and log-normal distributions, it is possible to investigate the robustness of the joint tests PAH, PHOV and PAHOV to asymmetry.

3.2 Vague tests

It is now assumed that there is some uncertainty about the specification error and that this uncertainty leads to overparameterised tests being employed.

Omitted variables

The vague test for omitted variables, denoted by VOV, is of the type proposed by Thursby and Schmidt (1977), being the F-test of the null model (11) against

$$y_t = \beta_1 + \beta_2 x_t + \gamma_1 x_t^2 + \gamma_2 x_t^3 + \gamma_3 x_t^4 + u_t.$$

Since the OLS predicted values from (11) are given by $\hat{y}_t = \hat{\beta}_1 + \hat{\beta}_2 x_t$, VOV can also be interpreted as a RESET test using the test variables $\hat{y}_t^2$, $\hat{y}_t^3$ and $\hat{y}_t^4$.
[F(3, n - 5)]

Autocorrelation

The test statistic, denoted by VA, is derived for an unrestricted AR(4) alternative and is calculated as the F-statistic for testing (11) against

$$y_t = \beta_1 + \beta_2 x_t + \delta_1 \hat{u}_{t-1} + \delta_2 \hat{u}_{t-2} + \delta_3 \hat{u}_{t-3} + \delta_4 \hat{u}_{t-4} + u_t.$$

This type of procedure is discussed by Godfrey and Tremayne in Section 5 of Chapter 1.
[F(4, n - 6)]

Heteroskedasticity

White's (1980) general check is used to obtain the statistic for detecting heteroskedasticity. This statistic, denoted by VH, is calculated as the F-statistic for testing $\alpha_1 = \alpha_2 = 0$ in the artificial regression

$$\hat{u}_t^2 = \alpha_0 + \alpha_1 x_t^2 + \alpha_2 x_t + \text{residual}.$$

[F(2, n - 4)]

As with the procedure based upon precise information, joint tests are constructed and are as follows.

Autocorrelation and omitted variables

The test statistic, denoted by VAOV, is the F-test of (11) against the expanded model

$$y_t = \beta_1 + \beta_2 x_t + \sum_{j=1}^{3} \gamma_j x_t^{j+1} + \sum_{j=1}^{4} \delta_j \hat{u}_{t-j} + u_t.$$

[F(7, n - 9)]

Autocorrelation and heteroskedasticity

The analogue of the precise joint test PAH is calculated as

VAH = (4VA + 2VH)/6.

[F(6, n - 8)]

Heteroskedasticity and omitted variables

The test statistic, VHOV, is obtained using

VHOV = (2VH + 3VOV)/5.

[F(5, n - 7)]

Autocorrelation, heteroskedasticity and omitted varibles

The statistic VAHOV, which can be regarded as a fairly general check for specification errors, is defined by

VAHOV = (3VOV + 4VA + 2VH)/9.

[F(9, n - 11)]

The above tests are used to study cases that generalise the design of BJ by allowing for a serially correlated regressor and overparameterisation of a relevant alternative. It has been argued that it is also important to extend the analysis by considering lagged dependent variables and the wrong choice of test variable in the check for omitted variables; see Chapter 3 and the remarks above. In order to evaluate the effects of having a lagged dependent variable as a test variable, the null model (11) is tested against

$$y_t = \beta_1 + \beta_2 x_t + \gamma_1 y_{t-1} + u_t, \tag{18}$$

using the standard F-test. The F-statistic relevant to testing $\gamma_1 = 0$ will be denoted YLAG. The problems associated with using an incorrect test variable are illustrated by considering the F-test of (11) against

$$y_t = \beta_1 + \beta_2 x_t + \gamma_1 w_{t2} + u_t, \tag{19}$$

in which w_{t2} is defined by (16). The F-statistic derived from OLS estimation of (11) and (19) is denoted by OVW.

3.3 Details of experiments for estimating significance levels

Estimates of significance levels are obtained by generating data from (11) with some specified error distribution, so that the parameters β_3, ρ and α_1 in (12), (13) and (14), respectively, are all set equal to zero. The parameters β_1 and β_2 of (11) are both set equal to one, but these two restrictions involve no loss of generality because Breusch's (1980) results on invariance apply.

The AR(1) regressor process of (15) is employed with $\sigma_a^2 = 9.0$ and ρ_x equal to either 0.75 or 0.95. Given values of ρ_x and σ_a^2, the variance of the independent and homoskedastic disturbances is selected to give an asymptotic R^2 for the null model of 0.75. Routines from the NAG library are used to generate (n + 50) observations for the regressor with the starting value x_0 set equal to zero. The first 50 observations are discarded to reduce the impact of using a fixed initial value for the regressor process. A sequence of disturbances u_t, $t = 1,...,n$, is then obtained from a NAG routine corresponding to one of the five distributions under consideration. The sample size n takes on the values of 40, 60 and 80.

For each replication of n artificial observations, the sample values of the test statistics described in the previous two subsections can be calculated and compared to suitable critical values. The nominal significance level is 5 per cent in all cases and 5000 replications are used to estimate the finite sample probabilities of rejecting a true null hypothesis. Consequently, values outside the range 4.4 to 5.6 per cent can be regarded as evidence against the assumption that the actual significance level equals 5 per cent (using $1.96[(5)(95)/5000]^{1/2} \simeq 0.6$ and the standard test for a sample proportion).

3.4 Details of experiments for estimating power

The behaviour of tests under data processes that are more general than (11) is studied by employing nonzero values for one or more of the parameters β_3, ρ and α_1 in (12)-(14). The values of the other parameters (β_1, β_2, σ_v^2, ρ_x, σ_a^2) are the same as those used when estimating significance levels, as are the values of the sample size n. Note, however, that the variance of u_t need not equal σ_v^2 and may even vary from one observation to another.

Having specified the values of all the parameters of (12)-(14) and selected an error distribution, data can be simulated and power estimates can be obtained for all the tests being employed. In order to obtain an interesting range of results, a set of target powers is used to guide the choice of data generation process. The target powers are 25, 50, 75 and 90 per cent. For each value of n, four processes are found (by trial and error) that produce rejection frequencies for the asymptotically optimal test that are reasonably close to these values. The estimates of power are based upon 1000 replications.

4. Some Monte Carlo results

The purpose of this section is to provide a representative sample of the Monte Carlo results that have been obtained, and to link our findings to those reported in the study by BJ and to the discussion of large sample properties contained in Chapter 3. The discussion will be divided into two parts: the first part covers estimates of significance levels; and the second part consists of a number of studies of power.

4.1 Estimates of significance levels

For any given test procedure, the finite sample significance level may vary with the parameter ρ_x that determines the autocorrelation structure of the regressor, the error distribution and the sample size n. The Monte Carlo results, however, suggest that the first of these influences does not merit lengthy

Table 2
Estimated significance levels with n = 40 and $\rho_x = 0.95$

Test	Distribution				
	D1	D2	D3	D4	D5
PA	4.5	4.5	2.5	3.4	4.9
PH	4.3	4.3	5.5	5.2	4.8
POV	4.6	5.2	5.1	4.7	4.6
PAH	4.4	4.7	4.9[a]	5.1[a]	4.6
PAOV	4.9	5.4	5.0	4.5	5.2
PHOV	4.4	5.0	6.6[a]	5.5[a]	4.4
PAHOV	4.6	5.7	6.4[a]	5.9[a]	4.7
VA	5.3	4.3	2.7	3.5	5.2
VH	4.8	5.4	7.2	6.5	4.7
VOV	4.5	6.1	6.2	5.8	3.8
VAH	4.7	4.8	5.2[a]	4.8[a]	4.5
VAOV	5.4	5.7	6.1	6.2	5.7
VHOV	4.8	5.9	7.8[a]	6.9[a]	4.0
VAHOV	5.2	6.4	7.6[a]	6.6[a]	5.0

Key: D1 ... Normal
D2 ... Student t(5)
D3 ... Log-normal
D4 ... $\chi^2(2)$
D5 ... Uniform

Note: [a] denotes an estimate for an asymptotically invalid test.

discussion. Changing ρ_x from 0.75 to 0.95 produces small changes with fluctuating signs; so that the effects are neither substantial nor systematic.

The effects of nonnormality are more interesting and are illustrated by the estimates of Table 2. While the estimated significance levels of most tests are not greatly affected by nonnormality, the autocorrelation tests appear to be sensitive to skewness, being undersized for the log-normal and $\chi^2(2)$ distributions. These findings are consistent with an estimate reported by BJ that indicates a low rejection frequency of their test T_I when errors are log-normal; see the second row of results in Table 1 of BJ (1982). The skewness associated with log-normal and $\chi^2(2)$ distributions does not always lead to poor results for joint tests against alternatives including heteroskedasticity, even though these tests require symmetry for their asymptotic validity. Table 2 contains results for cases with ρ_x = 0.95 and n = 40 that illustrate these general findings. All estimates in Table 2 are reported as percentages.

The figures given in Table 2 show that both autocorrelation tests are undersized for log-normal and $\chi^2(2)$ disturbances, but are fairly well behaved for the other distributions. The vague heteroskedasticity test VH is rather more sensitive to nonnormality than the precise form PH, with the former test being oversized for the skewed distributions. Both tests for omitted variables have estimated significance levels that are reasonably close to the nominal value, as do the two joint tests for autocorrelation and omitted variables (the vague forms having the larger values). The results for joint tests involving checks for heteroskedasticity are good for symmetric distributions, but, not surprisingly, are more variable in quality when the disturbances are drawn from skewed distributions - the tests being asymptotically invalid in the latter type of situation.

The final factor influencing the probability of rejecting a true null model is the sample size. When the errors come from a symmetric distribution, the results for all tests are good, even for the smallest value of n; see Table 2 for estimates with n = 40 and ρ_x = 0.95. We will, therefore, concentrate on the effects of increasing n when the disturbances are from skewed distributions. Table 3 is derived from results obtained using the log-normal distribution; the estimates for the $\chi^2(2)$ case are quite similar.

Table 3

Estimated significance levels with log-normal errors

	$\rho_x = 0.75$			$\rho_x = 0.95$		
Test	n = 40	60	80	40	60	80
PA	2.7	2.8	3.2	2.5	2.9	3.1
PH	5.2	5.6	4.8	5.5	5.7	4.8
POV	5.2	5.3	4.9	5.1	5.2	4.7
PAH	4.6[a]	5.2[a]	4.9[a]	4.9[a]	4.9[a]	4.6[a]
PAOV	5.0	4.7	4.6	5.0	4.5	4.3
PHOV	6.4[a]	6.8[a]	5.5[a]	6.6[a]	6.5[a]	5.7[a]
PAHOV	5.9[a]	6.2[a]	5.8[a]	6.4[a]	5.7[a]	5.9[a]
VA	3.0	2.8	3.3	2.7	2.8	3.6
VH	7.6	7.6	6.1	7.2	6.9	6.1
VOV	7.0	7.3	6.1	6.2	6.7	6.3
VAH	5.6[a]	5.7[a]	5.4[a]	5.2[a]	5.3[a]	5.4[a]
VAOV	5.9	6.0	5.7	6.1	6.4	6.7
VHOV	8.0[a]	8.7[a]	7.3[a]	7.8[a]	8.6[a]	7.6[a]
VAHOV	7.2[a]	7.7[a]	6.7[a]	7.6[a]	8.2[a]	7.7[a]

Note: [a] denotes that the test is asymptotically invalid

The use of the log-normal distribution implies that $E(u_t^3) \neq 0$ and, therefore, the only asymptotically valid joint tests are PAOV and VAOV. Although estimates in Table 3 corresponding to invalid joint tests should not be expected to tend to the value of 5 per cent as n increases, several of these estimates are quite close to this value. For the asymptotically valid tests, there is no clear evidence that the estimates corresponding to n = 40, 60, 80 form a sequence approaching the nominal value. Results for n = 40 are sometimes in better agreement with asymptotic theory than those for n = 80, e.g. see the estimates for PAOV. The changes in estimated significance levels associated with increasing the sample size are frequently quite small (only one exceeds 1 per cent) and are sometimes in a direction away from the nominal value. Thus we find no clear evidence that increasing the sample size leads to improved performance. However, it should be noted that, while these results are derived from fairly precise estimates, it would be unwise to extrapolate by claiming relevance to either much larger samples (e.g. n = 200), or much smaller samples (e.g. n = 10).

4.2 Estimates of power

The critical values of the approximate distributions of the test statistics under the null model (11) are used when estimating power. (This should not cause any important problems because variations of estimated significance levels about the nominal value are usually reasonably small for asymptotically valid tests; see Table 2 above. The exceptions are the low estimated significance levels of the autocorrelation tests in the cases with disturbances generated from either the log-normal or $\chi^2(2)$ distribution.) It is, therefore, more accurate to use "rejection frequency" rather than "power estimate" when discussing the Monte Carlo results. It will sometimes be convenient to use rf(.) to denote the rejection frequency of a test.

The departures from the null model that are used to investigate the finite sample properties of tests are as follows:

(a) autocorrelation;

(b) heteroskedasticity;

(c) omitted variables;

(d) a combination of autocorrelation and omitted variables; and

(e) errors in variables.

The errors in variables case is used to illustrate the behaviour of tests when the data are generated by an unconsidered alternative, i.e. it is an example of Case (i) of Section 3 of Chapter 3.

(a) Autocorrelation

The regressor of (11) is serially correlated and so, in contrast to the case studied by BJ, OLS inference will not be asymptotically valid under fixed autocorrelation alternatives with ρ of (13) being O(1). Similarly, as we follow BJ by using the squared value of a regressor in the omitted variables test (and higher powers in VOV), autocorrelation of the disturbances will invalidate checks of the adequacy of the regression function of (11) because the regressor and the test variables x_t^j are not serially independent. The effects of autocorrelated disturbances on tests for omitted variables will depend upon the pattern of autocorrelations of the regressors; see, for example, Porter and Kashyap (1984). Thus, in the context of the Monte Carlo design of Section 3, the effects of nonzero values of ρ in (13) will depend, in part, upon ρ_x in (15).

Although the omitted variables tests POV and VOV are not robust to autocorrelation when $\rho_x \neq 0$, they are both inconsistent under AR(1) disturbances because the regressor x is strictly exogenous. Consequently there may be some scope for isolating autocorrelation when it is the only departure from the null model. The situation is different if a test for omitted variables uses lagged values of the dependent variable. A combination of lagged dependent variables in the extended regressor set and AR(1) errors will cause OLS estimators to be inconsistent. Consequently a true assumption that the terms y_{t-i} are irrelevant will be rejected with probability tending to one as $n \longrightarrow \infty$. The YLAG procedure based upon (18) is introduced to provide an example of this situation.

Before including the test YLAG in power comparisons, it is,

of course, important to examine its behaviour under the null model (11). Such analysis reveals that YLAG is well behaved. For symmetric distributions, every estimate is in the range 4.4 to 5.6 per cent, i.e. every estimate is consistent with the hypothesis that the actual significance level equals the nominal value. For the skewed distributuions, there is a tendency for YLAG to be a little undersized, with an average estimated significance level of about 4 per cent. On the basis of these results, there seems to be no reason not to include YLAG in comparisons of power.

Table 4 contains a selection of the results for n = 60. Rejection frequencies are reported for both values of ρ_X because of the potential importance of the autocorrelation structure of the regressor. The disturbance distributions used in Table 4 are as follows: the normal, as a benchmark; the t(5), as an example of a symmetric nonnormal distribution; and the log-normal, as an example of a skewed distribution. The results of Table 4 are obtained using $\rho = 0.465$ and this value of the AR coefficient yields rejection frequencies of about 90 per cent for the asymptotically optimal test PA. The columns of Table 4 corresponding to the log-normal distribution, denoted by D3, only have estimates for asymptotically valid tests. (If the errors have an asymmetric distribution, then the joint tests involving a check for heteroskedasticity are not asymptotically valid.)

Topics of interest that can be addressed using the results of Table 4 include: the effects of autocorrelation on tests for irrelevant alternatives, viz. PH, POV, PHOV, VH, VOV, VHOV and YLAG; the consequences of using overelaborate alternatives either by overparameterising the autocorrelation model, or by considering irrelevant alternatives as well as autocorrelation; and the impact of the form of the error distribution on the behaviour of the tests. The first two topics correspond to cases (i) and (ii), respectively, of Section 3 in Chapter 3. (The results for PA correspond to case (iii) because the selected alternative is the actual data process.)

Table 4

Rejection frequencies under autocorrelation with n = 60

	$\rho_x = 0.75$			$\rho_x = 0.95$		
Test	D1	D2	D3	D1	D2	D3
PA	88	89	93	89	89	92
PH	5	5	7	6	7	8
POV	10	10	10	15	15	14
PAH	83	84	a	83	83	a
PAOV	83	85	86	83	82	85
PHOV	8	7	a	13	13	a
PAHOV	77	78	a	78	77	a
VA	74	72	70	73	71	69
VH	6	7	9	8	8	10
VOV	8	10	10	17	16	16
VAH	66	66	a	66	64	a
VAOV	63	60	58	64	62	57
VHOV	8	11	a	15	15	a
VAHOV	57	55	a	58	57	a
YLAG	57	60	50	80	79	76

Key: D1 ... Normal
D2 ... t(5)
D3 ... Log-normal

Note: [a] denotes that the rejection frequency has been omitted because the test is asymptotically invalid

Behaviour of tests for irrelevant alternatives

Tests for heteroskedasticity do not appear to be very sensitive to autocorrelation and no rejection frequency exceeds 10 per cent - this value being obtained for VH with $\rho_x = 0.95$ and log-normal disturbances. This degree of robustness is consistent with the theoretical discussion provided in Chapter 3. The checks for omitted variables are more sensitive and, as anticipated, the degree of sensitivity depends upon the value of ρ_x: the rejection frequencies for $\rho_x = 0.95$ being greater than those for $\rho_x = 0.75$, *ceteris paribus*. However, no form of an omitted variables test has a rejection frequency exceeding 17 per cent, so that the probability of adding irrelevant powers of x_t as a result of using the MCP is not likely to be high. The results for the joint tests PHOV and VHOV are quite similar to those for POV and VOV, respectively.

The tests discussed in the previous paragraph are all inconsistent when the only departure from the null specification is that ρ is not equal to zero. The procedure YLAG can, however, be consistent in this situation and, despite the fact that this test is designed for an irrelevant alternative, it is quite powerful in the presence of AR(1) errors. The performance of YLAG in the experiments depends upon the value of ρ_x with rejection frequencies increasing with ρ_x, *ceteris paribus*. This feature of the results may be explained as follows. The alternative model for YLAG is inadequate relative to the true process because it omits the term $-\rho\beta_2 x_{t-1}$ from the regression function, and most of the influence of this term can be picked up by the included regressor x_t when $\rho_x = \text{corr}(x_t, x_{t-1})$ is close to unity.

The application of an MCP involving YLAG and PA when the regressors are strongly serially correlated might quite frequently lead to the addition of the irrelevant regressor y_{t-1}. This error will produce estimator inefficiency and have effects on the small sample distributions of esimators. The situation would, of course, be worse if the MCP involved YLAG, but not a check for autocorrelation. Adding the irrelevant regressor y_{t-1} to a model with autocorrelated disturbances would cause the OLS estimators of

β_1 and β_2 in the expanded model to be inconsistent. Thus the asymptotic properties of OLS estimators would be poorer for the revised model than for the original specification.

The above results indicate the sensitivity of a test for the omission of y_{t-1} when the errors are AR(1). Similar results can be observed when the roles of these alternatives are reversed, i.e. PA may be significant because y_{t-1} has been incorrectly excluded; see Davidson and MacKinnon (1985, p. 45).

Behaviour of tests for overspecified alternatives

The asymptotically optimal test PA requires only one test variable, viz. the lagged OLS residual $\hat{u}_{t-1}$. Table 4 contains results for several tests that use not only $\hat{u}_{t-1}$, but also unnecessary indicators, for example, the vague autocorrelation test VA uses $\hat{u}_{t-i}$, i = 1, ..., 4. The asymptotic theory for local alternatives set out in the previous chapter predicts that power will be reduced by using alternatives that are more general than the true process. More precisely, under a sequence of local alternatives such that the statistic PA is asymptotically distributed as $\chi^2(1, \kappa^2)$, the asymptotic distributions of the tests for overspecified alternatives, when the errors come from a symmetric distribution, are as follows : 2PAH $\underset{a}{\sim}$ $\chi^2(2, \kappa^2)$; 2PAOV $\underset{a}{\sim}$ $\chi^2(2, \kappa^2)$; 3PAHOV $\underset{a}{\sim}$ $\chi^2(3, \kappa^2)$; 4VA $\underset{a}{\sim}$ $\chi^2(4, \kappa^2)$; 6VAH $\underset{a}{\sim}$ $\chi^2(6, \kappa^2)$; 7VAOV $\underset{a}{\sim}$ $\chi^2(7, \kappa^2)$; and 9VAHOV $\underset{a}{\sim}$ $\chi^2(9, \kappa^2)$, in which $\underset{a}{\sim}$ denotes "is asymptotically distributed as". Since all these distributions have the same noncentrality parameter, asymptotic local power will fall as the degrees of freedom parameter increases.

Inspection of Table 4 reveals that, for every combination of symmetric error distribution and ρ_x, the figures satisfy the inequality rf(PA) > rf(PAOV), rf(PAH) > rf(PAHOV) > rf(VA) > rf(VAH) > rf(VAOV) > rf(VAHOV), with rf(PAOV) and rf(PAH) being very similar. Thus the simulation results are in agreement with the predictions of asymptotic theory and provide several examples of the value of accurate information when choosing the alternative specification. It may be worth noting that, for cases with

ρ_x = 0.95, the inappropriate procedure YLAG rejects the null model more frequently than any of the tests using vague information, and is similar in performance to the procedure PAHOV.

Effects of the error distribution

The estimates reported in Table 4 do not suggest that variations in power over error distributions are likely to be substantial. In many cases, differences in rejection frequencies between D1, D2 and D3 are very small. The largest difference (10 per cent) is for the test YLAG with ρ_x = 0.75, and the t(5) and log-normal distributions being compared.

(b) Heteroskedasticity

The Monte Carlo evidence on the behaviour of tests under heteroskedasticity reveals two interesting features. First, while BJ report that their check for omitted variables is very robust to heteroskedasticity with power estimates close to (but smaller than) the significance level, we find that POV and VOV are moderately sensitive to nonzero values of α_1 in the variance model of (14). Second, power is only slightly affected by combining the heteroskedasticity tests with tests for irrelevant alternatives to obtain overelaborate joint tests. These two features can be observed in the estimates contained in Table 5.

Table 5 is derived from results for all three sample sizes with ρ_x = 0.75 and either normally or log-normally distributed disturbances. (The value of ρ_x does not appear to play a major role in determining the behaviour of tests and so little is lost by reporting the results for only one value of this parameter.) The target power for the optimal test PH is 25 per cent under log-normal errors and 75 per cent under normal errors. As with Table 4, estimates are only provided for asymptotically valid tests.

Behaviour of tests for irrelevant alternatives

The large sample properties of tests for autocorrelation and omitted variables under heteroskedasticity were examined in the previous chapter. This examination revealed that these tests for irrelevant alternatives should not be expected to have high rejection probabilities in the presence of heteroskedastic disturbances. The estimates of Table 5 suggest that both forms of autocorrelation test are robust to nonzero values of α_1, but that the two tests for omitted variables are more sensitive with rf(VOV) > rf(POV) > max[rf(PA),rf(VA)] in every column of results.

The degree of sensitivity of POV and VOV contrasts with the findings reported by BJ, and also with the results for PA and VA. The differences in results may be partly explained by the differences in the relationships between the test variables used to detect autocorrelation or omitted regressors, and the variances $Var(u_t)$, $t = 1, \ldots, n$. These relationships play a role in determining the directions of the biases of the invalid OLS estimators of sampling variances and covariances; see Judge et al. (1980, pp.127-8) for a simple example. In the experiments of this chapter, $Var(u_t)$ is a linear function of x_t^2 and so is positively and perfectly correlated with the test variable x_t^2 used in POV. The experiments carried out by BJ, however, use a design in which the variances are linear functions of a variable that is drawn independently of the regressors. The sensitivity of POV and VOV should not be overstated, but clearly there is the possibility that irrelevant regressors might be added to an adequate regression function after applying the MCP to a model with heteroskedastic disturbances.

Behaviour of tests for overspecified alternatives

It can be seen from Table 5 that constructing joint tests that check for irrelevant alternatives, as well as for heteroskedasticity, leads to unimportant changes in rejection frequencies. It is also clear that using VH with the additional, but irrelevant, test variable x_t produces only small variations in results. These findings support the conclusion of BJ that overtesting results in little loss of power relative to the asymptotically optimal procedure; see BJ (1982, p.71).

Table 5
Rejection frequencies under heteroskedasticity with ρ_x = 0.75 and either normal or log-normal errors

	Normal			Log-normal		
Test	n = 40	60	80	40	60	80
PA	8	6	6	4	4	4
PH	69	76	70	23	34	20
POV	18	19	18	14	21	14
PAH	62	68	64	a	a	a
PAOV	17	16	17	12	17	12
PHOV	65	71	66	a	a	a
PAHOV	60	66	63	a	a	a
VA	7	7	5	4	5	4
VH	66	72	67	24	34	22
VOV	28	28	25	16	25	16
VAH	50	59	56	a	a	a
VAOV	23	25	20	15	20	13
VHOV	60	67	62	a	a	a
VAHOV	51	59	54	a	a	a

Note: [a] denotes that the rejection frequency has been omitted because the test is asymptotically invalid

(c) Omitted variables

The examples provided in Section 2.2 of the previous chapter suggest that, when relevant regressors are excluded, autocorrelation tests and heteroskedasticity tests may both lead to significant outcomes, with the behaviour of the former type of check being related to the pattern of serial correlation of the omitted variables. It, therefore, seems worthwhile to present results for both values of ρ_x, and a representative sample of such results is provided in Table 6. This table is obtained using the same combinations of sample size and error distributions as Table 4. Two values of β_3, the coefficient of x_t^2 in (12), are used - one for each value of ρ_x. Given the value of ρ_x, β_3 is selected (by trial and error) to yield rejection frequencies for the asymptotically optimal POV test that are close to or inside the range 70-75 per cent.

In addition to the usual collection of tests for misspecification, the procedure OVW is included in Table 6. It will be recalled that OVW is a test of $\gamma_1 = 0$ in

$$y_t = \beta_1 + \beta_2 x_t + \gamma_1 w_{t2} + u_t,$$

where w_{t2} is generated according to (16), with the squared correlation between the random variables x_t^2 and w_{t2} being equal to 0.5. Thus the correct and incorrect test variables are not very strongly correlated.

Behaviour of tests for irrelevant alternatives

The check for autocorrelation was found to be robust to omitted variables in the study by BJ. However, as observed by BJ, their use of serially independent regressors limits the generality of their results. The regressor x_t of the experiments reported in this chapter is generated by the AR(1) scheme of (15) with parameter ρ_x. The estimates of Table 6 suggest that autocorrelation tests are only slightly affected in cases with $\rho_x = 0.75$, but are rather more sensitive when $\rho_x = 0.95$. For example, the average rejection frequency of PA is about 7 per cent when $\rho_x = 0.75$ and is 21 per cent when $\rho_x = 0.95$.

The estimates for heteroskedasticity tests in Table 6 reflect a moderate degree of sensitivity for both values of ρ_x. Several of these estimates are above 20 per cent and this sort of behaviour suggests the asymptotic local theory concerning heteroskedasticity tests under omitted variables provides a very poor approximation in the cases under consideration; see Section 2.3 of the previous chapter.

The final test for a single irrelevant alternative is OVW. Although the consequences of using an incorrect test for omitted variables were not considered by BJ, it seems worthwhile to include OVW because precise information about specification errors is not always available and mistakes may be made. It is clear from Table 6 that a test based upon an incorrect test variable may be quite powerful. Consequently an MCP using both POV and OVW could lead to the inclusion of the irrelevant variable w_{t2}.

Behaviour of tests for overspecified alternatives

As in the discussion of behaviour under autocorrelation, the asymptotic theory for the distributions of test statistics under local alternatives can be used to obtain a ranking by predicted power. If the errors have a symmetric distribution (such as D1 or D2 in Table 6), then the predicted ranking of rejection frequencies is

rf(POV) > rf(PAOV) = rf(PHOV) > rf(PAHOV) = rf(VOV) >
rf(VHOV) > rf(VAOV) > rf(VAHOV).

The corresponding predicted ranking of asymptotically valid tests in the case of asymmetric error distributions (such as D3 in Table 6) is

rf(POV) > rf(PAOV) > rf(VOV) > rf(VAOV).

Inspection of the estimates of Table 6 indicates that they are in close agreement with these rankings. It is clear that consideration of irrelevant alternatives or overparameterised models tends to reduce the probability of rejecting an inadequate model. The estimates for overspecified alternatives are, however, sometimes greater than would be expected from asymptotic local powers; see Table 1 of Chapter 3.

Table 6

Rejection frequencies under omitted variables with n = 60

	$\rho_x = 0.75$			$\rho_x = 0.95$		
Test	D1	D2	D3	D1	D2	D3
PA	6	6	8	19	19	25
PH	23	20	20	19	17	16
POV	68	71	75	76	74	78
PAH	20	17	[a]	25	22	[a]
PAOV	60	64	67	70	68	72
PHOV	61	64	[a]	71	69	[a]
PAHOV	56	60	[a]	66	65	[a]
VA	4	4	4	13	12	18
VH	21	19	21	24	23	23
VOV	54	58	60	66	64	68
VAH	15	15	[a]	22	21	[a]
VAOV	41	44	48	54	54	57
VHOV	50	51	[a]	61	59	[a]
VAHOV	41	40	[a]	52	52	[a]
OVW	50	57	59	40	44	45

Key: D1 ... Normal
D2 ... t(5)
D3 ... Log-normal

Note [a] denotes that the rejection frequency has been omitted because the test is asymptotically invalid.

(d) Autocorrelation and omitted variables

Ramsey and Kmenta (1980, p.11) have remarked that the chief difficulty in the use of specification error tests is the isolation and identification of separate effects in the presence of more than one error. In order to obtain some evidence on the behaviour of tests in such a situation, data are generated with nonzero values of both β_3 and ρ. The null model is, therefore, inadequate because it fails to allow for autocorrelation and also omits a relevant regressor.

Two tables are used as a basis for discussion and to illustrate the general findings of the larger body of results. Table 7 contains a sample of estimates for $\rho_x = 0.75$ covering all three values of n and processes with either normal or log-normal errors. For each value of n, values of β_3 and ρ are selected to obtain a rejection frequency of about 75 per cent for the asymptotically optimal test PAOV. Table 8 is derived by considering cases with $\rho_x = 0.95$, n = 60 and a target power for PAOV of about 50 per cent: rejection frequencies are reported for all five distributions. These two tables reflect the general finding that the variations in the error distribution do not lead to large changes in rejection frequencies.

As in previous cases, the discussion will cover tests for both irrelevant and overspecified alternatives. However, the joint presence of two misspecifications implies that a third category of tests requires attention. This category consists of procedures designed to detect only one of the two types of error: such tests will be termed "tests for underspecified alternatives".

Behaviour of tests for irrelevant alternatives

Tables 7 and 8 contain results for three tests for irrelevant alternatives - the two checks for heteroskedasticity, viz. PH and VH, and YLAG. (The procedure YLAG is included because of its sensitivity to autocorrelation.) While PH and VH are not completely robust, the rejection frequencies of Tables 7 and 8 do not suggest that the MCP will frequently lead to the unnecessary

Table 7

Rejection frequencies under autocorrelation and omitted variables with ρ_x = 0.75 and either normal or log-normal errors

		Normal			Log-normal	
Test	n = 40	60	80	40	60	80
PA	69	64	62	69	64	61
PH	12	13	14	16	14	13
POV	45	45	52	54	55	56
PAH	62	58	56	a	a	a
PAOV	75	74	72	81	77	75
PHOV	38	38	44	a	a	a
PAHOV	70	67	67	a	a	a
VA	45	41	39	37	36	32
VH	14	14	14	18	16	13
VOV	33	34	36	42	42	42
VAH	42	39	39	a	a	a
VAOV	52	50	52	53	52	50
VHOV	30	31	32	a	a	a
VAHOV	48	46	48	a	a	a
YLAG	41	34	34	34	30	30

Note: [a] denotes that the rejection frequency has been omitted because the test is asymptotically invalid.

specification of heteroskedasticity. The test YLAG is rather more sensitive, as might be anticipated from the earlier discussion of behaviour under autocorrelation. This sensitivity could cause problems. If the error autocorrelation model of the revised model is incorrect - for example, because the wrong member of the correct family of LEAs has been selected - then the presence of the irrelevant variable y_{t-1} will lead to estimator inconsistency.

Behaviour of tests for overspecified alternatives

The tests PAHOV, VAOV and VAHOV correspond to overspecified alternatives. The sets of super-models associated with PAHOV and VAOV are nonnested, but both sets are included in the super-models of VAHOV. As expected, the use of tests for overspecified alternatives leads to a reduction of power relative to the asymptotically optimal procedure PAOV. Consideration of the estimates in Tables 7 and 8 reveals that it is the use of vague, as opposed to precise, information about the nature of autocorrelation and omitted variables that leads to most of the reduction. Allowing for the irrelevant alternative of heteroskedasticity produces quite small changes (compare the figures for PAOV and VAOV with those for PAHOV and VAHOV, respectively). These features of the results are what would be expected on the basis of the numbers of irrelevant test indicators being employed in the tests.

The rankings of rejection frequencies are again in agreement with predictions derived from consideration of asymptotic local power since

rf(PAOV) > rf(PAHOV) > rf(VAOV) > rf(VAHOV)

for the symmetric distributions (normal in Table 7, and D1, D2 and D5 in Table 8) and

rf(PAOV) > rf(VAOV)

for the asymmetric distributions (log-normal in Table 7, and D3 and D4 in Table 8). The differences between rf(PAOV) and rf(VAOV) are sometimes quite large, reflecting the cost of having only vague information.

Table 8

Rejection frequencies under autocorrelation and omitted variables with ρ_x = 0.95 and n = 60

Test	Distribution D1	D2	D3	D4	D5
PA	47	46	44	46	43
PH	8	8	10	9	10
POV	42	47	50	47	42
PAH	37	40	a	a	35
PAOV	53	58	58	55	53
PHOV	36	39	a	a	36
PAHOV	48	51	a	a	46
VA	27	28	25	26	26
VH	10	9	12	12	12
VOV	35	38	42	39	34
VAH	25	25	a	a	24
VAOV	37	38	39	38	37
VHOV	30	35	a	a	32
VAHOV	35	35	a	a	35
YLAG	38	38	34	37	34

Key: D1 ... Normal D2 ... Student t(5)
D3 ... Log-normal D4 ... $\chi^2(2)$
D5 ... Uniform

Note: [a] denotes that the rejection frequency has been omitted because the test is asymptotically invalid

Behaviour of tests for underspecified alternatives

Underspecified alternatives fail to take one of the misspecifications into account. The tests that are not designed to detect autocorrelation are POV, PHOV, VOV and VHOV. The tests that are not designed to detect omitted variables are PA, PAH, VA and VAH. It will be convenient to start the discussion of the performance of tests for underspecified alternatives by considering PA and POV. It was found in the experiments carried out by BJ that undertesting could lead to a considerable loss of power. The estimates for POV in Table 7 provide evidence of this kind of loss with rf(POV) being, on average, about two-thirds of rf(PAOV). However, the use of underspecified alternatives does not always lead to such substantial reductions. The ratio of min[rf(PA), rf(POV)] to rf(PAOV) is, on average, about 0.8 in Table 8 and the ratio of rf(PA) to rf(PAOV) is always greater than this value in Table 7.

Entertaining the irrelevant alternative of heteroskedasticity by combining PA and POV with PH to obtain PAH and PHOV, respectively, leads to further reductions in power estimates relative to those of PAOV. Performances that are clearly inferior to those of PA and POV are also observed when the error of ignoring one of the two misspecifications that are present is combined with the error of overparameterisation by using either VA or VOV. For example, in the first column of results in Table 8, the power estimates for PA, POV, VA and VOV are 47, 42, 27 and 35 per cent, respectively.

LEAs and the identification of specification errors

Section 3 of Chapter 3 contains a description of a method for detecting irrelevant alternatives. This method is based upon an artificial regression equation that is locally equivalent to the super-models associated with the battery of tests being employed. Comparison of the rejection frequencies of YLAG with those of PA and POV in Tables 7 and 8 indicates that the former test (which is designed for an irrelevant alternative) appears to produce significant outcomes with a probability that is not relatively small. It may, therefore, be of interest to investigate the usefulness of a suitable LEA in identifying the specification errors that are present.

In order to simplify the analysis, it is assumed that the disturbances are taken to be homoskedastic under null and alternative hypotheses. (The modifications required to take account of heteroskedasticity are outlined below.) A suitable LEA for the super-models associated with PA, POV and YLAG can then be written as

$$y_t = \beta_1 + \beta_2 x_t + \beta_3 x_t^2 + \gamma_1 \hat{u}_{t-1} + \delta_1 y_{t-1} + u_t,$$

in which $\hat{u}_{t-1}$ is the lagged value of the residual obtained by OLS estimation of the null model. An attempt to assess the relevance of the three separate alternatives can then be based upon checks of the corresponding separate restrictions $\beta_3 = 0$, $\gamma_1 = 0$ and $\delta_1 = 0$. Let the F-tests for these three single-parameter restrictions be denoted by LEAPOV, LEAPA and LEAYLAG, respectively; so that, e.g., LEAPOV is the square of the t-ratio of x_t^2 obtained by OLS estimation of the LEA.

Including the three tests derived from the LEA in the experiments produces some interesting results. Whatever the error distribution or sample size, the power estimates for LEAPA, LEAPOV and LEAYLAG display two clear features: the true assumption that y_{t-1} is irrelevant is not often rejected; and the behaviour of LEAPA is greatly influenced by the value of ρ_x. These findings are illustrated by the estimates of Table 9. This table is constructed from results for cases with n = 60 and either normal or log-normal error distributions. In order to assist comparisons, Table 9 also contains results for the asymptotically optimal test PAOV and the three separate tests PA, POV and YLAG.

It is clear from Table 9 that the strategy of using the LEA is successful in terms of detecting the irrelevance of y_{t-1}. The identification of the irrelevant alternative has, however, been bought at a price. The tests for autocorrelation and omitted variables both suffer reductions in rejection frequencies with rf(PA) > rf(LEAPA) and rf(POV) > rf(LEAPOV). The reductions for the omitted variables test are not small and those for the autocorrelation test are very large when $\rho_x = 0.95$. These results suggest that, when the regressors are strongly serially correlated, the procedure LEAPA may fail to detect autocorrelation, leading to inadequate revisions of the original specification. Loosely speaking, there is a problem of multicollinearity present in the regressors of the LEA when $\rho_x = 0.95$ and it is difficult to estimate all the individual coefficients with a useful degree of precision. When the information content of the LEA is low, a researcher might well be faced by a significant joint test of $\gamma_1 = \delta_1 = 0$, but insignificant values of LEAPA and LEAYLAG.

Table 9

Rejection frequencies under autocorrelation and omitted variables with n = 60 and either normal or log-normal errors

	Normal		Log-normal	
Test	$\rho_x = 0.75$	0.95	0.75	0.95
PAOV	74	53	77	58
PA	64	47	64	44
POV	45	42	55	50
YLAG	34	38	30	34
LEAPA	36	11	34	12
LEAPOV	39	34	51	43
LEAYLAG	7	6	7	7

Note: The values of β_3 and ρ in the data generation process vary with ρ_x.

The results of Table 9, and those of other experiments, indicate that inference based upon an appropriate LEA has both costs and benefits. Such inference will, of course, require modification in the presence of heteroskedasticity. As suggested in Section 3 of Chapter 3, a convenient approach to modifying the procedure consists of estimating the LEA by OLS and then constructing the required significance tests using White's heteroskedasticity-consistent covariance matrix estimate, rather than the standard estimate derived under the assumption of homoskedastic errors. The results of Chesher and Austin (1987) indicate that it may be useful to examine the leverage values for the LEA before implementing this modified procedure.

(e) Errors in variables

The final misspecification to be considered is that the regressor is measured with an error. None of the tests under examination is specifically designed for this alternative and so they do not generate a super-model that includes the actual data process. The data on y_t are generated by the null model (11), but the "observed" values of the regressors (as opposed to the "true" values) are obtained from (16), being given by

$$w_{t1} = x_t + z_{t1},$$

where the z_{t1} are NID(0, Var(x_t)) variates that are distributed independently of the x_t, t = 1, ... , n. The true regressor values x_t are also normally distributed, but are autocorrelated. The fitted model can therefore be written as

$$y_t = \beta_1^e + \beta_2^e w_{t1} + u_t^e, \ t = 1, \dots , n$$

and the expressions for the tests for misspecification are derived from those for previous cases by replacing x_t with w_{t1}. Thus, for example, POV is the F-test for testing $\beta_3^e = 0$ in

$$y_t = \beta_1^e + \beta_2^e w_{t1} + \beta_3^e w_{t1}^2 + u_t^e, \ t = 1, \dots , n.$$

The rejection frequencies for cases with n = 80 and $\rho_x = 0.75$ are given in Table 10 - these results are representative of the full set.

As illustrated by the figures of Table 10, the variations in results between the different error distributions are quite small. Indeed, if the log-normal distribution is excluded, the remaining four sets of results are very similar. The main features of the results are clear. Tests designed to detect heteroskedasticity and/or omitted variables, but not autocorrelation, are insensitive to the specification error that has been made. Such tests have rejection frequencies under errors in variables that are close to the nominal size. Tests involving a check for autocorrelation are much more sensitive with rejection frequencies increasing as the number of test indicators falls. For example, the tests PA and VAHOV use 1 and 9 indicators, respectively, and, under symmetric distributions, rf(PA) ≥ 46 per cent = 2rf(VAHOV).

It is, however, the procedure YLAG that yields the highest rejection frequencies with rf(YLAG) being close to 90 per cent in all cases. The use of an MCP based upon, say, PA, PH, POV and YLAG might well lead a researcher to reformulate the model as

$$y_t = \beta_1^e + \beta_2^e w_{t1} + \gamma_1^e y_{t-1} + u_t^e, \; t = 1, \ldots , n,$$

possibly allowing for an AR(1) structure for u_t^e, rather than to re-estimate the original model using the method of instrumental variables. Given the specification of the data process for y_t and w_{t1}, the lagged variable $y_{t-1} = \beta_1 + \beta_2 x_{t-1} + u_{t-1}$ is a valid instrument. If y_{t-1} were used as the instrument, then the procedure YLAG would be equivalent to a form of Hausman's (1978) test for errors in variables. This interpretation provides an explanation of the performance of YLAG, as reflected by the estimates in Table 10. The autocorrelation test PA can also be regarded as a Hausman-type test - it is derived using the instrument $\hat{u}_{t-1}^e$.

It might be thought that POV could also be regarded as a Hausman-type test, but this is not the case. The test variable used in this procedure, viz. w_{t1}^2, is not a valid instrument. The covariance between w_{t1}^2 and the true value x_t is

$$E(w_{t1}^2 x_t) = E(x_t^3 + 2x_t^2 z_{t1} + x_t z_{t1}^2) = 0,$$

Table 10
Rejection frequencies under errors-in-variables with ρ_x = 0.75 and n = 80

Test	Distribution				
	D1	D2	D3	D4	D5
PA	46	47	52	47	46
PH	4	4	4	4	4
POV	6	6	6	7	7
PAH	36	38	[a]	[a]	37
PAOV	37	39	44	38	38
PHOV	5	5	[a]	[a]	6
PAHOV	30	32	[a]	[a]	32
VA	34	34	40	34	34
VH	5	4	6	5	5
VOV	5	6	6	6	6
VAH	29	29	[a]	[a]	29
VAOV	26	27	32	28	27
VHOV	6	6	[a]	[a]	6
VAHOV	23	23	[a]	[a]	23
YLAG	90	90	88	89	89

Key: D1 ... Normal D2 ... Student t(5)
D3 ... Log-normal D4 ... $\chi^2(2)$

Note: [a] denotes that the rejection frequency has been omitted because the test is asymptotically invalid.

since x_t and z_{t1} are independent random variables, both having zero mean, and the distribution of x_t is symmetric.

The results of our experiment and, in particular, the estimates in Table 10 should not be interpreted as implying that RESET-type tests are always robust to the errors in variables problem. If the regressor x_t is a random variable with $E(x_t^3) \neq 0$ and the errors of observation z_{t1} are such that $E(z_{t1}^3) = 0$, then the covariance between w_{t1}^2 and x_t is not equal to zero and w_{t1}^2 is a valid instrument.

5. Conclusions

Several topics concerning the finite sample behaviour of test procedures have been considered in this chapter. The central issue, however, has been the usefulness of separate (one-directional) tests in identifying the specification errors that are present and eliminating irrelevant generalisations of the null model. If tests for individual alternatives, such as autocorrelation, heteroskedasticity and omitted variables, do not provide accurate information about the nature of specification errors, there is little point in following current practice by calculating them. Instead a single joint test can be derived that has the advantage of having a known significance level, at least in large samples.

The study by Bera and Jarque (1982) provides an important starting point for assessing the constructive value of misspecification tests. These authors examine a method in which four separate tests are obtained: one is designed as a check of the adequacy of the regression function, and the others are derived as tests of the three classical assumptions concerning the disturbance term (serial independence, homoskedasticity and normality). This method, termed the "multiple comparison procedure" (MCP), involves comparing each of the four statistics to its corresponding marginal significance level. If a test produces a significant outcome, then it is assumed that the alternative for which it is derived forms part of the true data process.

It has been suggested that the original MCP should be modified. Bera and Jarque use the Breusch-Pagan (1979) test for heteroskedasticity in their MCP. This procedure is sensitive to nonnormality and, in order to avoid false signals of heteroskedasticity, a variant of Koenker's (1981) Studentised test is recommended. The Monte Carlo results of Section 4 indicate that this robust form performs well under correct specification for several distributions. If an MCP is to be used, this test should be used in preference to the Breusch-Pagan procedure.

The simulation evidence adduced by Bera and Jarque to support the use of the MCP has been summarised. On the basis of results obtained from experiments with data processes involving one or more extensions of the null model, Bera and Jarque conclude that, although the MCP can lead to overadjustment, its success in identifying specification errors should make it a valuable aid for applied research; see Bera and Jarque (1982, p. 76). The generality of these conclusions is, however, limited by certain features of the experimental design adopted by Bera and Jarque, and by the range of cases that they study. More specifically, the following points have received attention in this chapter.

(i) As noted by Bera and Jarque (1982, p. 71), their evidence concerning the good performance of tests for autocorrelation and omitted variables in the MCP may reflect the serial independence of the regressors of their experiments. The regressors for the experiments of this chapter are serially correlated, as are most economic time series variables. In these experiments, the effect of strong serial correlation in the regressors is to make autocorrelation (resp. omitted variables) tests more sensitive to omitted variables (resp. autocorrelation). Consequently tests are less specific and the MCP is more likely to lead to overadjustment.

(ii) By using only strictly exogenous variables as regressors, Bera and Jarque restrict attention to cases in which tests for autocorrelation and omitted variables are asymptotically independent under correct specification. The implications of such independence for the construction of joint tests and the behaviour of tests under alternatives are discussed in Chapter 3. The assumption of strictly exogenous variables clearly excludes the consideration of a model with lagged values of the dependent variable in the regressor set. However, this type of model is quite commonly used in time series econometrics. It is, therefore, useful to introduce lagged dependent variables and this has been done in some of our experiments. The tests for omitted variables and autocorrelation are asymptotically dependent under correct specification in such experiments and the analysis of Chapter 3 suggests that the MCP may lead to overadjustment. The results of Table 4 provide some support for this prediction and indicate how the introduction of lagged dependent variables can reduce the effectiveness of the MCP; also see the example given by Davidson and MacKinnon (1985, p. 45).

(iii) Bera and Jarque assume that very precise information about alternative hypotheses is available, or at least that very precise beliefs are held. Their Monte Carlo experiments have the following characteristics: all separate tests have just one degree of freedom (so that joint tests for overelaborate alternatives do not use many irrelevant indicators); the check for omitted variables uses the correct variable in cases with this specification error; and the actual data process is always included in the super-models generated by the tests of the MCP.

In practice, applied workers may have much weaker information and beliefs about the nature of specification errors. One way in which the problem of weak information has been tackled is the development of fairly general tests involving quite large numbers of degrees of freedom for the associated χ^2 distributions, e.g. the portmanteau tests for autocorrelation discussed in Chapter 1 and White's (1980) test for heteroskedasticity. We have, therefore, considered the performance of tests using only vague information. The estimated rejection frequencies of Section 4 suggest that, under the alternative for which they are designed, such tests are rather less powerful than the corresponding one degree of freedom tests which are asymptotically optimal. The loss of power associated with overparameterisation of a true alternative is sometimes substantial. These effects do not imply anything about the relative merits of the MCP and a suitable joint test because the latter procedure would also be based upon vague information and would have several irrelevant terms.

The use of weak information in an MCP can, however, produce invalid inferences when the poor quality of the information leads, for example, to the use of an incorrect test variable for checking the adequacy of the regression function. Section 4 contains some results that illustrate this problem. The MCP using a test with the wrong variable, denoted by OVW above, often leads to the inclusion of this irrelevant variable, even though it is only moderately correlated with the correct omitted regressor.

If weak information leads to a situation in which the researcher does not carry out a test against a relevant alternative, then the MCP may produce quite misleading results. This possibility has been examined in the context of an errors in variables problem in Section 4. The results show that the MCP cannot point to the correct reformulation and may strongly suggest incorrect revisions. It is, of course, inevitable that an MCP will point to an inadequate specification when its tests do not cover all the specification errors that have been made. Consequently it is very important to use a wide range of tests in empirical work if the MCP approach is to be employed.

As a result of extending the Monte Carlo analysis of Bera and Jarque (1982) in the ways described above, we think that their conclusions concerning the accuracy of the MCP may be too optimistic. Alternative models that are themselves inconsistent with the data can lead to the rejection of inadequate null models with high probability. Moreover the MCP is by design incapable of distinguishing between omitted regressors and incorrect functional form and provides only a general check of the regression function.

Evidence obtained by Bera and Jarque suggests that the power of this general check may be adversely affected by the presence of heteroskedasticity and the estimates of Section 4 above provide examples of how autocorrelation can lead to significant values of tests for omitted variables. Given the need for a reliable test of the regression function, it is natural to consider the possibility of constructing procedures that are robust to autocorrelation and heteroskedasticity of unspecified form. This approach to the derivation of robust tests has been outlined in Section 5 of Chapter 3, but it would be very useful to have some Monte Carlo results concerning the finite sample performance of these procedures and those recently proposed by Wooldridge (1990).

References

Bera, A.K. and C.M. Jarque (1982). Model specification tests: a simultaneous approach, *Journal of Econometrics*, 20, 59-82.

Bera, A.K. and C.R. McKenzie (1986). Alternative forms and properties of the Score test, *Journal of Applied Statistics*, 13, 13-25.

Breusch, T.S. (1980). Useful invariance results for generalized regression models, *Journal of Econometrics*, 13, 327-430.

Breusch, T.S. and A.R. Pagan (1979). A simple test for heteroskedasticity and random coefficient variation, *Econometrica*, 47, 1287-94.

Chesher, A.D. and G. Austin (1987). Finite sample behaviour of heteroskedasticity robust Wald statistics, unpublished paper, University of Bristol (DP87/187).

Davidson, R. and J.G. MacKinnon (1985). The interpretation of test statistics, *Canadian Journal of Economics*, 18, 38-57.

Epps, T.W. and M.L. Epps (1977). The robustness of standard tests for autocorrelation and heteroskedasticity when both problems are present, *Econometrica*, 45, 745-53.

Hausman, J. (1978). Specification tests in econometrics,

Econometrica, 46, 1251-71.

Jarque, C.M. and A.K. Bera (1987). An efficient large sample test for normality of observations and regression residuals, *International Statistical Review*, 55, 163-72

Judge, G.G, W.E. Griffiths, R.C. Hill and T.-C. Lee (1980). *The Theory and Practice of Econometrics*. Wiley: New York.

Koenker, R. (1981). A note on studentizing a test for heteroskedasticity, *Journal of Econometrics*, 17, 107-12.

Pagan, A.R. and A.D. Hall (1983). Diagnostic tests as residual analysis, *Econometric Reviews*, 2, 159-218.

Porter, R.D. and A.K. Kashyap (1984). Autocorrelation and the sensitivity of RESET, *Economics Letters*, 14, 229-33.

Ramsey, J.B. (1983). Comments on diagnostic checks as residual analysis, *Econometric Reviews*, 2, 241-8.

Ramsey, J.B. and J. Kmenta (1980). Problems and issues in evaluating econometric models, in J. Kmenta and J.B. Ramsey (eds.) *Evaluation of Econometric Models*, 1-11. Academic Press: New York.

Shapiro, S.S. and R.S. Francia (1972). An approximate analysis of variance test for normality, *Journal of the American Statistical Association*, 67, 215-16.

Thursby, J.G. and P. Schmidt (1977). Some properties of tests for specification error in a linear regression model, *Journal of the American Statistical Association*, 72, 635-641.

White, H. (1980). A heteroskedasticity-consistent covariance matrix and a direct test for heteroskedasticity, *Econometrica*, 48, 421-48.

White, H. and G.M. MacDonald (1980). Some large sample tests for non-normality in the linear regression model, *Journal of the American Statistical Association*, 75, 16-28.

Wooldridge, J. M. (1990). A unified approach to robust, regression-based specification tests, *Econometric Theory*, 6, 17-43.

S. P. BURKE, L. G. GODFREY AND MICHAEL MCALEER

5. MODIFICATIONS OF THE RAINBOW TEST

1. Introduction

Many empirical analyses involve the ordinary least squares (OLS) estimation of a linear regression model. Applied workers involved in such analyses will often wish to test the consistency with the data of the assumptions that are required to justify the use of standard theoretical results concerning the properties of OLS estimators and associated test statistics. It is particularly important to check the assumptions that the functional form of the regression model is correctly specified and that no relevant regressors have been omitted because, in general, these assumptions are necessary for the consistency of OLS estimators and the validity of the usual methods of inference.

In the previous two chapters, it has been stressed that testing for misspecification, as for any other statistical procedure, benefits from the availability of accurate information. For example, overparameterisation of the alternative model will lead to loss of power of any test of a false null hypothesis. Thus, in the context of testing the hypothesis that the mean function of a regression equation is correctly specified against the alternative that relevant regressors have been excluded, it would be simple to carry out the appropriate F-test if there were precise information of the omitted variables. Unfortunately there is sometimes only vague information about the nature of specification errors and, in such cases, Ramsey's (1983, p. 244) comment that "one needs a general, moderately robust, information parsimonious procedure for a specification error test for omitted variables or incorrect functional form" will be very pertinent.

Several general checks of the type described by Ramsey (1983) have been proposed; see, for example, Ramsey (1969), Thursby and Schmidt (1977), White (1980), and Plosser, Schwert and White (1982). These tests have the useful property that they can be implemented by what Pagan (1984) calls the "method of variable addition". More precisely, the test statistics can be calculated by adding certain variables to the original regressor set and then testing their joint significance; see Godfrey (1988, Sections 4.2 and 4.10). The expanded regression model corresponding to a given general check of the null model can be regarded as the implicit alternative hypothesis.

Thursby (1989) has recently examined the small sample performance of several tests that are frequently used as general checks, such as Ramsey's (1969) RESET test and the differencing test proposed by Plosser, Schwert and White (1982). He reports that his results are disappointing in so far as the procedures that are designed to be powerful against general misspecification tend to perform well only against their own "specific" alternative; see Thursby (1989, pp. 229-30) for a useful summary. The main purpose of this chapter is to derive new general checks that satisfy the criteria proposed by Ramsey (1983) and can be easily calculated in applied econometric studies.

The starting point for our analysis is an article by Utts (1982) whose "rainbow" test is intended to accommodate a broad range of specification errors. The Utts rainbow test is pedagogically appealing and computationally simple. It was employed in a study of diagnostic checking in practice by Krämer, Sonnberger, Maurer and Havlik (1985) and, as explained by Kmenta (1986), can be calculated by the method of variable addition. Although Utts's approach is, at first sight, attractive, it is argued below that the rainbow test suffers from the serious drawback that it is not even asymptotically robust to nonnormality of the regression disturbances. More robust modifications of the rainbow test are therefore developed. Also, following a suggestion made by Pagan and Hall (1983), the use of a check for heteroskedasticity as a general test is examined.

The plan of this chapter is as follows. In Section 2, the rainbow test is described, its sensitivity to nonnormality is explained, and the modified procedures are proposed. Some Monte Carlo experiments are described in Section 3 and the results derived from these experiments are discussed in Section 4. In order to implement either the original rainbow test or the modified versions derived in Section 2, it is necessary to reorder the data according to the leverage values. Many other systems of reordering are available and the impact of the basis for reordering on the power of tests when the model is misspecified is examined in Section 5. Section 6 contains some concluding comments.

2. The model and test procedures

Consider the linear regression model

$$y = X\beta + u \ , \ u \sim D(0, \sigma_u^2 I_n), \tag{1}$$

in which y is an n by 1 vector of observations on the dependent variable, X is an n by k nonstochastic matrix with full column rank, β is a k by 1 vector of unknown coefficients, and u is an n by 1 vector of independently and identically distributed random disturbances with zero mean and variance σ_u^2. The elements of X are uniformly bounded with $n^{-1}(X'X)$ tending to a finite nonsingular matrix and the first four moments of the disturbance term are assumed to be finite.

Utts's rainbow test involves partitioning the total sample into three subsamples and comparing the sum of squared residuals derived from estimation using all n observations to the corresponding quantity obtained using only the "central" subsample. Utts recommends that the central subsample be defined after reordering the data by leverage values, i.e. by the values of the diagonal elements of the hat matrix $H = X(X'X)^{-1}X'$. Let the model for data after such reordering be written as

$$\bar{y} = \bar{X}\beta + \bar{u}\ ,\ \bar{u} \sim D(0,\ \sigma_u^2 I_n), \tag{2}$$

and the numbers of observations for the first, second and third subsamples be n_1, n_2 and n_3 respectively, with $n_2 > k$. The rainbow test statistic is then

$$R = \frac{(S - S_2)}{S_2} \cdot \frac{(n_2 - k)}{(n_1 + n_3)}, \tag{3}$$

in which S and S_2 are the sums of squared residuals based upon the complete sample and the central subset of n_2 observations respectively. Under the null hypothesis of correct specification, R is distributed as $F(n_1 + n_3,\ n_2 - k)$ if the disturbances are normally distributed. Thursby's (1989) results suggest that a rule of the form "reject the null hypothesis if $R > c$" is generally more useful than a two-sided test. The power of the test will depend upon, in part, the way in which the reordered sample is split into subsamples. Utts (1982, p.2806) recommends that, in general, n_2 should be about n/2 and n_1 and n_3 should both be about n/4.

Equation (3) indicates that the rainbow test can be interpreted as a variant of Chow's (1960) prediction error test in which the original data have been reordered. Consequently, using the ideas of Salkever (1976), it can be shown that the Utts test can be implemented as a test of $\alpha = 0$ in the expanded model

$$\bar{y} = \bar{X}\beta + A\alpha + \bar{u},\ \bar{u} \sim D(0,\ \sigma_u^2 I_n), \tag{4}$$

in which A is an n by $(n_1 + n_3)$ matrix of dummy variables defined by

$$A = \begin{bmatrix} I_{n_1} & 0 \\ 0 & 0 \\ 0 & I_{n_3} \end{bmatrix}$$

and $\alpha = (\alpha_1', \alpha_3')'$ is an $(n_1 + n_3)$ by 1 vector of parameters. The use of the dummy variable matrix A may also be interpreted as corresponding to a test of suitably reordered outliers; see McAleer and Tse (1988).

As well as illustrating how the rainbow test can be carried out by the method of variable addition, equation (4) can also be used to examine the robustness of this procedure to nonnormality. A consideration of the robustness of the test is important because the asymptotic properties of OLS estimators are unaffected by many types of nonnormality. It would clearly be unhelpful if nonnormality caused the rainbow test to signal that an adequate model was data inconsistent.

Standard theory concerning significance tests for regression models implies that when the errors are not normally distributed, the statistic R (which is simply the usual F-criterion for testing $\alpha = 0$ in (4)) will not be distributed as $F(n_1 + n_3, n_2 - k)$ under the null hypothesis in finite samples. Thus the rainbow test lacks finite sample robustness, and some evidence on finite sample effects of nonnormality is provided in Section 4.

Turning to asymptotic robustness, it will be convenient to introduce some additional notation by partitioning the reordered data matrices as follows: $\bar{y}' = (\bar{y}_1', \bar{y}_2', \bar{y}_3')$, where $\bar{y}_i$ is n_i by 1, and $\bar{X}' = (\bar{X}_1', \bar{X}_2', \bar{X}_3')$, where $\bar{X}_i$ is n_i by k, for i = 1, 2, 3. Also it will be useful to consider two cases that are differentiated by the assumptions made about the orders of magnitude of n_1 and n_3.

Case (i): n_1, n_2 and n_3 are all O(n)

This case includes Utts's suggested combination, viz. $n_1 = n/4$, $n_2 = n/2$ and $n_3 = n/4$. Let $\tilde{\alpha} = (\tilde{\alpha}_1', \tilde{\alpha}_3')'$ and $\tilde{\beta}$ denote the OLS estimators for the parameters of (4). Salkever's (1976) results imply that $\tilde{\beta} = (\bar{X}_2'\bar{X}_2)^{-1}\bar{X}_2'\bar{y}_2$, the OLS estimator for (2) based upon the central subset of observations, and that $\tilde{\alpha}_1$ and $\tilde{\alpha}_3$ are the prediction error vectors

$$\tilde{\alpha}_1 = \bar{y}_1 - \bar{X}_1\tilde{\beta}$$

and

$$\tilde{\alpha}_3 = \bar{y}_3 - \bar{X}_3\tilde{\beta},$$

respectively. Under the assumption of correct specification, plim $\tilde{\beta} = \beta$ and R is asymptotically equivalent to

$$R' = (n_1 + n_3)^{-1}(\bar{u}_1'\bar{u}_1 + \bar{u}_3'\bar{u}_3)/\sigma_u^2,$$

in which $\bar{u}_i$ is a n_i-dimensional subvector of $\bar{u}' = (\bar{u}_1', \bar{u}_2', \bar{u}_3')$, i = 1, 3. The random variable R′ has probability limit equal to unity. Hence, if (1) is correctly specified, the rainbow statistic R has a degenerate asymptotic distribution. Loosely speaking, the standard central limit theorems used to establish the asymptotic validity of many variable addition tests are not applicable in this case because the dimension of α, i.e. the number of test variables, is increasing at the same rate as n.

Case (ii): n_1 and n_3 are O(1); n_2 is O(n)

Suppose now that n_1 and n_3 are both fixed and finite. After suitable reordering, this case corresponds to the one for which Hendry (1979, p.222) developed his z_4 test. If $F(v_1, v_2)$ denotes an F variate with degrees of freedom parameters v_1 and v_2, then $F(v_1,v_2) \rightarrow \chi^2(v_1)/v_1$ as $v_2 \rightarrow \infty$. It follows that the asymptotic validity of the rainbow test requires that the test statistic R of (3) converge to $\chi^2(n_1 + n_3)/(n_1 + n_3)$. Under the hypothesis that (1) is correctly specified, $S_2/(n_2 - k)$ is consistent for σ_u^2 and R is again asymptotically equivalent to R′. It follows that the rainbow test is not asymptotically robust even when it uses a finite number of test variables because the scaled sums of squared errors $\sum_i \bar{u}_t^2/\sigma_u^2$ will not, in general, be distributed as $\chi^2(n_i)$ when the errors $\bar{u}_t$ are nonnormal, i = 1, 3.

Modified rainbow tests

It seems worthwhile to consider modifications of the rainbow test that are asymptotically robust to nonnormality when n_i is O(n), i = 1, 2, 3. The assumption that $n_i = O(n)$ for i = 1, 2, 3 is needed to ensure that the differences between OLS estimators in each subsample and the corresponding true values are all $O_p(n^{-1/2})$, rather than varying in asymptotic order of magnitude from one subsample to another. Given the interpretation of Utts's original procedure as a prediction error test, it seems natural to examine the corresponding analogues of the analysis of covariance test described by Chow (1960). Consider, therefore, the k-dimensional estimator contrast vector

$$q = b_2 - \hat{\beta},$$

in which b_2 is the OLS estimator of β based upon the central n_2 observations and $\hat{\beta}$ is the OLS estimator using all n observations. The estimator b_2 can be regarded as being derived by applying OLS to transformed data $(\bar{y}_*, \bar{X}_*)$, where $\bar{y}_* = T\bar{y} = \bar{y}_2$, $\bar{X}_* = T\bar{X} = \bar{X}_2$, and T is the n_2 by n matrix $[0 : I_{n_2} : 0]$, in which the first and second null matrices are of dimensions n_2 by n_1 and n_2 by n_3 respectively. The results of Breusch and Godfrey (1986), therefore, imply that a test of the significance of $q = b_2 - \hat{\beta}$ can be performed as the F-test of the hypothesis $\theta = 0$ in

$$\bar{y} = \bar{X}\beta + [T'T\bar{X}]\theta + \bar{u}, \ \bar{u} \sim D(0, \sigma_u^2 I_n), \tag{5}$$

in which $[T'T\bar{X}]' = [0: \bar{X}_2': 0]$.

The test of $\theta = 0$ in (5) will be denoted the modified rainbow test - version 1 (or MR1). Under the null hypothesis, the usual

F-statistic will be exactly distributed as F(k, n - 2k) when disturbances are normal and will be asymptotically distributed as $\chi^2(k)/k$ for many types of nonnormal disturbances. The reason why MR1, in contrast to the original rainbow test, is asymptotically robust to nonnormality can be seen by considering the vectors of cross-products between disturbances and the regressors of (4) and (5). For the tests to be robust, these vectors (and hence the coefficient estimators) should be asymptotically normally distributed. The vectors of cross-products are $\bar{X}'\bar{u}$, $[T'T\bar{X}]'\bar{u} = \bar{X}_2'\bar{u}_2$, and $A'\bar{u} = (\bar{u}_1, \ldots, \bar{u}_{n_1}, \bar{u}_{n_1+n_2+1}, \ldots, \bar{u}_n)'$. Clearly, while central limit theorems are applicable to the elements of the first two vectors with $n_2 = O(n)$, the third vector contains only individual elements of u and it will only be normally distributed if the disturbances are normal.

The form of equation (5) indicates that, even under the alternative hypothesis, it is being assumed that the regression parameter vector is the same in the first and third subsamples. This assumption is also revealed by the following formula for MR1 which can be regarded as a modification of expression (3) for R:

$$MR1 = \frac{S - (S_{13} + S_2)}{(S_{13} + S_2)} \cdot \frac{(n - 2k)}{k}, \tag{6}$$

in which S and S_2 are as previously defined, and S_{13} is the residual sum of squares from the regression of the reordered dependent variable on the reordered regressors using only the first n_1 and last n_3 observations.

Relaxing the assumption that the same parameter vector applies to the first and third subsamples will produce a test that is consistent against a wider range of specification errors than MR1. If $n_i > k$ for i = 1, 2, 3, then an appropriate test can be based upon the 2k by 1 contrast vector

$$d = \begin{bmatrix} b_1 - \hat{\beta} \\ b_2 - \hat{\beta} \end{bmatrix},$$

or equivalently

$$d = \begin{bmatrix} b_1 - b_3 \\ b_2 - b_3 \end{bmatrix},$$

where b_i denotes the OLS estimator for the regression of $\bar{y}_i$ on $\bar{X}_i$, for i = 1, 2, 3. This test is simply a generalisation of the standard analysis of covariance test and is denoted the modified rainbow test - version 2 (or MR2). It can be computed as the F-test of $\delta = 0$ in the expanded model

$$\bar{y} = \bar{X}\beta + \bar{Z}\delta + \bar{u}, \; \bar{u} \sim D(0, \sigma_u^2 I_n), \tag{7}$$

where: $\delta' = (\delta_1', \delta_2')$; $\bar{Z} = (\bar{Z}_1, \bar{Z}_2)$; $\bar{Z}_1' = (\bar{X}_1', 0, 0)$; $\bar{Z}_2' = (0, \bar{X}_2', 0)$; and, to economise on notation, 0 is used to denote any matrix or vector with every element equal to zero (the various dimensions are obvious from the context).

The F-statistic for testing $\delta = 0$ in (7) can be calculated as

$$MR2 = \frac{S - (S_1 + S_2 + S_3)}{(S_1 + S_2 + S_3)} \cdot \frac{(n - 3k)}{2k} \tag{8}$$

in which S_i denotes the residual sum of squares for estimation using only the ith subsample of the reordered data, i.e. $S_i = (\bar{y}_i - \bar{X}_i b_i)'(\bar{y}_i - \bar{X}_i b_i)$ for i = 1, 2, 3. The test statistic in (8) is distributed as F(2k, n - 3k) under the null hypothesis when the errors are normal and is asymptotically distributed as $\chi^2(2k)/2k$ under many departures of the error distribution from normality.

Equation (8) makes it clear that MR2 can be computed by estimating (2) using different sets of reordered observations, so that it is not necessary to estimate (7). Equation (7) is, however, not without interest because it leads to a useful theoretical result. More precisely, (7) permits the derivation of the class of implicit alternative hypotheses associated with MR2; see Davidson and MacKinnon (1987) for a definition and discussion of such hypotheses. If the observations on the variables of (7) are transformed to the original ordering, then we obtain a model that bears the same relationship to (7) as (1) does to (2). This transformation of (7) can be written as

$$y = X\beta + Z\delta + u,$$

in which the rows of Z are derived by rearranging those of $\bar{Z}$, and the class of implicit alternative hypotheses for MR2 consists of all models that are locally equivalent to this expanded version of (1); see Godfrey (1988, Chapter 3) for a discussion of locally equivalent alternative models. The corresponding result for MR1 is obtained by imposing the restriction that the first k elements of δ equal zero, i.e. $\delta_1 = 0$ in (7).

The MR2 test is, of course, consistent against a range of fixed alternatives that is more general than the class of implicit alternative hypotheses. Under standard assumptions, the statistic MR2 will be $O_p(n)$ if either $(b_1 - b_3)$ or $(b_2 - b_3)$ is $O_p(1)$. Consequently the probability of rejecting the model (1) when it is inadequate will tend to unity as $n \longrightarrow \infty$ for all data processes with the property that the OLS estimators b_i, $i = 1, 2, 3$, calculated from subsets of reordered observations do not have the same probability limit. Since the MR2 test is proposed as a general check to be used when there is no clear idea about the nature of the specification error, it is desirable that it should be consistent against such a wide range of alternative models.

It is natural to ask why it is worth reordering the original observations (y, X) to obtain $(\bar{y}, \bar{X})$ before carrying out tests. Conventional Chow tests for parameter constancy using the original data are clearly easier to implement than the MR1 and MR2 procedures. The justification for reordering is based upon considerations of power. If there were no reordering, then, for stationary processes, the OLS estimator of β for each subsample would have the same probability limit as $n_i \rightarrow \infty$, n_i denoting the number of observations in a typical subsample. Consequently the estimator contrast vector would tend to a null vector, implying the inconsistency of the associated test in the presence of misspecification. This result suggests that conventional Chow tests may have low power against general specification error in moderately large samples. Thursby (1989) finds results that support this conjecture in his Monte Carlo experiments; also see Thursby (1982). The impact of the basis for reordering on power is discussed in Section 5.

There are other ways in which the sensitivity of the rainbow test to nonnormality might be tackled. For example, as explained above, Utts's original procedure can be regarded as a prediction error test with the number of prediction errors being O(n). Rather than attempting to test the joint significance of all these prediction errors, a finite number of linear combinations could be constructed in such a way that a central limit theorem could be applied to them. A χ^2 test with degrees of freedom equal to the number of linear combinations could then be derived. Thus a one degree of freedom test could be used to check the significance of the sum of the prediction errors and such a procedure would be robust to nonnormality. However, the choice of the number of linear combinations to be tested and of the weights used to form such combinations is somewhat arbitrary. The approach of this paper in which tests are derived by considering alternative estimates of the parameters of interest seems preferable.

3. Monte Carlo experiments

In the Monte Carlo experiments considered here, the assumed model is

$$y_i = a + bx_i + u_i, \; u_i \text{ iid}(0, \sigma_u^2), \tag{9}$$

and the actual data generation process is given by

$$y_i = a + bx_i + cx_i^2 + u_i, \; u_i \text{ iid}(0, \sigma_u^2) \tag{10}$$

in which x_i is an autoregressive process defined by

$$x_i = \rho x_{i-1} + v_i \; , \; v_i \text{ NID}(0, \sigma_v^2),$$

for $i = 1, 2, \ldots , n$. The first observation for the regressor, x_1, is taken from the $N(0, \sigma_v^2/(1 - \rho^2))$ distribution. The following parameter values are fixed for all experiments: $a = 1.0$; $b = 1.0$; $\rho = 0.5$; $\sigma_u^2 = 4$; and $\sigma_v^2 = 9$. On the basis of these values, it follows that $\sigma_x^2 = 12$ and, under the null hypothesis $c = 0$, $R^2 \rightarrow 0.75$ as $n \rightarrow \infty$. The case in which the errors u_i are normally distributed is examined and, in order to investigate the effects of nonnormality (such as skewed and thick-tailed distributions) on the properties of tests, the log-normal, chi-squared with two degrees of freedom ($\chi^2(2)$), t distribution with five degrees of freedom (t(5)) and the uniform distribution are also considered. All pseudo-random number generators are taken from the NAG library. After being drawn from the specified distribution, the u_i are transformed to have zero mean and variance $\sigma_u^2 = 4$. As an index of the extent of departure from normality, the rejection frequency of the Jarque-Bera (1987) test, denoted by LM(N), is calculated in each case.

Having selected an error distribution, it only remains to choose the pair (n, c) to complete the specification of a data generation scheme. The sample sizes considered are n = 20, 40, 60. In order to avoid using uninformative extreme values, the parameter c is selected by reference to the performance of a test of (9) against (10), i.e. the test derived from perfect information about the nature of the specification error. The F-test of c = 0 in (10) will be denoted by FX2 and the results in Table 2 are for combinations of n and c that yield power estimates for FX2 of about 75 per cent under normality. Some results for more severe misspecifications are reported in Table 3.

In addition to R, MR1, MR2, FX2 and the Jarque-Bera test, four other procedures are used. Three of these procedures are standard Chow-type tests obtained by applying R, MR1 and MR2 to the data before reordering by leverage values. These procedures are included in the study in order to illustrate the usefulness of reordering. The remaining procedure is an LM test for heteroskedasticity of the errors u_i; Pagan and Hall (1983) have suggested that such checks may be effective general tests. In the experiments, the LM test is implemented in the form proposed by Koenker (1981), so that it is asymptotically robust to nonnormality. More precisely, the test statistic, denoted by LM(H), is calculated as n times the coefficient of determination in the auxiliary regression of the squared OLS residuals from (9) on an intercept and the squared OLS predicted values from (9); see Chapter 4 for some evidence on an F-variant of this procedure.

For each combination of n, c and error distribution, 1000 replications are obtained and used to estimate rejection probabilities. The nominal level of significance for all tests is five per cent. The critical values for R, MR1 and MR2 (and the corresponding Chow tests derived from the original data) are taken from F distributions with $n_1 = n/4$, $n_2 = n/2$, $n_3 = n/4$, and $k = 2$; the F(1, n - 3) distribution provides critical values for the FX2 test; the critical value for the heteroskedasticity test LM(H) is taken from the $\chi^2(1)$ distribution; and the LM(N) test for nonnormality uses the critical values proposed by Jarque and Bera (1987), rather than those for the $\chi^2(2)$ distribution.

4. Monte Carlo results

The results are presented in three tables. Table 1 relates to the behaviour of tests when the regression function is correctly specified. Table 2 contains estimates under misspecifications that lead to power estimates for FX2 in the region of 75 per cent when the u_i are normally distributed: such cases are regarded as representing important, but not overwhelming, errors. Estimates of rejection probabilities for more severe misspecifications are given in Table 3.

4.1 Correct specification

The nominal level of significance for all tests is 5 per cent, so that, with 1000 replications, estimates outside the range $5 \pm 2\ [5(95)/1000]^{1/2} = 5 \pm 1.4$ can be viewed as being inconsistent with the hypothesis that the actual finite sample significance level equals the nominal value.

Estimates of significance levels of the test statistics under the null hypothesis $c = 0$ are given in Table 1. It should be noted that entries for LM(N) for the four nonnormal distributions correspond to rejection probabilities under alternative hypotheses, rather than significance levels. The following comments apply to the results of this table.

(i) The Utts rainbow R test is satisfactory for the normal distribution, but is inadequate for the four other distributions with every estimate for these distributions differing significantly from the nominal value. The significance level estimates are too high under the log-normal, $\chi^2(2)$ and t(5) distributions and, as noted by Godfrey, McAleer and McKenzie (1988), these estimates tend to increase with the sample size. The Utts test is, however, undersized when the disturbances are drawn from a uniform distribution.

(ii) In contrast to Utts's original procedure, the modified tests MR1 and MR2 are always satisfactory and seem to be robust to nonnormality, even for $n = 20$.

Table 1

Rejection frequencies under correct specification of the regression function, i.e. c = 0

		distribution of disturbances				
n	Test	normal	log-normal	$\chi^2(2)$	t(5)	uniform
20	R	4.4	22.2[b]	13.6[b]	9.2[b]	2.9[b]
	MR1	5.7	5.2	4.0	4.6	4.9
	MR2	4.7	6.4	5.9	5.4	5.6
	LM(H)	3.1[b]	6.4	6.4	4.5	4.1
	FX2	4.3	5.4	4.7	5.9	4.6
	LM(N)[a]	6.1	79.4[b]	62.9[b]	20.5[b]	0.3[b]
40	R	4.3	26.9[b]	17.4[b]	12.0[b]	1.8[b]
	MR1	4.6	4.2	4.6	4.8	4.6
	MR2	5.1	6.2	4.8	5.7	5.5
	LM(H)	3.5[b]	5.1	5.7	5.0	5.1
	FX2	4.5	5.0	4.5	5.0	3.8
	LM(N)[a]	5.7	99.0[b]	93.5[b]	39.6[b]	1.0[b]
60	R	5.3	29.9[b]	17.7[b]	12.1[b]	2.0[b]
	MR1	4.7	4.5	5.5	4.7	5.7
	MR2	4.0	5.9	4.9	5.2	4.4
	LM(H)	3.8	5.6	5.1	4.2	4.5
	FX2	5.8	6.7[b]	4.1	4.9	4.1
	LM(N)[a]	7.2[b]	100.0[b]	99.6[b]	49.2[b]	23.0[b]

Notes:

a) Rejection frequencies for LM(N) given in the first column are estimates of significance levels. Estimates for LM(N) in all other columns are of power under nonnormal distributions.

b) Denotes values which lead to the rejection of the hypothesis that the actual significance level equals the nominal value of 5 per cent.

(iii) The LM(H) test behaves reasonably well, especially for the two larger sample sizes. Koenker's (1981) approach appears to be effective in making the test robust to nonnormality. Indeed, his variant performs better with nonnormal distributions than with normal errors.

(iv) The test based upon perfect information, viz. FX2, has estimated significance levels that are satisfactory, except for a value of 6.7 per cent in the case with n = 60 and log-normal disturbances.

(v) The estimated significance levels of LM(N) based upon critical values suggested by Jarque and Bera (1987) are a little high, especially for n = 60. The LM(N) test is effective in detecting the log-normal and $\chi^2(2)$ distributions, even for n = 20. It has a reasonable performance against the t(5) distribution, but has almost no power against the uniform distribution, except when n = 60.

The significance level estimates for Chow tests applied to data before reordering by leverage are quite similar to those of the corresponding tests evaluated after reordering and are not reported in detail. The similarity can be illustrated using the first row of results in Table 1. When n = 20, the estimated significance levels of Chow's prediction error test corresponding to Utts's test are 4.5, 22.3, 14.5, 9.3 and 2.2 per cent.

4.2 Moderate misspecification

Table 2 contains estimated rejection frequencies for "moderate" misspecification, loosely defined to be cases with nonzero values of c that lead to FX2 having power close to 0.75 when the errors are normally distributed. One issue to be examined in these cases is the importance of reordering the data before carrying out the tests of Section 2. Entries corresponding to the original and modified rainbow tests are, therefore, given in the form $\bar{e}/e$, where $\bar{e}$ denotes an estimate after reordering and e denotes an estimate before reordering. For example, with n = 60

Table 2
Rejection frequencies under moderate misspecification of the regression function

n	c	Test	distribution of disturbances normal	log-normal	$\chi^2(2)$	t(5)	uniform
20	0.125	R[b,c]	31/12	54/22	38/16	34/15	24/ 8
		MR1[b]	32/19	43/24	30/20	34/22	28/19
		MR2[b]	55/23	67/28	56/25	56/25	55/22
		LM(H)	22	28	24	21	21
		FX2	76	82	76	77	76
		LM(N)	13	56	38	17	6
40	0.07	R[b,c]	22/ 9	47/26	36/18	32/13	18/ 5
		MR1[b]	36/16	46/20	37/17	35/18	36/16
		MR2[b]	58/20	67/25	63/18	61/21	57/19
		LM(H)	26	27	25	28	35
		FX2	75	80	81	80	78
		LM(N)	11	94	82	34	4
60	0.05	R[b,c]	19/ 7	44/29	33/18	24/14	12/ 4
		MR1[b]	31/14	43/16	37/12	30/13	31/14
		MR2[b]	54/19	66/23	60/14	53/17	58/15
		LM(H)	26	24	22	20	35
		FX2	77	82	77	74	77
		LM(N)	10	100	99	44	7

Notes:

a) Combinations of n and c are selected so that, for normally distributed errors, the estimated power of the test derived for the correct alternative (namely, FX2) is approximately 75 per cent.

b) Entries corresponding to the original and modified rainbow tests are given as estimates after/before reordering on the basis of leverage.

c) Apart from the normal and uniform distributions, the Utts R test is oversized (see Table 1) and this feature should be borne in mind when comparing rejection frequencies for cases with $c \neq 0$.

and normal errors, the MR2 test carried out after reordering by leverage values has an estimated power of 54 per cent, while the corresponding analysis of covariance test using the original sequence of observations has an estimate of only 19 per cent. The estimates given in Table 2 suggest the following comments.

(i) Of the modified rainbow tests, MR2 consistently outperforms MR1. It is difficult to compare these tests to the original rainbow test because the latter test is oversized for nonnormal distributions, except for the uniform distribution for which it is undersized. However, the estimates for the original rainbow test are always lower than those for MR2 and it seems reasonable to conjecture that MR2 is the preferred procedure of the three rainbow tests. Further, the MR2 test nearly always has in excess of two-thirds of the estimated power of the FX2 test based upon perfect information about the specification error. This is a creditable performance for an information-parsimonious test.

(ii) The entries for the original and modified rainbow tests also reveal the value of reordering by leverage. Application of Chow tests to the original sequence of data points yields much lower estimates of power, with the proportional reductions being greatest for the MR2 test.

(iii) Although the omission of a quadratic term in the regressor might be expected to lead to significant evidence of heteroskedasticity in the OLS residuals, the LM(H) test is generally not very powerful. The estimate for MR2 is usually about twice the estimate for LM(H).

(iv) The power estimates for FX2 are not greatly affected by nonnormality.

(v) As indicated by the results of Table 1, the LM(N) test cannot be recommended as a general check for omitted variables or incorrect functional form. Even when neither of these errors had been committed, the use of this test could often lead to the rejection of models as a result of nonnormal disturbances. For the cases of Table 2 with normal errors, LM(N) is poor at rejecting false null hypotheses. It is interesting to note that in several cases rejection frequencies for LM(N) in Table 2 are lower than the corresponding estimates in Table 1, so that the joint presence of nonnormality and incorrect functional form can reduce the sensitivity of LM(N) to the error for which it was designed.

4.3 Severe misspecification

While the results of Table 2 are encouraging, it seems worthwhile to investigate the performance of tests in situations in which there is "severe" misspecification, that is, when the FX2 procedure rejects the restriction c = 0 with very high probability. Table 3 contains results for three representative cases. The LM(N) test is no longer included for the reasons given above.

The following features can be seen by considering the estimates of Table 3.

(i) As in the cases of Table 2, MR2 dominates the original Utts R, MR1 and LM(H) tests. Also reordering the data by leverage values again produces substantial increases in estimated powers for all three rainbow tests.

(ii) The performance of MR2 relative to the correct specific test, viz. FX2, has improved as the degree of specification error has increased. Indeed, the estimates for MR2 are usually within 10 per cent of those for FX2 and, in about half of the cases, the power estimates differ by less than 5 per cent.

5. The effects of alternative orderings

Implementing the modified rainbow tests MR1 and MR2 will not be computationally convenient if the program being used to estimate a linear regression model such as (1) does not provide leverage values. The importance of reordering the data has been stressed above and so, if the leverage values are not readily available, other bases for reordering should be considered. One alternative to the leverage approach is to reorder the data

Table 3
Rejection frequencies under severe misspecification of the regression function

			distribution of disturbances				
n[a]	c[a]	Test	normal	log-normal	$\chi^2(2)$	t(5)	uniform
40	0.10	R[b,c]	41/12	58/26	50/20	49/15	39/10
		MR1[b]	56/22	65/25	58/24	56/25	57/24
		MR2[b]	82/29	85/33	84/29	82/30	83/30
		LM(H)	39	40	39	44	46
		FX2	93	93	94	93	93
40	0.125	R[b,c]	58/15	66/27	64/22	64/18	58/14
		MR1[b]	68/26	77/29	70/30	70/29	71/30
		MR2[b]	92/35	91/39	92/37	92/37	91/37
		LM(H)	45	47	46	51	50
		FX2	97	97	97	96	97
60	0.10	R[b,c]	61/15	70/29	63/21	63/17	64/12
		MR1[b]	76/27	82/30	78/26	77/28	74/29
		MR2[b]	96/39	95/42	95/34	96/35	96/35
		LM(H)	64	53	55	54	68
		FX2	100[d]	98	99	99	99

Notes:
a) Combinations of n and c are selected so that FX2 has high estimated power.

b) Entries corresponding to the original and modified rainbow tests are given as estimates after/before reordering on the basis of leverage.

c) Apart from the normal and uniform distributions, the Utts R test is oversized (see Table 1) and this feature should be borne in mind when comparing rejection frequencies for cases with $c \neq 0$.

d) The actual estimate in this case is 99.5 per cent.

according to the values of one of the regressors - this type of reordering is sometimes used when implementing the Goldfeld-Quandt test for heteroskedasticity. Suppose, for example, that there is some concern about the modelling of the impact of the last regressor, with the null model being

$$y_t = \sum_{i=1}^{k-1} x_{ti}\beta_i + x_{tk}\beta_k + u_t, \; u_t \text{ iid}(0, \sigma_u^2),$$

and the incompletely specified alternative being

$$y_t = \sum_{i=1}^{k-1} x_{ti}\beta_i + h(x_{tk}; \xi) + u_t, \; u_t \text{ iid}(0, \sigma_u^2),$$

in which h(.) is an unspecified function of x_{tk} and unknown parameters ξ. The original data (y, X) can then be reordered according to the values of x_{tk}, t = 1, ..., n, to obtain $(\dot{y}, \dot{X})$ with $\dot{x}_{1k} \leq \dot{x}_{2k} \leq \ldots \leq \dot{x}_{nk}$.

An alternative to the MR2 test (which emerged as the preferred test according to the results in the previous section) can then be derived by testing the 2k restrictions $\pi = (\pi_1', \pi_2')' = 0$ in the expanded model

$$\dot{y} = \dot{X}\beta + \dot{V}\pi + \dot{u}, \; \dot{u} \sim D(0, \sigma_u^2 I_n), \tag{7A}$$

where the matrix $\dot{V}$ is defined by

$$\dot{V} = \begin{bmatrix} \dot{X}_1 & 0 \\ 0 & \dot{X}_2 \\ 0 & 0 \end{bmatrix}.$$

Equation (7A) should be compared with equation (7). Let the F-statistic for testing $\pi = 0$ in (7A) be denoted the alternative modified rainbow test - version 2 (or AMR2), so that AMR2 is distributed as F(2k, n-3k) in large samples when (1) is correctly specified. Since MR2 and AMR2 test the same number of restrictions, it is fairly straightforward to compare these procedures and hence to examine the effects of alternative schemes for reordering the original data.

As MR2 and AMR2 have the same asymptotic distribution (and the same finite sample distribution under normality and nonstochastic regressors) when the null hypothesis is true, the effects of alternative schemes for reordering are confined to power. If asymptotic local powers of MR2 and AMR2 are to be compared, then the results provided in Chapter 3 by Eastwood and Godfrey can be employed. More precisely, suppose that power comparisons are to be made under the sequence of local alternatives

$$y = X\beta + F(n^{-1/2}\eta) + u,$$

in which F is an appropriate matrix of estimated first order partial derivatives and η is a vector with elements that are O(1); see equation (15) of Section 2.3 of Chapter 3. Let the noncentrality parameters of the MR2 and AMR2 tests in χ^2 form be denoted by ν^2_{MR2} and ν^2_{AMR2}, respectively. Equation (17) of Section 2.3 of Chapter 3 then implies that

$$\nu^2_{MR2} - \nu^2_{AMR2} = (\Delta_{AMR2} - \Delta_{MR2})/\sigma^2_u,$$

in which Δ_{AMR2} (resp. Δ_{MR2}) is the probability limit of the residual sum of squares from the OLS regression of $n^{-1/2}\dot{F}\eta$ (resp. $n^{-1/2}\bar{F}\eta$) on $(\dot{X}, \dot{V})$ of (7A) (resp. $(\bar{X}, \bar{Z})$ of (7)).

The sign of $(\Delta_{AMR2} - \Delta_{MR2})$ is indeterminate and so it is not possible to obtain a generally valid ranking of MR2 and AMR2 in terms of asymptotic local power. Similarly a comparison of approximate slopes (see Geweke, 1981) yields inconclusive results because the probability limit of $\dot{y}'M(\dot{X}, \dot{V})\dot{y}/\bar{y}'M(\bar{X}, \bar{Z})\bar{y}$ can be greater than or less than one under fixed alternatives, where M(B) denotes $[I_n - B(B'B)^{-1}B']$ for a n by b matrix B with full column rank. Given precise information about the true data generating process, it may be possible to rank the powers of MR2 and some alternative AMR2-type procedures. However, if such information were available, then there would be little incentive to use such general checks and a specific test would presumably be preferred.

In the absence of information about the form of the misspecification, it would be possible to use each regressor in turn to reorder the data and then to select the maximum of the associated test statistics, as well as MR2 if leverage points are used. This strategy is similar to one discussed by White (1980) in the context of general checks derived by comparing OLS and weighted least squares estimators of β. The significance level associated with this method is unknown, even asymptotically, but a joint test with known asymptotic size could be constructed using variable addition techniques; see Godfrey (1988, pp.155-7).

6. Conclusions

The rainbow test of Utts (1982) for testing a linear regression model against a broad range of alternative hypotheses has been examined in detail. Since the rainbow test is sensitive to nonnormality of regression disturbances, even in large samples, it cannot be recommended as a robust general check for omitted variables or incorrect functional form. We therefore propose modifications of the rainbow test. Utts's original test may be interpreted, after suitable reordering of the data, as Chow's (1960) prediction error test. The modifications proposed can be similarly regarded as analysis of covariance tests for two or three different structures, being based upon alternative estimates of the parameters of interest. The small sample properties of the original rainbow test, the two modifications derived above, and other diagnostic checks are investigated using Monte Carlo experiments, with special attention being paid to the effects of nonnormal error distributions. The modified rainbow tests are found to be robust to nonnormality and one of them, viz. the MR2 procedure, emerges as the preferred test, being quite powerful in rejecting false models.

The modified rainbow tests, like the procedure originally proposed by Utts, rely upon the leverage values to provide a basis for reordering the data in order to achieve power in the presence of specification errors. Since other schemes for reordering observations could also be adopted, the impact of the choice of reordering scheme has been examined. Perhaps not surprisingly, it has been shown that it is difficult to rank general checks derived from different orderings unless there is precise information about the data generation process.

There are several topics of interest that might be pursued in future work. The experiments employed by Thursby (1989) to compare different specification error tests for linear regression models could be generalised to include the modified rainbow test MR2. The extension of the modified rainbow tests beyond the OLS estimation of a multiple regression model also merits consideration, but will not be without difficulties.

In order to illustrate these difficulties, consider the problem of extending the results above to be relevant to relationships with limited or qualitative dependent variables. (Widely used examples of such relationships include the logit and probit models for binary choice, the truncated regression model, and the tobit model; see Maddala (1983) for a discussion of estimation methods and test procedures for these and other models.) An important feature of models involving limited or qualitative dependent variables is that incorrect specification of the error distribution usually leads to estimator inconsistency; see, for example, Arabmazar and Schmidt (1982) and Robinson (1982). This effect is in marked contrast to the case of the linear regression model of Section 2 in which nonnormality is asymptotically irrelevant, provided conditions for suitable central limit theorems are satisfied. There is, therefore, no prospect of generalising a test such as MR2 to obtain a procedure that is robust to the incorrect specification of the distribution of the disturbances. Moreover, it is difficult to devise an appropriate analogue of the hat matrix upon which to base reordering of data. The general idea of reordering data and then carrying out tests of parameter constancy can, however, be used.

The reordering of the data can be based upon the values of a particular regressor, as described in Section 5 above. Alternatively, if the data being used for estimation and testing are derived from some survey of households or individuals, it may be useful to partition the sample according to socio-demographic criteria. Utts's proposal for the division of the data will, in general, be inappropriate if the second approach is adopted. Let the number of subsamples (socio-demographic subgroups) be denoted by G, and the numbers of observations in these subsamples be denoted by $n_g = O(n)$, $g = 1, \ldots, G$. Checks for parameter constancy using the selected partitioning of the data can then be carried out by means of the likelihood ratio (LR) test. The assumption of independent observations implies that an LR test statistic is defined by

$$LR_G = -2[\hat{\ell}_0 - \sum_{g=1}^{G} \hat{\ell}_g],$$

in which $\hat{\ell}_0$ is the maximised log-likelihood using the full sample of n observations and $\hat{\ell}_g$ $(g = 1, \ldots, G)$ denotes the corresponding function obtained by estimation using only the gth subsample of reordered data. The statistic LR_G is likely to be easy to implement in applied work because estimation programs usually provide maximised log-likelihood functions and allow subsample estimation. Moreover some programs, e.g. LIMDEP (Greene, 1989), permit reordering according to the values of a specified regressor. As an alternative to a check of parameter constancy, a general test for the misspecification of models with limited or qualitative dependent variables can be obtained using White's

(1982) information matrix test. The sampling performance of different forms of the information matrix test is discussed in Chapter 7.

References

Arabmazar, A. and P. Schmidt, (1982), An investigation of the robustness of the Tobit estimator to non-normality, *Econometrica*, 50, 1055-63.

Breusch, T.S. and L.G. Godfrey, (1986), Data transformation tests, *Economic Journal*, 96, 45-58.

Chow, G.C., (1960), Tests of equality between sets of coefficients in two linear regressions, *Econometrica*, 28, 591-605.

Davidson, R. and J.G. MacKinnon, (1987), Implicit alternatives and the local power of test statistics, *Econometrica*, 55, 1305-29.

Geweke, J., (1981), The approximate slopes of econometric tests, *Econometrica*, 49, 1427-42.

Godfrey, L.G., (1988), *Misspecification Tests in Econometrics: the Lagrange Multiplier Principle and other Approaches*. Cambridge University Press: Cambridge.

Godfrey, L.G., M. McAleer and C.R. McKenzie, (1988), Variable addition and Lagrange multiplier tests for linear and logarithmic regression models, *Review of Economics and Statistics*, 70, 492-503.

Greene, W.H., (1989), *LIMDEP*, Version 5.1, Econometric Software Inc., New York.

Hendry, D.F., (1979), Predictive failure and econometric modelling in macroeconomics: the transactions demand for money, in *Economic Modelling*, P. Omerod (ed.), London: Heinemann Educational Books, 217-42.

Jarque, C.M. and A.K. Bera, (1987), An efficient large-sample test for normality of observations and regression residuals,

International Statistical Review, 55, 163-72.

Kmenta, J., (1986), *Elements of Econometrics*, second edition, New York: Macmillan.

Koenker, R., (1981), A note on studentizing a test for heteroscedasticity, *Journal of Econometrics*, 17, 107-12.

Krämer, W., H. Sonnberger, J. Maurer and P. Havlik, (1985), Diagnostic checking in practice, *Review of Economics and Statistics*, 67, 118-23.

Maddala, G.S., (1983), *Limited-dependent and Qualitative Variables in Econometrics*. Cambridge: Cambridge University Press.

McAleer, M. and Y.K. Tse, (1988), A sequential testing procedure for outliers and structural change, *Econometric Reviews*, 7, 103-11.

Pagan, A.R., (1984), Model evaluation by variable addition, in *Econometrics and Quantitative Economics*, D.F. Hendry and K.F. Wallis (eds.), Oxford: Blackwell, 103-33.

Pagan, A.R. and A.D. Hall, (1983), Diagnostic tests as residual analysis, *Econometric Reviews*, 2, 159-218.

Plosser, C.I., G.W. Schwert and H. White, (1982), Differencing as a test of specification, *International Economic Review*, 23, 535-52.

Ramsey, J.B., (1969), Tests for specification error in classical least-squares regression analysis, *Journal of the Royal Statistical Society*, Series B, 31, 350-71.

Ramsey, J.B., (1983), Comment on diagnostic tests as residual analysis, *Econometric Reviews*, 2, 241-8.

Robinson, P.M., (1982), On the asymptotic properties of estimators of models containing limited dependent variables, *Econometrica*, 50, 27-41.

Salkever, D.S., (1976), The use of dummy variables to compute predictions, prediction errors, and confidence intervals, *Journal of Econometrics*, 4, 393-7.

Thursby, J.G., (1982), Misspecification, heteroscedasticity, and the Chow and Goldfeld-Quandt tests, *Review of Economics and Statistics*, 64, 314-21.

Thursby, J.G., (1989), A comparison of several specification error tests for a general alternative, *International Economic Review*, 30, 217-30.

Thursby, J.G. and P. Schmidt, (1977), Some properties of tests for specification error in a linear regression model, *Journal of the American Statistical Association*, 72, 635-41.

Utts, J.M., (1982), The rainbow test for lack of fit in regression, *Communications in Statistics - Theory and Methods*, 11, 2801-15.

White, H., (1980), Using least squares to approximate unknown regression functions, *International Economic Review*, 21, 147-70.

White, H., (1982), Maximum likelihood estimation of misspecified models, *Econometrica*, 50, 1-25.

C. D. ORME

6. THE DISTRIBUTION OF MAXIMUM LIKELIHOOD CORRECTIONS UNDER LOCAL MISSPECIFICATION [1]

1. Introduction

It is now well established in the literature that maximum likelihood (ML) estimation of a model which is misspecified (by the imposition of incorrect restrictions on the parameters, say, or by incorrect distributional assumptions in general) can yield estimates which are inconsistent (see, for example, White (1982)). A potential source of misspecification when estimating models using cross-section data is neglected heterogeneity; that is, incorrectly maintaining that certain parameters are constant across observations when in fact they are not. The consequences of such misspecification in the context of a normal linear model where the dependent variable is grouped or censored (e.g., the tobit or probit models) have been highlighted by Chesher, Lancaster and Irish (1985).

For general models, Chesher (1984) developed a score test of the hypothesis that parameters have zero variance.[2] This, rather interestingly, turned out to be an information matrix (IM) test which was originally introduced by White (1982) as a general test for misspecification, and which is discussed at some length in

---

[1] A large part of the material presented in this chapter is drawn from Orme (1989 a,b).

[2] See also Chesher (1983).

Chapter 7. The test derived by Chesher does not depend upon the specification of a particular distribution for the parameters under the alternative. Thus, although parameter variation may be the specification error, the precise form of the misspecification will in general remain unknown.

Lancaster (1985) derived expressions giving linear approximations to the local asymptotic bias of maximum likelihood estimates (MLEs) in the presence of neglected heterogeneity of unknown form.[3] Drawing these two strands of the literature together Lancaster and Chesher (1985) (in the context of duration models) and Chesher, Lancaster and Irish (1985) (in the context of general models) proposed a simple correction procedure to remove the local asymptotic bias from MLEs due to parameter variation when departures from the null are detected, but believed to be small. These corrections are based on the IM test statistic and require no knowledge of the precise form of the misspecification. Their work was incomplete, however, in that the asymptotic distribution of these corrections was not established.

Independently of Lancaster (1985), Kiefer and Skoog (1984) gave linear approximations to the local asymptotic bias of MLEs, in general, when subject to the imposition of incorrect restrictions. Orme (1986), applying the reasoning of Chesher, Lancaster and Irish to this general case, provided corrections and their asymptotic distribution, based on the score test which examined the validity of the imposed restrictions (see Breusch and Pagan (1980)). It is assumed in the analysis of Orme (1986) that there is a fully specified alternative model and the corrections produced are asymptotically equivalent to the unrestricted MLEs obtained from estimating the alternative model.

This chapter discusses four issues. Firstly the construction and asymptotic distribution of maximum likelihood corrections in the presence of local parameter variation of unknown form. This would appear to be an interesting question, particularly with

---

[3] That is when the misspecification is local or small.

regard to models of cross-section data where such misspecification can have serious consequences for parameter estimator consistency, but where (maximum likelihood) estimation of an appropriate alternative becomes intractable since higher order integrals enter the likelihood specification as more parameters are allowed to vary.

Secondly, the question arises concerning the extent to which ML corrections, based on the IM test statistic, will be useful in the presence of general unknown local specification error. This essentially draws on the work of Chesher, Lancaster and Irish (1985) who demonstrated that "minimal parameter variation" induces heteroskedasticity, skewness and super-normal kurtosis in the normal linear model and that a score test for parameter constancy (IM test) will examine such departures from standard assumptions even when the dependent variable is grouped or censored. One example considered in this chapter is a probit model in which the unknown specification error is one of parameter variation, heteroskedasticity or non-normality. A second example follows up the investigations of Chesher, Lancaster and Irish (1985) by considering the method of correcting tobit regression parameter estimates for neglected heteroskedasticity of unknown form. Such a misspecification arises when a subset of the regression parameters vary randomly in the population but are, mistakenly, assumed constant. This procedure, which uses a double length artificial regression in order to calculate the efficient version of the IM test and associated corrections, readily generalises to linear models with different forms of censoring. In the case of the probit model, only a single length regression is required.

As a measure of relative sampling performance, corrections are compared with uncorrected MLEs by considering an 'asymptotic' mean square error criterion. The above three issues are dealt with in Sections 2, 3 and 4.

Finally, in Section 5, the idea of local corrections is used to construct tests of parameter consistency. This is a straightforward procedure which contrasts the maximum likelihood parameter estimator with the correction, along the lines of

Hausman (1978).

Some Monte Carlo evidence on the sampling performance of these corrections in probit and tobit models is presented in Section 6.

2. Specification tests and maximum likelihood corrections

We shall consider the situation in which an econometrician is interested in modelling one or more endogenous variables, y_i, as a function of exogenous variables, x_i', within a general stochastic framework, where $u_i = (y_i', x_i')'$ form a sequence of independently and identically distributed random variables, $i = 1,...,n$.[4] The joint density of u is assumed, up to a knowledge of the marginal density of x, to have parametric form $f(u;\theta) = f_1(y|x;\theta).f_2(x)$, i.e., the product of the conditional density of y given x and the marginal density of x which, since x is exogenous, does not depend upon θ. The "true" value of the $(k \times 1)$ parameter vector θ will be denoted θ° and is estimated, under the maintained assumption, by the MLE, $\hat{\theta}$, which is given by the solution to

$$\max_{\theta} \sum_{i=1}^{n} \log\{f_1(y_i|x_i;\theta)\} \tag{1}$$

based on the available independent observations on y and x (i.e., $\hat{\theta}$ maximises the conditional log-likelihood). In what follows, unless stated otherwise, summations will be over the sample observations, $i = 1,...,n$.

Recent work (e.g. White (1982)) has shown that the properties of $\hat{\theta}$ are sensitive to correctly specifying the conditional density $f_1(y|x;\theta)$.[5] Testing the model specification is thus desirable and

---

[4] This assumption is somewhat strong in the context of time series models but is often assumed when analysing cross-section data. I shall later refer to Newey (1985) who also makes this assumption.

[5] Misspecifying $f_2(x)$ is inconsequential (in general, it is unknown anyway) provided that in truth it does not depend upon θ.

a specification test examines the statistical significance of some chosen statistic which shall be denoted $D_n(\mathbf{u};\hat{\theta}) = n^{-1}\Sigma\ d(u_i;\hat{\theta})$, where $\mathbf{u} = (u_1,u_2,...,u_n)'$, and the $(q \times 1)$ vector $d(u;\theta°)$ satisfies

$$E[d(u;\theta°)\,|\,x] = 0 \tag{2}$$

and $E[.\,|\,x]$ denotes expectation conditional on x taken at $f(u;\theta°)$. Thus, by the law of iterative expectations, $d(u;\theta°)$ is uncorrelated with x and $E[d(u;\theta°)] = 0$, where $E[.]$ denotes expectation (unconditional) taken at $f(u;\theta°)$. The idea is that $D_n(\mathbf{u};\hat{\theta})$ should not differ significantly from zero, in large samples, if (2) is satisfied. Significant departures from zero may indicate model misspecification. Such specification tests have been examined by Newey (1985) who called them "conditional moment" (CM) tests.

For the econometrician, the ability of a model to pass a specification test is reassuring. However, its inability to do so poses a difficult problem: if the assumed specification is incorrect what might the correct specification be ? If the answer is not known precisely then it would seem that little else can be achieved, within a parametric framework, and the researcher is left with estimates which may be inconsistent for the parameters of interest. Expressions for the inconsistency of MLEs are difficult to obtain, in general, even when the true model is known. If, however, the true model represents a local departure from the maintained specification (to be defined formally below for the present situation; but see also, for example, Godfrey (1988)) then expressions for the "local" inconsistency of MLEs have been established by Kiefer and Skoog (1984) (see also Yatchew and Griliches (1985)). But even so, these depend upon the specification of the true model.

In an attempt to address these issues, the purpose of this section is

(i) to extend the work of Chesher, Lancaster and Irish (1985) (and also Orme (1986)) by showing how, without being precise about the true model specification, MLEs can be updated to remove the local inconsistency due to local specification error;

and

(ii) to obtain the asymptotic distribution of these corrected MLEs.

In what follows we adopt Newey's (1985) assumptions and appeal to his results in order to derive the asymptotic distribution of these corrected estimates in the presence of local specification error of unknown form.

The first assumption concerns the true data generation process (DGP) which corresponds to local, unknown, departures from the maintained specification $f(u;\theta)$. The true joint density of u is $p(u;\theta^\circ,\alpha_n)$ where $\alpha_n = \alpha^\circ + \delta/\sqrt{n}$, $\delta'\delta < \infty$.[6] The parametric form of p(.) is unknown, but the maintained model is correctly specified at α° which assumes that, for all θ in Θ, $p(u;\theta,\alpha^\circ) = f(u;\theta)$. Thus $c_n = \alpha_n - \alpha^\circ$ measures the extent of the misspecification. Now $c_n = O(n^{-1/2})$ is a convenient mathematical device but is regarded as a metaphor for "$c = \alpha - \alpha^\circ$" small.

Testing for parameter variation

For the example of neglected heterogeneity, misspecification arises because the parameter vector θ (or a certain sub-vector of θ) is assumed fixed when in fact it is the realisation of a random variable Θ. The maintained specification is correct when $\mathrm{var}(\Theta) = 0$. Chesher (1984) very elegantly derives a specification error test statistic (examining whether $\mathrm{var}(\Theta) = 0$) without precisely specifying a particular distribution for Θ. The resulting test statistic is the information matrix (IM) test (see White (1982)) which is a CM test based on the following q distinct quantities

---

[6] For instance, $p(u;\theta,\alpha_n) = p_1(y|x;\theta;\alpha_n)f_2(x)$ or $p(u;\theta;\alpha_n) = p_1(y|x;\theta,\alpha_n)p_2(x;\alpha_n)$. However, $p(u;\theta,\alpha_n) = f_1(y|x;\theta)p_2(x;\alpha_n)$ is inconsequential for reasons outlined in footnote 5.

$$d_h(u;\theta) = \partial \log\{f_1(y|x;\theta)\}/\partial\theta_j \cdot \partial \log\{f_1(y|x;\theta)\}/\partial\theta_g + \partial^2 \log\{f_1(y|x;\theta)\}/\partial\theta_j \partial\theta_g \ ,$$

$$(h = 1,\ldots,q; \quad j = 1,\ldots,g; \quad g = 1,\ldots,k)$$

and the test examines the statistical significance of $D_n(\mathbf{u};\hat{\theta})$, where $\hat{\theta}$ solves (1).

We briefly summarise Chesher's results since they will be of use later. Chesher assumes that Θ is a continuous variate with elements being twice differentiable functions, $\Theta(\Gamma)$, of a $(k \times 1)$ random vector Γ. Since we are only interested in local misspecification, without loss of generality, we can assume that $\Theta - \theta_0 = Q(\Gamma - \gamma_0)$, $E(\Gamma) = \gamma_0$, i.e., $\theta_0 = Q\gamma_0$, (expectations taken with respect to the probability density function (*pdf*) of Γ) where Q, $(k \times k)$, is finite with elements not all equal to zero but generally unknown. The vector Γ has a proper absolutely continuous *pdf* defined over $\mathbb{R}^k$, $s(\gamma)$, which can be written for all γ as

$$s(\gamma) = \eta^{-k/2} h\{(\gamma-\gamma_0)'(\gamma-\gamma_0)/\eta\}$$

where $h\{.\}$ does not depend upon $\eta > 0$ and such that the elements of the random vector $W = (\Gamma - \gamma_0)/\sqrt{\eta}$, $\eta > 0$, are uncorrelated with zero means and finite variances equal to $\lambda^2 = \int w_j^2 h\{w'w\}dw$, $j = 1,\ldots,k$.

Thus $\text{var}(\Theta) = \eta\lambda^2 QQ'$ and the marginal *pdf* of y with respect to Θ (conditional on x) is then

$$p_1(y|x;\theta_0,\eta) = \int_{\mathbb{R}^k} f_1(y|x;\theta(\gamma))s(\gamma)d\gamma \ ,$$

which is unknown because the precise form of $s(\gamma)$ is not specified.

At this point it might be beneficial to draw the reader's attention to the notation and its implications. Maximum likelihood estimation of the maintained model yields $\hat{\theta}$ as an estimate θ_0, the "mean" of Θ, which is unknown but has "true" value θ_0°.

Essentially, the parameter of interest is θ_0 whereas, in general, it would usually be written θ. It is hoped that this will not cause too much confusion.

Writing $\Omega = \text{var}(\Theta)$, a test of $\Omega = 0$ was derived by Chesher as a score test of $\eta = 0$. The important point here is that, although the parametric form of $p_1(.)$ is not specified precisely, Chesher nevertheless shows that the mean score for η evaluated at $\theta_0 = \hat{\theta}$, $\eta = 0$ is [7]

$$\frac{\lambda^2}{2} \frac{1}{n} \sum_i \text{trace}\left\{QQ'D(u_i;\hat{\theta})\right\} \tag{3}$$

where

$$D(u_i;\theta) = \frac{\partial \log f_1}{\partial \theta} \cdot \frac{\partial \log f_1}{\partial \theta'} + \frac{\partial^2 \log f_1}{\partial \theta \partial \theta'}$$

and $f_1(y|x;\theta_0) = p_1(y|x;\theta_0,0)$.

Now,

$$\text{tr}\{QQ'D(u_i;\hat{\theta})\} = \{\text{vec}(QQ')\}'\text{vec}(D(u_i;\hat{\theta})) = r'd(u_i;\hat{\theta})$$

where $r = (M'M)\text{vech}(QQ')$ and $d(u_i;\theta_0) = \text{vech}\{D(u_i;\theta_0)\}$.[8]

Thus the mean score for η, (3), can be written

---

[7] See Appendix A. Although the test statistic is derived for local specification error, Chesher (1984) in actual fact derives the same test statistic without recourse to this local approximation.

[8] (i) vec(A) column stacks the matrix A;
(ii) vech(A) eliminates supra-diagonal elements from vec(A) for a symmetric matrix A. Further vec(A)=Mvech(A) where M is the duplication matrix and A is symmetric.
(iii) M'M is a diagonal matrix with non-zero elements equal to 1 or 2; see Magnus and Neudecker (1980).

$$\frac{\lambda^2}{2} r' \frac{1}{n} \sum_i d(u_i;\hat{\theta}) \tag{4}$$

which equals zero for all $r \neq 0$ if and only if $n^{-1}\Sigma\, d(u_i;\hat{\theta})$ equals zero. So the test examines the statistical significance of $D_n(\mathbf{u};\hat{\theta}) = n^{-1}\Sigma\, d(u_i;\hat{\theta})$, which is an IM test.

Although his analysis only considers symmetric distributions for the varying parameters, Chesher points out in a footnote that the same test statistic emerges from a variety of non-symmetric distributions as well. Thus the IM test is sensitive to neglected heterogeneity for a realistically wide range of alternative distributional assumptions about Θ.

General conditional moment tests and corrections

Proceeding generally, for the moment, suppose the CM test $D_n(\mathbf{u};\hat{\theta})$, not necessarily an IM test, is employed to examine the correctness of the maintained model specification; where $E[d(u;\theta^\circ)\,|\,x] = 0$, under the maintained model specification, but where the true DGP is given by $p(u;\theta^\circ,\alpha_n)$. To facilitate the forthcoming exposition the following matrices are defined :

$$\mathcal{I}_{\theta\theta} = E\left[\frac{\partial \log\{f_1(y\,|\,x;\theta^\circ)}{\partial\theta} \cdot \frac{\partial \log\{f_1(y\,|\,x;\theta^\circ)}{\partial\theta'}\right],$$

$$\mathcal{I}_{\theta\alpha} = E\left[\frac{\partial \log\{f_1(y\,|\,x;\theta^\circ)}{\partial\theta} \cdot \frac{\partial \log\{p(u;\theta^\circ)}{\partial\alpha'}\right],$$

$$\mathcal{I}_{d\theta} = E\left[d(u_i;\theta^\circ)\,\frac{\partial \log\{f_1(y\,|\,x;\theta^\circ)}{\partial\theta'}\right] = -\,E\left[\frac{\partial d(u\,;\,\theta^\circ)}{\partial\theta'}\right],$$

$$\mathcal{I}_{d\alpha} = E\left[d(u;\theta^\circ)\,\frac{\partial \log\{p(u;\theta^\circ,\alpha^\circ)\}}{\partial\alpha'}\right],$$

$$\mathcal{I}_{dd} = E\left[d(u;\theta^\circ)\ d(u;\theta^\circ)'\right],$$

and

$$\mathcal{H}' = \left[\mathcal{I}_{\alpha\theta}, \mathcal{I}_{\alpha d}\right]$$

where expectations are taken conditionally on x, then unconditionally with respect to the maintained model $p(u;\theta°,\alpha°) = f(u;\theta°)$.[9] The equality for $\mathcal{I}_{d\theta}$ ($= \mathcal{I}_{\theta d}'$) is a generalisation ofthe information matrix equality, for any zero mean random variable $d(u;\theta°)$, as pointed out by Newey (1985). Then under the regularity assumptions of Newey (1985), his Lemma 2.1 can be modified for the present purposes to give [10]

$$\sqrt{n}(\hat{\theta} - \theta°) \cong \mathcal{I}_{\theta\theta}^{-1} \frac{1}{\sqrt{n}} \sum_i \frac{\partial \log\{f_1(y|x;\theta°)\}}{\partial\theta} \tag{5}$$

and

$$\sqrt{n}\ D_n(\mathbf{u};\hat{\theta}) \cong \frac{1}{\sqrt{n}} \sum_i d(u_i;\theta°) - \mathcal{I}_{d\theta}\sqrt{n}(\hat{\theta} - \theta°). \tag{6}$$

Substituting (5) into (6) gives

$$\sqrt{n}\ D_n(\mathbf{u};\hat{\theta}) = \left\{ -\mathcal{I}_{d\theta}\mathcal{I}_{\theta\theta}^{-1},\ I_q\right\} \frac{1}{\sqrt{n}} \sum_i g(u_i;\theta°) \tag{7}$$

where $g(u;\theta°) = (\partial \log f_1(y|x;\theta°)/\partial\theta', d(u;\theta°)')'$ and I_q is the $(q \times q)$ identity matrix. We assume that there are no linear dependencies among the elements of $g(u;\theta°)$ so that the matrix $\mathcal{I}^*$ is positive definite, where

---

[9] $f_2(x)$ does not depend upon θ, so $\partial \log f_1(.)/\partial\theta = \partial \log f(.)/\partial\theta$.

[10] In what follows $\cong$ denotes "asymptotically equivalent to". The two sides of $\cong$ have the same probability limit. (5) is achieved by expanding the function $n^{-1/2}\sum \partial \log f_1(y|x;\theta)/\partial\theta$ (evaluated at $\hat{\theta}$) about $\theta°$ and ignoring asymptotically negligible terms. Similarly, $\sqrt{n}\ D_n(\mathbf{u};\hat{\theta})$ is expanded about $\theta°$.

$$\mathcal{I}^* = \begin{bmatrix} \mathcal{I}_{\theta\theta} & \mathcal{I}_{\theta d} \\ \mathcal{I}_{d\theta} & \mathcal{I}_{dd} \end{bmatrix}.$$

Newey further proves that

$$\sqrt{n}\ G_n(\mathbf{u};\theta^\circ) \xrightarrow{d} N(\mathcal{H}\delta,\ \mathcal{I}^*) \tag{8}$$

where $G_n(\mathbf{u};\theta^\circ) = n^{-1}\Sigma\ g(u_i;\theta^\circ)$ and $\delta = \sqrt{n}(\alpha_n - \alpha^\circ)$. This result together with (5) implies that the asymptotic distribution of $\hat{\theta}$ is

$$\sqrt{n}(\hat{\theta} - \theta^\circ) \xrightarrow{d} N(\mathcal{I}_{\theta\theta}^{-1}\mathcal{I}_{\theta\alpha}\delta,\ \mathcal{I}_{\theta\theta}^{-1}) \tag{9}$$

which is the familiar result (see, for example, Newey (1985)). Notice that if $p(u;\theta,\alpha) = f_1(y|x;\theta)p_2(x;\alpha)$ then, as suggested in footnote 5, the misspecification is inconsequential in that $\mathcal{I}_{\theta\alpha} = 0$ (using iterative expectations under the maintained model) and the usual asymptotic distribution results are valid. Upon substitution of (8) into (7) and noting that

$$(-\mathcal{I}_{d\theta}\mathcal{I}_{\theta\theta}^{-1},\ I_q)\mathcal{I}^*(-\mathcal{I}_{d\theta}\mathcal{I}_{\theta\theta}^{-1},\ I_q)' = \mathcal{I}_{dd} - \mathcal{I}_{d\theta}\{\mathcal{I}_{\theta\theta}^{-1}\}\mathcal{I}_{\theta d} \equiv \{\mathcal{I}^{dd}\}^{-1},$$

where $\mathcal{I}^{dd}$ is the bottom right $(q \times q)$ partition in the inverse of $\mathcal{I}^*$, we see that

$$\sqrt{n}\ D_n(\mathbf{u};\hat{\theta}) \xrightarrow{d} N\left(\{\mathcal{I}_{d\alpha} - \mathcal{I}_{d\theta}\mathcal{I}_{\theta\theta}^{-1}\mathcal{I}_{\theta\alpha}\}\delta,\ \{\mathcal{I}^{dd}\}^{-1}\right). \tag{10}$$

Adopting the assumption of a sequence of local departures which converges to the maintained specification at a rate of $n^{-1/2}$ is a mathematical device that enables us to derive the asymptotic distribution of, say, $\sqrt{n}(\hat{\theta} - \theta^\circ)$ and $\sqrt{n}\ D_n(\mathbf{u};\hat{\theta})$, as above. However, for a fixed departure these could also be regarded as giving the approximate limiting distributional results when the specification error is small, i.e., $\alpha - \alpha^\circ = c$ small. For example, for small c, the approximate sampling distribution of $\hat{\theta}$ (in large samples) could be regarded as $N\left(\theta^\circ + \mathcal{I}_{\theta\theta}^{-1}\mathcal{I}_{\theta\alpha}c,\ \frac{1}{n}\mathcal{I}_{\theta\theta}^{-1}\right)$. In particular, denoting the probability limit of $\hat{\theta}$ to be $\theta^*(c)$ (with $\theta^\circ(0) = \theta^\circ$), and in the spirit of Kiefer and Skoog (1984) or Lancaster (1985), a "small c" approximation to this probability limit is obtained by taking a Taylor series expansion of $\theta^*(c)$ about c = 0. This yields

$$\theta^*(c) = \theta^\circ + \left\{ \frac{\partial\theta^*(0)}{\partial c'} \right\} c + o(c).$$

The expression for $\partial\theta^*(0)/\partial c'$ can be obtained by totally differentiating the "expected" likelihood equations (which solve to yield θ^*), i.e.,

$$E_p\left\{ \frac{\partial \log\{f_1(y|x;\theta^*)}{\partial\theta} \right\} = \int \left\{ \frac{\partial \log\{f_1(y|x;\theta^*)}{\partial\theta} \right\} p(u;\theta,\alpha)du = 0$$

$$\text{for all } \theta,\ \alpha,$$

which yields (Gourieroux, Monfort and Trognon (1983), Lancaster (1985))

$$\int \left\{\partial^2 \log\{f_1(y|x;\theta^*)\}/\partial\theta\partial\theta'\right\} p(u;\theta,\alpha)du \left\{\frac{\partial\theta^*(0)}{\partial c'}\right\}$$

$$+ \int \left\{\partial \log\{f_1(y|x;\theta^*\})/\partial\theta\right\} \left\{\partial \log\{p(u;\theta,\alpha)\}/\partial\alpha'\right\} p(u;\theta,\alpha)du = 0.$$

Evaluating the above at $\theta = \theta^\circ$, $\alpha = \alpha^\circ$ (i.e., c = 0) and substituting into the previous expression yields

$$\theta^*(c) = \theta^\circ + \{\mathcal{I}_{\theta\theta}\}^{-1}\mathcal{I}_{\theta\alpha}c + o(c).$$

Chesher and Lancaster (1985) and Chesher, Lancaster and Irish (1985) suggested "updating" $\hat{\theta}$ to remove this local asymptotic bias by forming $\bar{\theta} = \hat{\theta} + \hat{\varphi}$, where $\hat{\varphi}$ is the update and, to a linear approximation, plim $\hat{\varphi} = -\{\mathcal{I}_{\theta\theta}\}^{-1}\mathcal{I}_{\theta\alpha}c$. Formally, in the presence of a sequence of local departures, $c_n = \delta/\sqrt{n}$, this sort of updating procedure removes the asymptotic bias from the limiting distribution of $\sqrt{n}(\hat{\theta} - \theta°)$. For practical purposes, however, we interpret this as removing the approximate inconsistency from $\hat{\theta}$ when departures from the null specification are small.

The choice of the appropriate update, $\hat{\varphi}$, is difficult, it appears, since it depends upon the specification of p(.) which is unknown. In the analysis of Orme (1986), p(.) was completely specified. However, if

$$\frac{\partial \log\{p(u;\theta;\alpha°)\}}{\partial\alpha} = R\ d(u;\theta) \qquad (11)$$

for some finite matrix R, of full row rank and not depending on u, then useful corrections can be obtained based on $D_n(\mathbf{u};\hat{\theta})$ even though R is unknown (since $p_1(.)$ is unknown). These corrections are constructed as

$$\bar{\theta} = \hat{\theta} + \hat{\varphi}\ ;\ \varphi = -\{\mathcal{I}_{\theta\theta}\}^{-1}\mathcal{I}_{\theta d}\mathcal{I}^{dd}D_n(\mathbf{u};\theta) \qquad (12)$$

where $\hat{}$ denotes functions evaluated at $\hat{\theta}$. These are similar to the "corrections" discussed in Orme (1986), which were formulated using a fully specified alternative model. The important distinction here is that (12) does not depend upon the specification of an alternative model.[11]

The restriction embodied in (11) requires further comment, perhaps. If the alternative model was known then "asymptotically efficient" corrections could be constructed based on the score test indicator vector; i.e., using the statistic $D_n(\mathbf{u};\hat{\theta})$ defined by (11) but where R is the identity matrix. These corrections are asymptotically efficient in the sense that they have the same limiting distribution, under the sequence of local alternatives, as the maximum likelihood estimator for $\theta°$ obtained under the specified alternative (and are similar in spirit to the

---

[11] The rank condition on R ensures that the true score contains no linear dependencies.

"linearised" maximum likelihood estimator of Rothenberg and Leenders (1964)). In particular, the score contains the information that is required to enable these corrections to remove the asymptotic bias from the limiting distribution of $\sqrt{n}(\hat{\theta} - \theta°)$. In one sense, then, the restriction (11) says that the statistic $D_n(\mathbf{u};\hat{\theta})$ contains all the necessary information needed to produce useful corrections, but that this information, as it stands, is not in its most "efficient form".

Another way of interpreting (11) is to regard it as implicitly defining a family of locally equivalent alternatives. For example, (11) arises if a set of globally distinct alternatives each has a locally equivalent alternative of the form

$$p(u;\theta,\alpha) = f(u;\theta)\ g(a(\alpha)'d(u;\theta))$$

where (locally, at least) g(.) is a positive scalar random function having unit expectation under the null hypothesis, a(.) is a vector function where $a(\alpha°) = 0$, $g(0) = 1$, $\partial a(\alpha°)/\partial\alpha' \neq 0$, and $g'(0) \neq 0$ (the first derivative of g(.)). Then $\partial\log\{p(u;\theta;\alpha°)\}/\partial\alpha' = \{g'(0)\}^{-1}\{\partial a(\alpha°)/\partial\alpha'\}d(u;\theta)$, which is of the form (11). Again, however, the statistic that is employed to update the MLE will not be the true score (associated with the correct alternative).

Thus, we might conjecture that corrections based on $D_n(\mathbf{u};\hat{\theta})$ will remove the asymptotic bias from the limiting distribution of $\sqrt{n}(\hat{\theta} - \theta°)$, but that they will have larger sampling variances than both the constrained MLE, $\hat{\theta}$, and the MLE obtained under the specified alternative. This is, in fact, the case as is demonstrated below.

The asymptotic distribution of maximum likelihood corrections

Under a sequence of local alternatives, and given Newey's regularity conditions, $\mathcal{I}^*(\hat{\theta})$ consistently estimates $\mathcal{I}^*$, so that (from (12))

$$\sqrt{n}(\bar{\theta} - \theta°) \cong \sqrt{n}(\hat{\theta} - \theta°) - \{\mathcal{I}_{\theta\theta}\}^{-1}\mathcal{I}_{\theta d}\mathcal{I}^{dd}\ \sqrt{n}D_n(\mathbf{u};\hat{\theta}). \qquad (13)$$

Substituting (5)-(7) into the left hand side of (13) yields

$$\sqrt{n}(\bar{\theta} - \theta^{\circ}) \cong \left(\mathcal{I}_{\theta\theta}^{-1} + \mathcal{I}_{\theta\theta}^{-1}\mathcal{I}_{\theta d}\mathcal{I}^{dd}\mathcal{I}_{d\theta}\mathcal{I}_{\theta\theta}^{-1},\ -\mathcal{I}_{\theta\theta}^{-1}\mathcal{I}_{\theta d}\mathcal{I}^{dd}\right)\sqrt{n}G_n(\mathbf{u};\theta^{\circ})$$

$$= \left(\mathcal{I}^{\theta\theta},\ \mathcal{I}^{\theta d}\right)\sqrt{n}G_n(\mathbf{u};\hat{\theta})\ ,$$

where $\mathcal{I}^{\theta\theta}$ and $\mathcal{I}^{\theta d}$ are respectively the top left and top right partitions of $\{\mathcal{I}^{*}\}^{-1}$. This means that $\sqrt{n}(\bar{\theta} - \theta^{\circ})$ has the same limiting sampling distribution as the random variable $\left(\mathcal{I}^{\theta\theta},\ \mathcal{I}^{\theta d}\right)\sqrt{n}G_n(\mathbf{u};\theta^{\circ})$. Thus a simple application of Slutsky's theorem and appealing to the result given in (8) gives the asymptotic distribution of these (ML) corrections to be

$$\sqrt{n}(\bar{\theta} - \theta^{\circ}) \xrightarrow{d} N\left(\left(\mathcal{I}^{\theta\theta},\ \mathcal{I}^{\theta d}\right)\mathcal{H}\delta,\ \left(\mathcal{I}^{\theta\theta},\ \mathcal{I}^{\theta d}\right)\mathcal{I}^{*}\left(\mathcal{I}^{\theta\theta},\ \mathcal{I}^{\theta d}\right)'\right). \quad (14)$$

However if (11) is true then

$$\mathcal{H}' = \left(\mathcal{I}_{\alpha\theta},\ \mathcal{I}_{\alpha d}\right) = R\left(\mathcal{I}_{d\theta},\ \mathcal{I}_{dd}\right)$$

so that

$$\left(\mathcal{I}^{\theta\theta},\ \mathcal{I}^{\theta d}\right)\mathcal{H} = \left(\mathcal{I}^{\theta\theta}\mathcal{I}_{\theta d} + \mathcal{I}^{\theta d}\mathcal{I}_{dd}\right)R' = 0 \quad \text{for all } R,$$

by the rules for the partitioned inversion of the matrix $\mathcal{I}^{*}$. Using these same rules it is easily demonstrated that $\left(\mathcal{I}^{\theta\theta},\ \mathcal{I}^{\theta d}\right)\mathcal{I}^{*}\left(\mathcal{I}^{\theta\theta},\ \mathcal{I}^{\theta d}\right)' = \mathcal{I}^{\theta\theta}$, which implies that

$$\sqrt{n}(\bar{\theta} - \theta^{\circ}) \xrightarrow{d} N\left(0,\ \mathcal{I}^{\theta\theta}\right), \quad (15)$$

which is the principal result of this section; that is, (15) gives the limiting distribution of (ML) corrections in the face of local misspecification of unknown form.

Assumption (11), on which the result given in (15) rests, is relevant as the example of neglected heterogeneity illustrates. It can be seen from (4) that contributions to the score for η are of the form (11). Thus, even though the precise form for the distribution of the varying parameters, under the alternative, is unknown, ML corrections can be obtained (based on the IM test statistic) that remove the asymptotic bias from the limiting distribution of $\sqrt{n}(\hat{\theta} - \theta^{\circ})$; or, to a local approximation, these corrections remove the asymptotic bias from $\hat{\theta}$.

Indeed, as indicated previously, if there is a family of alternative models, p(.), for which (11) is true, given $d(u;\theta)$, then corrections based on $D_n(\mathbf{u};\hat{\theta})$ have the asymptotic distribution given in (15) for each (unknown) member of this family (provided the departure from the maintained specification, in the direction of this family, is local). In particular, this is true of a family of locally equivalent alternative models (see Godfrey (1988), who discusses in some detail the concept of local equivalence).

For the case of parameter variation of unknown form, Chesher (1983) derived the locally equivalent alternative model by taking a two term Taylor series expansion of $p_1(y|x;\theta_0,\Omega)$ about $\Omega = \text{var}(\Theta) = 0$, in order to give an approximation to the true marginal *pdf* of y valid for small Ω. This same approximation can be obtained by taking a one term Taylor series expansion of $p_1(y|x;\theta_0,\eta)$ about $\eta = 0$ and using $\Theta = Q\Gamma$; for a simple review of these details see Appendix B.

However, corrections based on $D_n(\mathbf{u};\hat{\theta})$ will not in general remove the local asymptotic bias from $\hat{\theta}$ if (11) is not satisfied. An important example of such a situation arises when the statistic $m(u;\theta)$ satisfies (11) (where $m(u;\theta^{\circ})$ has expectation zero under the null hypothesis of no model misspecification) with $M_n(\mathbf{u};\hat{\theta}) = n^{-1}\Sigma\, m(u_i;\hat{\theta}) = D_n(\mathbf{u};\hat{\theta})$ but where $m(u;\theta) \neq d(u;\theta)$. This can occur because $m(u;\theta)$ takes the form

$$m(u;\theta) = d(u;\theta) + B\,\frac{\partial \log\{f_1(y|x;\theta)}{\partial\theta} \tag{16}$$

for some finite matrix B not depending on u. Then, by the first order conditions for the MLE $\hat{\theta}$, $M_n(\mathbf{u};\hat{\theta}) = D_n(\mathbf{u};\hat{\theta})$, but $m(u;\theta) \neq d(u;\theta)$.

Suppose, then, that corrections are based on $D_n(\mathbf{u};\hat{\theta})$ and assume (15) holds where $m(u;\theta)$ (not $d(u;\theta)$) satisfies (11). For this case the analysis is quite simply extended because all the asymptotic distribution results obtained above remain valid up to the point where the assumption embodied in (11) is used. That is, (8)-(10) and (14) remain true, since $D_n(\mathbf{u};\hat{\theta})$ is a valid CM test. In particular, notice that the variance of these corrections is still correctly specified as $\mathcal{I}^{\theta\theta}$. However, the move from (14) to (15) regarding the mean in the asymptotic distribution of $\sqrt{n}(\bar{\theta} - \theta°)$ is now incorrect. When $m(u,\theta)$ satisfies (11) the matrix $\mathcal{H}$ takes the form

$$\mathcal{H} = \begin{bmatrix} \mathcal{I}_{\theta m} \\ \mathcal{I}_{dm} \end{bmatrix} R' = \begin{bmatrix} \mathcal{I}_{\theta d} + \mathcal{I}_{\theta\theta}B' \\ \mathcal{I}_{dd} + \mathcal{I}_{d\theta}B' \end{bmatrix} R' \tag{17}$$

where $\mathcal{I}_{\theta m}$ and $\mathcal{I}_{dm}$ are defined in exactly the same way as $\mathcal{I}_{\theta d}$ and $\mathcal{I}_{dd}$ and we have used (16). Then some relatively straightforward matrix algebra shows that

$$\left(\mathcal{I}^{\theta\theta},\mathcal{I}^{\theta d}\right)\mathcal{H} = \left(\mathcal{I}^{\theta\theta}\mathcal{I}_{\theta d} + \mathcal{I}^{\theta\theta}\mathcal{I}_{\theta\theta}B' + \mathcal{I}^{\theta d}\mathcal{I}_{dd} + \mathcal{I}^{\theta d}\mathcal{I}_{d\theta}B'\right)R' = B'R',$$

which gives the result that

$$\sqrt{n}(\bar{\theta} - \theta°) \xrightarrow{d} N\left(B'R'\delta, \mathcal{I}^{\theta\theta} \right). \tag{18}$$

Thus, in such a situation, even though the test indicator vectors are equivalent if $M_n(\mathbf{u};\hat{\theta})$ is employed rather than $D_n(\mathbf{u};\hat{\theta})$, $\bar{\theta}$ will not in general remove the asymptotic bias from $\hat{\theta}$.

3. Specification error in censored normal linear models

The IM test may be a useful test of general misspecification in many micro-econometric models, which are based on censored regressions. The apparent reason for this is that it can be interpreted as a score test of random parameter variation which is a specification error inducing heteroskedasticity, skewness and non-normal kurtosis.

To see why this is so, consider the latent normal linear model

$$y^* = x'\beta + \varepsilon$$

where conditional on the $(1 \times k)$ vector x', the maintained distribution of y^* is $N(x'\beta, \tau)$. Now suppose that, in fact, β and τ are random variables with means β_0 and τ_0 and variances and covariances $\Omega_{\beta\beta}$, $\Omega_{\tau\tau}$ and $\Omega_{\beta\tau}$, distributed independently of x. Since the effect of minimal parameter variation is of concern it is assumed that higher order central moments of β and τ are negligible and as such will be ignored.

Given this misspecification, we shall focus on the true conditional distribution of y^* given x. It is evident that the mean of y^* given x, but marginal with respect to the random parameters, is $E(y^*|x) = x'\beta_0$. Higher order moments, which examine the variance, skewness and kurtosis in the conditional distribution of y^* given x, can be obtained as follows (see Chesher, Lancaster and Irish (1985)), where third and higher order central moments of β and τ are assumed so small as to be taken as zero:

$$E[(y^* - x'\beta_0)^2|x] = \tau_0 + x'\Omega_{\beta\beta}x$$

$$E[(y^* - x'\beta_0)^3|x] = 3\Omega_{\beta\tau}x$$

$$E[(y^* - x'\beta_0)^4|x] - 3E[(y^* - x'\beta_0)^2|x] = 3\Omega_{\tau\tau} .$$

From the above expressions it can thus be seen that minimal parameter variation induces heteroskedasticity in the conditional distribution of y^* given x, where the variance in this distribution is a function of the squares and cross products of the elements in x. Further, if β and τ are correlated the distribution will exhibit skewness depending linearly on x. The final expression reveals that random variation in τ induces super-normal kurtosis.

Therefore, not surprisingly perhaps, Hall (1987) showed that the IM test for the fully observed normal linear model is sensitive to all these specification errors. This property of the IM test has also been pointed out by Chesher, Lancaster and Irish (1985) in the context of the tobit model (which is, of course, a censored normal linear model).

Expanding on these (and other) themes Chesher and Irish (1987) give an excellent discussion of specification tests, based on generalised residuals, in the context of the normal linear model which is subject to a more general scheme of censoring. In such models, which include as special cases the tobit and probit models, the range of y^* (which is $(-\infty,\infty)$) is partitioned into J groups or classes. Let these groups be denoted $g_j = (m_{j-1}, m_j]$, $j = 1,\ldots,J-1$ with $m_0 = -\infty$ and the Jth group being open-ended with $m_J = \infty$, where the class boundaries, m_j, are known constants. If y^* falls in certain groups, then only the interval (group) in which it falls is known. Let this subset of groups be defined as $C = \{g_j;\ j \in J^C\}$, where J^C is simply a subset of the integers $j = 1,\ldots,J$. (Intervals of the real line outside this set are g_j, $j \notin J^C$.) Any realisation of y^* must fall in a group g_j, for some j. There is thus only one of two possibilities: either $g_j \notin C$ so that $y = y^*$ is fully observed, or $g_j \in C$ so that only the fact that $m_{j-1} < y^* \leq m_j$ is revealed. For all observations it is always known into which group y^* has fallen.

The probit model is a special case of the above scheme where there are only two classes with $m_1 = 0$ and where $C = \{(-\infty,0],(0,\infty)\}$, so that y^* is never fully observed, only its sign is recorded. For the tobit model also the real line is divided into the same two intervals but $C = \{(-\infty,0]\}$, so that y^* is only fully observed when positive; if y^* is negative only this fact will be revealed.

Chesher and Irish (1987), for the general case, show that the IM test statistic, in terms of the parameter vector $\theta' = (\beta', \tau)$, can be expressed as

$$D_n(\mathbf{u};\hat{\theta}) = \frac{1}{n}\sum_{i=1}^{n}\left\{ \frac{\hat{v}_i^{(2)} - 1}{\hat{\tau}}\, z_i',\ \frac{\hat{v}_i^{(3)}}{2\hat{\tau}^{3/2}}\, x_i',\ \frac{\hat{v}_i^{(4)} - 3}{4\hat{\tau}^2} \right\}'$$

where $\hat{\theta}' = (\hat{\beta}', \hat{\tau})$, z_i contains non-redundant terms of $\text{vech}(x_i x_i')$ and the $\hat{v}_i^{(r)}$ are "moment residuals" which correspond to the following "moment errors" (r = 1,2,3,4)

$$v^{(r)} = \begin{cases} \left(\dfrac{y - x'\beta}{\sqrt{\tau}}\right)^r, & \text{if } g_j^* \notin C, \text{ i.e., if } y^* \text{ is observed,} \\ E\left[\left(\dfrac{y^* - x'\beta}{\sqrt{\tau}}\right)^r \,\middle|\, y^* \in g_j^*\right], & \text{if } g_j^* \in C, \end{cases}$$

and g_j^* is the group into which y^* has fallen.

In particular, given the above definitions, it can be seen that the elements in the IM test indicator vector are, in fact, the direct analogues of well known test statistics for heteroskedasticity, skewness and kurtosis for the fully observed linear model. Indeed, these latter statistics can readily be obtained from the above when y^* is fully observed by defining the set C to be empty. Thus, as pointed out by Chesher, Lancaster and Irish (1985), the IM test could provide useful diagnostics sensitive to such departures from the standard assumptions that are typically made in models of this kind.

Specification error in a probit model

Recent work has shown that misspecification in a binary probit model such as heteroskedasticity (Yatchew and Griliches (1985), Kiefer and Skoog (1984)), non-normality (Arabmazar and Schmidt (1982)) and parameter variation (Hausman and Wise (1978)) can lead to inconsistent estimates of the parameters of interest (see, however, Ruud (1983) who gives sufficient conditions for parameters to be consistently estimated up to a scale factor). Thus, as noted by Newey (1985) and Chesher and Irish (1987), the IM test would be a good test of general misspecification since it would appear to be sensitive to all three of these departures. The

IM test (using the full set of indicators) would not, however, be informative about the precise nature of the misspecification. However, if misspecification is detected but believed to be small then, as is shown below, ML corrections based on the IM test statistic will remove the local asymptotic bias from the estimated regression parameter vector even when the precise form of the misspecification (be it heteroskedasticity, non-normality or parameter variation) is not known.

Consider the latent model, $y^* = x_1'\theta + \varepsilon$; where the maintained specification is that, conditional on x_1, y^* is normally distributed with mean $x_1'\theta$ and unit variance. We assume throughout that $x_1'\theta$ includes an intercept term.[12] The observed response, however, is the sign of y^* indicated by the variable y which equals one if $y^* > 0$ and zero otherwise. Then $f_1(y|x_1;\theta) = \{\Phi(x_1'\theta)\}^y.\{\Phi(-x_1'\theta)\}^{1-y}$, where $\Phi(.)$ is the cumulative distribution function for a standard normal variate; $\phi(.)$ will denote the standard normal *pdf* and, also, it will be useful to define $\lambda(.) = \phi(.)/\{\Phi(.)(1-\Phi(.))\}$.

A parameterisation of heteroskedasticity is obtained by assuming that, conditional on $x' = (x_1', x_2')$, y^* is normally distributed with mean $(x_1'\theta)$ and standard deviation $h(x_2'\alpha) > 0$, where $h(x_2'\alpha)$ does not include a constant term and h(.) is twice differentiable, with respect to its argument, with $h(0) = 1$ and $h'(0) \neq 0$.[13] This yields

$$\log\{p(u;\theta,\alpha)\} = y\log[\Phi(x_1'\theta/h(x_2'\alpha))] + (1-y)\log[\Phi(-x_1'\theta/h(x_2'\alpha))] + \log\{f_2(x)\}.$$

---

[12] This is the usual normalising restriction. We have thus used θ instead of β since this is the only unknown parameter vector which is identified.

[13] This assumption ensures that estimation is possible under the null (see Godfrey (1988), for a discussion).

The maintained specification is correct at $\alpha^\circ = 0$ and the score for α is given by

$$\text{(H):} \quad \frac{\partial \log\{p_1(u;\theta,\alpha^\circ)\}}{\partial\alpha} = -\{y-\Phi(x_1'\theta)\}\cdot\lambda(x_1'\theta)\cdot(x_1'\theta)\cdot x_2. \tag{19}$$

As reported by Newey (1985) a parameterisation of non-normality as presented by Ruud (1981) is achieved by assuming that conditional on x, $\varepsilon = y^* - x_1'\theta$ has distribution function $G(\eta) = \Phi(\alpha_0+\eta+\alpha_1\eta^2+\alpha_2\eta^3)$ where α_1 and α_2 are parameters which satisfy $1+2\alpha_1\eta+3\alpha_2\eta^2 > 0$ for all η (since $G'(\eta) > 0$). This yields

$$\begin{aligned}\log\{p(u;\theta,\alpha)\} = {} & y\log[\Phi\{\alpha_0 + x_1'\theta + \alpha_1(x_1'\theta)^2 + \alpha_2(x_1'\theta)^3\}] \\ & + (1-y)\log[\Phi\{-\alpha_0 - (x_1'\theta) - \alpha_1(x_1'\theta)^2 - \alpha_2(x_1'\theta)^3\}] \\ & + \log\{f_2(x)\}.\end{aligned}$$

The maintained specification is correct at $\alpha_1^\circ = \alpha_2^\circ = 0$ and the score for $\alpha' = (\alpha_1,\alpha_2)$ is given by

$$\text{(NN):} \quad \frac{\partial \log\{p_1(u;\theta,\alpha^\circ)\}}{\partial\alpha'} = \{y-\Phi(x_1'\theta)\}\cdot\lambda(x_1'\theta)\cdot[(x_1'\theta)^2,(x_1'\theta)^3]. \tag{20}$$

Finally contributions to the IM test indicator vector come from

$$d(u;\theta) = - \{y - \Phi(x_1'\theta)\}\cdot\lambda(x_1'\theta)\cdot(x_1'\theta)\cdot z \tag{21}$$

(see, for example, Newey (1985)) where z contains distinct non-redundant elements of (x_1x_1').

Firstly, comparing (21) with (19), if x_2 contains some regressors included in x_1 and cross products of these regressors, then equation (19) can be written

(H): $\partial \log\{p_1(u;\theta,\alpha^\circ)/\partial\alpha = R_H d(u;\theta)$, for some selection matrix R_H.[14]

If the (unknown) misspecification present is non-normality then comparing (20) with (21) note that $(x_1'\theta)^2$ and $(x_1'\theta)^3$ are linear combinations of the elements of $(x_1'\theta)\cdot z$ (Newey, 1985,) so that equation (20) can be written

$$\text{(NN): } \partial \log\{p_1(u;\theta,\alpha^\circ)/\partial\alpha = R_{NN} d(u;\theta),$$

for some matrix R_{NN} not depending on u.

The above analysis shows that the IM test is sensitive to heteroskedasticity, non-normality and, of course, parameter variation. Moreover, since (11) is satisfied in all three cases, ML corrections, $\bar{\theta}$, of the form (12) based on the IM test statistic will remove the local asymptotic bias from $\hat{\theta}$ irrespective of whether the local misspecification is heteroskedasticity, non-normality or parameter variation. These corrections are particularly easy to compute by appealing to the results of Orme (1988). There I show that the asymptotically efficient variant of the IM test statistic can be calculated as the explained sum of squares from an artificial regression which has left hand side variable

$$\{y-\Phi(x_1'\hat{\theta})\}/\surd\{\Phi(x_1'\hat{\theta})(1-\Phi(x_1'\hat{\theta}))\}$$

and right hand side variables

$$\hat{w}' = [\phi(x_1'\hat{\theta})/\surd\{\Phi(x_1'\hat{\theta})(1-\Phi(x_1'\hat{\theta}))\}].(x_1', -(x_1'\hat{\theta})z') = (\hat{w}_1', \hat{w}_2');$$

see also Chapter 7 of this volume. The necessary corrections are also available from this regression: the estimated least squares coefficient on the "regressor" $\hat{w}_1$ is simply $\bar{\theta} - \hat{\theta}$. The estimated

---

[14] It is often assumed that heteroskedasticity is a function of the included regressors.

variance of $\bar{\theta}$ is the top left partition of $(\hat{W}'\hat{W})^{-1}$ where the $(n \times k+q)$ matrix $\hat{W}$ has rows $\hat{w}' = (\hat{w}_1', \hat{w}_2')$, i.e. the "regressor" matrix. Specifically, in the notation of Section 2 of this chapter, the "asymptotically efficient" estimator of $\mathcal{I}^*$, $\mathcal{I}^*(\hat{\theta})$, is simply $\hat{W}'\hat{W}/n$, (see Orme (1988) and also Orme (1989b) who establishes a similar calculation procedure for the asymptotically efficient variant of the IM test or, indeed, any conditional moment test for binary data models).

These corrections based on the IM test statistic may be somewhat lacking, however, if in reality the true distribution function of the errors, in the underlying latent model, is not well approximated by $G(\eta) = \Phi(\alpha_0 + \eta + \alpha_1\eta^2 + \alpha_2\eta^3)$. Suppose the true distribution function is nested in the Pearson family as proposed by Bera, Jarque and Lee (1984). The score test statistics for testing normality, in both cases, are identical numerically. That is, although contributions to the test indicator vector are different, the sums of these contributions coincide. The problem now, however, concerns the corrections based on the IM test statistic since it no longer satisfies (11).

When the error distribution is nested in the Pearson family the appropriate score is given by

$$\frac{\partial \log\{p_1(u;\theta,\alpha^\circ)\}}{\partial\alpha} =$$

$$\begin{bmatrix} -\frac{1}{3} & 0 \\ 3 & \frac{1}{4} \\ 0 & 4 \end{bmatrix} (y-\Phi(x_i'\theta))\lambda(x_i'\theta) \begin{bmatrix} (x_i'\theta)^2 - 1 \\ (x_i'\theta)^3 + 3(x_i'\theta) \end{bmatrix} \tag{22}$$

(see Bera, Jarque and Lee (1984)) which can be expressed as

$$\frac{\partial \log\{p_1(u;\theta,\alpha^\circ)\}}{\partial\alpha} = AR_{NN}\left\{ d(u;\theta) + B\,\frac{\partial \log\{f_1(y|x;\theta)\}}{\partial\theta} \right\},$$

where

$$A = \begin{bmatrix} -\frac{1}{3} & 0 \\ 0 & \frac{1}{4} \end{bmatrix},$$

the $(2 \times q)$ matrix R_{NN} is defined as before, and the $(q \times k)$ matrix B is such that

$$R_{NN}B = \begin{bmatrix} -1 & 0' \\ & 3\theta' \end{bmatrix} = C, \text{ say.}$$

The problem regarding the asymptotic distribution of corrections based on the IM test statistic can now be addressed quite simply since the above formulation corresponds exactly to the discussion given at the end of Section 2. Appealing to the result given in (18), where in the present situation $R = AR_{NN}$, we obtain

$$\sqrt{n}(\bar{\theta} - \theta^{\circ}) \xrightarrow{d} N\left(C'A'\delta,\ \mathcal{I}^{\theta\theta}\right)$$

where

$$C'A'\delta = \begin{bmatrix} \frac{1}{3} & \frac{3}{4}\theta \\ 0 & \end{bmatrix} \delta = \begin{bmatrix} \frac{1}{3}\alpha_1 + \frac{3}{4}\alpha_2\theta_1 \\ \frac{3}{4}\alpha_2\theta_2 \end{bmatrix},$$

and θ has been partitioned as $\theta' = (\theta_1, \theta_2')$, with θ_1 being the intercept term. Substituting θ° for θ in the above expressions gives the following approximations (following, for example, Kiefer and Skoog (1984))

$$\text{plim } \bar{\theta}_1 = \frac{1}{3}\alpha_1 + (\frac{3}{4}\alpha_2 + 1)\ \theta_1^{\circ},$$

$$\text{plim } \bar{\theta}_2 = (\frac{3}{4}\alpha_2 + 1)\ \theta_2^{\circ},$$

which indicate that, for local specification error, corrections based on the IM test statistic "consistently" estimate θ_2°, but only up to a scaling factor.

Specification error in a tobit model

The work of Arabmazar and Schmidt (1981), amongst others, has highlighted the fact that maximum likelihood estimates of the regression parameters in a tobit model will, in general, be inconsistent as a consequence of neglected heteroskedasticity. Thus testing for its presence has become more evident in empirical studies over recent years. Lee and Maddala (1985) provided the appropriate score indicator vector which enabled applied workers to construct the "Outer Product of the Gradient"(OPG) variant of the score test statistic quite easily (see Godfrey and Wickens (1982)). It has become apparent, from a variety of studies, that the OPG method yields a statistic which is disastrously over-sized (causing it to reject a correct null hypothesis far too often). Chapter 7 of this volume investigates this phenomenon and concludes that this particular method of calculation typically gives highly misleading inferences. On the other hand, other variants (which differ only in the calculation of the covariance matrix of the test indicator) can exhibit vast improvements over the OPG variant and, moreover, the efficient variant has much to recommend it. Motivated by these observations, Orme (1992) extends the work of Lee and Maddala by providing an expression for the asymptotically efficient covariance matrix of the score test statistic and showing how the efficient version of the score test statistic can be obtained as standard output from an artificial linear regression. Moreover, since the IM test (corresponding to the $\beta\beta'$ partition of the IM equality) can be interpreted directly as a score test for heteroskedasticity of unknown form, then the same algorithm can be used to construct the efficient version of the IM test.[15]

Chesher, Lancaster and Irish (1985) discussed the possibility of correcting tobit regression parameter estimates for

---

[15] The form of the heteroskedasticity is assumed only to depend in some way on the squares and cross products of the included regressors in the tobit specification.

heteroskedasticity. Their work, however, employed the Outer Product of the Gradient variant of the IM test. On the other hand, "efficient" corrections can be obtained from the estimated least squares slope coefficients in the artificial regression used to construct the efficient test statistic. This procedure is detailed below.

In the censored or tobit regression model, the observed limited dependent variable is a transformation of the latent variable y_i^*, such that

$$y_i = \begin{cases} y_i^*, & x_i'\beta + \varepsilon_i > 0,\ \varepsilon_i \ iid\ N(0,\sigma^2) \\ 0, & \text{otherwise.} \end{cases}$$

The conditional density of y_i given x_i is a continuous discrete mix defined by

$$f(y_i;a_i,\sigma) = \left[\frac{1}{\sigma}\phi(\{y_i - a_i\}/\sigma)\right]^{1-s_i}\left[\Phi(-a_i/\sigma)\right]^{s_i},$$

where $a_i = x_i'\beta$ and s_i is a censoring indicator which equals 0 if $y_i > 0$ and 1 if $y_i = 0$. The log-likelihood based on a sample of n observations is

$$\mathcal{L}(\theta',\sigma) = \sum_i \left((1-s_i)\{-\log(\sigma) - \frac{1}{2\sigma^2}\varepsilon_i^2\} + s_i\log(\Phi(-a_i/\sigma))\right),$$

giving the following likelihood equations

$$\sum_i \hat{\mu}_i x_i = 0, \quad \sum_i \hat{\gamma}_i = 0,$$

where

$$\mu_i = \{(1 - s_i)v_i - s_i m_i\}/\sigma$$

and

$$\gamma_i = \{(1 - s_i)(v_i^2 - 1) + s_i c_i m_i\}/\sigma,$$

in which the following definitions apply: $v_i = \varepsilon_i/\sigma$, $c_i = a_i/\sigma$ and $m_i = \phi(c_i)/\{1 - \Phi(c_i)\}$. Alternatively, μ_i and γ_i can be expressed as $v_i^{(1)}/\sigma$ and $(v_i^{(2)} - 1)/\sigma$, respectively, where

$$v_i^{(1)} = (1 - s_i)v_i + s_i E(v_i | y_i = 0)$$

and

$$v_i^{(2)} = (1 - s_i)v_i^2 + s_i E(v_i^2 | y_i = 0).$$

(Chesher and Irish (1987) refer to these as "generalised errors", and see above for a further discussion). From the preceding discussion, it can be seen that the IM test investigates the statistical significance of $G_n(\mathbf{u};\hat{\theta}) = n^{-1}\Sigma\hat{\sigma}^{-1}\{\hat{v}_i^{(2)} - 1\}z_i$, which is equivalent to $n^{-1}\Sigma\hat{\gamma}_i z_i$.

The following moments are defined conditional on x_i:

$$\text{var}(\mu_i) = \sigma^{-2}(1-\{1-\Phi(c_i)\}\{1-m_i(m_i-c_i)\}) > 0;$$

$$\text{cov}(\mu_i,\gamma_i) = \sigma^{-2}\phi(c_i)(1-c_i(m_i-c_i)) > 0;$$

$$\text{var}(\gamma_i) = \sigma^{-2}(2-\{1-\Phi(c_i)\}\{2+c_i m_i(1-c_i(m_i-c_i))\}) > 0.$$ [16]

Then, following the method of Orme (1992) various variables need to be constructed from the above (evaluated at maximum likelihood estimates). These are:

$$\upsilon_{1i} = \{\text{var}(\mu_i) - \{\text{cov}(\mu_i,\gamma_i)\}^2/\text{var}(\gamma_i)\}^{1/2};$$

$$\upsilon_{2i} = \text{cov}(\mu_i,\gamma_i)/\{\text{var}(\gamma_i)\}^{1/2}$$

and

---

[16] These results and associated inequalities are well established.

$$\upsilon_{3i} = \{\mathrm{var}(\gamma_i)\}^{1/2}.$$

From these expressions we obtain $\hat{\upsilon}_{1i}$, $\hat{\upsilon}_{2i}$ and $\hat{\upsilon}_{3i}$ where the hats indicate that unknown parameters have been replaced by maximum likelihood estimates. In particular, it can be shown that the variable $\hat{\upsilon}_{1i}$ is strictly positive. As before the $(q \times 1)$ vector z_i is defined to contain distinct non-constant terms of $\mathrm{vech}(x_i x_i')$.[17]

The artificial regression employed to calculate the efficient IM test statistic (and associated corrections) is of "double length" form. The first set of n observations on the left hand side variable are set equal to

$$\hat{r}_{1i} = - \{\hat{\gamma}_i/\hat{\upsilon}_{3i}\}\hat{\upsilon}_{2i}/\hat{\upsilon}_{1i},\ i = 1,...,n$$

and the second set of n elements are

$$\hat{r}_{2i} = \hat{\gamma}_i/\hat{\upsilon}_{3i},\ i = 1,...,n.$$

Correspondingly, the right hand side variables come in two sets of n observations. The first set are typically $\{\hat{w}_{1i}', \hat{w}_{1i}^{+\prime}\}$, where

$$\hat{w}_{1i}' = \{\hat{\upsilon}_{1i}x_i', 0\} \quad (1 \times k+1), \quad \hat{w}_{1i}^{+} = \{0'\} \quad (1 \times q),$$

and the second set are typically $\{\hat{w}_{2i}', \hat{w}_{2i}^{+\prime}\}$, where

$$\hat{w}_{2i}' = \{\hat{\upsilon}_{2i}x_i', \hat{\upsilon}_{3i}\} \quad (1 \times k+1), \quad \hat{w}_{2i}^{+} = \{\hat{\upsilon}_{3i}z_i'\} \quad (1 \times q).$$

Then stacking the observations into appropriate vectors and matrices the double length artificial regression can be expressed as

---

[17] As in the probit model, the first element in the IM test indicator vector is identically zero.

$$\begin{bmatrix} \hat{r}_1 \\ \hat{r}_2 \end{bmatrix} = \begin{bmatrix} \hat{W}_1 \\ \hat{W}_2 \end{bmatrix} b_1 + \begin{bmatrix} \hat{W}_1^+ \\ \hat{W}_2^+ \end{bmatrix} b_2 + \text{errors}$$

$$(2n \times 1) \quad (2n \times k+1) \quad (2n \times q)$$

where the matrix $\hat{W}_j$ has rows $\hat{w}'_{ji}$ and $\hat{W}_j^+$ has rows $\hat{w}^{+\prime}_{ji}$, $j = 1,2$, $i = 1,\ldots,n$. The efficient IM test statistic is simply the difference between the two residual sums of squares obtained from imposing the restriction that $b_2 = 0$ and unrestricted ordinary least squares estimation (see Orme (1992)). The corrected tobit regression parameter estimates (i.e., corrected estimates of β, $(k \times 1)$) are obtained by taking the first k elements of the ordinary least squares estimate of b_1, from the above regression, and adding them to the elements of $\hat{\beta}$, the tobit maximum likelihood estimate of β. The appropriate estimated covariance matrix of these corrections is the top left $(k \times k)$ partition of $(\hat{W}'\hat{W})^{-1}$, where $\hat{W}$ is the $(2n \times k+1+q)$ regressor matrix employed in the double length artificial regression and is defined as

$$\hat{W} = \begin{bmatrix} \hat{W}_1 & \hat{W}_1^+ \\ \hat{W}_2 & \hat{W}_2^+ \end{bmatrix}.$$

The sampling performance of such correction procedures, in both the probit and tobit models, is investigated in Section 6.

Specification error in censored normal linear models

The previous two subsections detail special cases of the more general censoring scheme that can occur. Further, the structure of the information matrix tests derived for these two special cases, and the associated interpretation of the artificial regressions employed, suggests that the procedure can be readily generalised. This, indeed, turns out to be the case and, for completeness, it is outlined here.

The general scheme of censoring is as described at the start of this section where, as in the tobit model, the latent variable y^* is normally distributed with mean $x'\beta$ and variance $\tau = \sigma^2$, and here we shall deal in terms of the parameter vector $\theta' = (\beta',\sigma)$.

Following Chesher and Irish (1987), the likelihood equations can be written $n^{-1}\Sigma\hat{\sigma}^{-1}\hat{v}_i^{(1)}x_i = 0$ and $n^{-1}\Sigma\hat{\sigma}^{-1}\hat{v}_i^{(2)} = 0$, where $\hat{v}_i^{(1)}$ and $\hat{v}_i^{(2)}$ are, respectively, the first and second order generalised residuals, as defined for the general censoring scheme. As in the tobit model, we define $\mu_i = v_i^{(1)}/\sigma$ and $\gamma_i = v_i^{(2)}/\sigma$ and consider a test for neglected heteroskedasticity of unknown form which is based on the information matrix test indicator given by $n^{-1}\Sigma\hat{\gamma}_i z_i$. Thus far, the algebraic formulation is identical to that of the tobit model and all that remains is to define the following variances and covariances. In order to do this we define $c_{ij} = (m_j - x_i'\beta)/\sigma$, where m_j is the upper limit of group g_j, with $m_0 = -\infty$ and $m_J = \infty$, and also a (censoring) indicator variable s_{ij}, $j = 1,...,J$, which take the value 1 if $j \in J^c$ and 0 otherwise. Further, for compact notation we have $\Phi_{ij} = \Phi(c_{ij})$, $\phi_{ij} = \phi(c_{ij})$, and $P_{ij} = \Phi_{ij} - \Phi_{i,j-1}$. Then

$$\sigma^2\text{var}(\mu_i) = \begin{cases} \sum_j (1-s_{ij})\left[P_{ij} - (c_{ij}\phi_{ij} - c_{i,j-1}\phi_{i,j-1})\right] \\ \qquad + s_{ij}\, P_{ij}^{-1}\, (\phi_{i,j-1} - \phi_{ij})^2, \end{cases}$$

$$\sigma^2\text{cov}(\mu_i,\gamma_i) = \begin{cases} \sum_j (1-s_{ij})\left[-(\phi_{ij}-\phi_{i,j-1})-(c_{ij}^2\phi_{ij}-c_{i,j-1}^2\phi_{i,j-1})\right] \\ + s_{ij}P_{ij}^{-1}\left[(\phi_{ij}-\phi_{i,j-1})(c_{ij}\phi_{ij}-c_{i,j-1}\phi_{i,j-1})\right], \end{cases}$$

$$\sigma^2\text{var}(\gamma_i) = \begin{cases} \sum_j (1-s_{ij})\Big[2P_{ij} - (c_{ij}\phi_{ij}-c_{i,j-1}\phi_{i,j-1}) \\ \qquad\qquad - (c_{ij}^3\phi_{ij}-c_{i,j-1}^3\phi_{i,j-1})\Big] \\ + s_{ij}P_{ij}^{-1}\left[(c_{ij}\phi_{ij}-c_{i,j-1}\phi_{i,j-1})\right]^2. \end{cases}$$

Observing that these variances and covariances have been multiplied by a factor of σ^2 (for ease of notation), then the definitions of υ_{1i}, υ_{2i} and υ_{3i} are as in the tobit model. The double length artificial regression required to calculate the IM test statistic and associated corrections follows immediately. The algebraic structure of the required regression was detailed for the tobit model and, given the above modified definitions, remains the same for the more general censoring scheme.

4. Asymptotic mean square error considerations

In the above example the corrections of the form (12) can be constructed from the same artificial regression that is used to compute the test statistic. Asymptotically valid corrections can be constructed in a similar fashion for more general models. Newey (1985) showed that an asymptotically valid CM test statistic can be constructed as n times the uncentred R^2 (alternatively, n minus the residual sum of squares) from the regression of a vector of ones, s (the sum vector), on a matrix which has rows $\hat{w}_i' = (\partial \log\{f_1(y_i|x_i;\hat{\theta})/\partial\theta', d(u_i;\hat{\theta})')$.[18] Asymptotically valid corrections are given by $\bar{\theta} = \hat{\theta} + \bar{b}_1$ where $\bar{b}_1$ is the least squares estimate of the coefficient vector associated with $\partial \log\{f_1(y_i|x_i;\hat{\theta})/\partial\theta'$ from this artificial regression. By "asymptotically" valid it is meant that its "$\sqrt{n}$-normed" limiting distribution is given by (15). The least squares estimate of the coefficient vector associated with $d(u_i;\hat{\theta})'$ will be denoted $\bar{b}_2$ and is given by $\bar{b}_2 = (\hat{W}'\hat{W})^{dd}\Sigma\, d(u_i;\hat{\theta})$, where $(\hat{W}'\hat{W})^{dd}$ is the bottom right partition of $(\hat{W}'\hat{W})^{-1}$ and the matrix $\hat{W}$ has rows $\hat{w}_i'$. Again, appealing to Newey's results, $\hat{W}'\hat{W}/n$ consistently estimates $\mathcal{I}^*$ so that by (7) $\sqrt{n}\bar{b}_2$ is asymptotically equivalent to $(\mathcal{I}^{d\theta}, \mathcal{I}^{dd})\sqrt{n}G_n(\mathbf{u};\theta^\circ)$; then by the result in (8), the limiting distribution of $\sqrt{n}\bar{b}_2$ is obtained as $N(R'\delta, \mathcal{I}^{dd})$. Thus $\bar{b}_2$ can be thought of as estimating the extent of the local specification error, since $R'\delta = \underline{0}$ for all $R \neq 0$ if and only if $\delta = 0$. Given this interpretation $\bar{b}_2$ is itself a corrected ML estimate: under

---

[18] Orme (1990a) (see also Chapter 7 of this volume) considers alternative nR^2 variants which do not appear to exhibit the severe size bias that is, typically, a feature of Newey's procedure.

the maintained specification the maximum likelihood estimate of b_2 is identically zero (i.e. no misspecification). What we have then are two competing estimators of $\theta°$: $\hat{\theta}$ and $\bar{\theta}$; and two competing "estimators" of the local specification error: $b_2 \equiv 0$ and $\hat{b}_2$. In most situations, however, we shall only be interested in comparing the performance of the alternative estimators for $\theta°$. This is an important consideration since although the corrections, described above, are constructed to remove the local asymptotic bias from $\hat{\theta}$ they are potentially very inefficient (asymptotically). This can be seen by comparing the covariance matrices in the limiting distributions of $\bar{\theta}$ and $\hat{\theta}$, given in (15) and (9) respectively. The difference between the two is

$$\mathcal{I}^{\theta\theta} - \mathcal{I}_{\theta\theta}^{-1} = \{\mathcal{I}_{\theta\theta}\}^{-1}\mathcal{I}_{\theta d}\mathcal{I}^{dd}\mathcal{I}_{d\theta}\{\mathcal{I}^{\theta\theta}\}^{-1},$$

which is a positive semi-definite matrix; it is positive definite if and only if the $(q \times k)$ matrix $\mathcal{I}_{d\theta}$ has rank k, in which case the number of parameters of interest (k) must be less than or equal to the number of moment restrictions (q) introduced to test the model specification. (This is in fact the case if the (full) information matrix test is used.)

In order to evaluate this asymptotic trade-off an "asymptotic mean square error" (AMSE) criterion will be used, defined as follows:

Asymptotic Mean Square Error Criterion: Suppose some estimator $\hat{\beta}$ *of* β *has a limiting distribution such that* $\sqrt{n}(\hat{\beta} - \beta) \xrightarrow{d} N(\mu,V)$, *then we define AMSE*$(\hat{\beta}) = \mu\mu' + V$. *If* $\bar{\beta}$ *is an alternative competing estimator then the "strong" criterion says that* $\hat{\beta}$ *is better than* $\bar{\beta}$ *if and only if AMSE*$(\bar{\beta})$ - *AMSE*$(\hat{\beta}) \geq 0$, *in the sense of being non-negative definite (nnd).*

These sorts of comparisons between $\hat{\theta}$ and $\bar{\theta}$ are valid only if $\hat{\theta}$ is employed "all the time" or if $\bar{\theta}$ is employed "all the time". A more obvious strategy, perhaps, would be to correct only when misspecification is detected. That is, one would usually first look for significant values of the test statistic

$$T_n = nD_n(\mathbf{u};\hat{\theta})'\left\{\mathcal{I}^*(\hat{\theta})\right\}^{dd} D_n(\mathbf{u};\hat{\theta}) \cong n\, R^2_{s.\hat{W}} \qquad (23)$$

which has a limiting non-central χ^2 distribution ($\chi^2(q,\upsilon)$), on q degrees of freedom with non-centrality parameter (*ncp*) $\upsilon = \delta' R\{\mathcal{I}^{dd}\}^{-1} R'\delta$. Corrections are then employed if T_n exceeds a pre-determined (asymptotic) critical value c, so that a "pre-test" estimator $\tilde{\theta}$ is achieved where

$$\tilde{\theta} = \begin{cases} \hat{\theta} \text{ if } T_n \leq c \\ \bar{\theta} \text{ if } T_n > c\,. \end{cases}$$

Since $\bar{\theta}$ will be used some of the time and $\hat{\theta}$ at other times neither of the limiting distributions described in (14) or (9) will be correct for the actual estimator employed.[19]

For the present though we shall proceed as if either $\hat{\theta}$ or $\bar{\theta}$ is to be employed all the time regardless of any inferences obtained from T_n so that comparisons can be made.[20] Now, applying the strong AMSE criterion we find that (see (14) and (19))

$$\text{AMSE}(\bar{\theta}) - \text{AMSE}(\hat{\theta}) = \mathcal{I}_{\theta\theta}^{-1}\mathcal{I}_{\theta d}\{\mathcal{I}^{dd} - R'\delta\delta' R\}\mathcal{I}_{d\theta}\mathcal{I}_{\theta\theta}^{-1} = \mathcal{A}\,.$$

---

[19] Many econometric texts gloss over a similar problem in the context of the general linear model; cf., Johnston (1984, p.204) who talks about estimation subject to linear restrictions and writes "If H_0 is not rejected, one may wish to re-estimate the model, incorporating the restrictions into the estimation process". He then proceeds to derive the distribution of the restricted least squares estimator as if it were employed "all the time".

[20] The analytical difficulties that arise from a pre-test solution could provide the focus of future work.

Hence $(\mathcal{I}^{dd} - R'\delta\delta'R)$ being *nnd* is sufficient to ensure that $\mathcal{A}$ is *nnd*. The converse is not necessarily true, however, even if the rank of $\mathcal{I}$ is k. With this in mind, a testable condition can be established which ensures the superiority of $\hat{\theta}$ over $\bar{\theta}$ in the sense that $\mathcal{A}$ is *nnd*. That is, although $\bar{\theta}$ achieves a reduction in bias it is potentially very inefficient and on the basis of an AMSE criterion for comparing $\bar{\theta}$ and $\hat{\theta}$, in the presence of local specification error, $\hat{\theta}$ is still to be preferred to $\bar{\theta}$ if $\lambda'(\mathcal{I}^{dd} - R'\delta\delta'R)\lambda \geq 0$ for all vectors $\lambda \neq 0$. Note that $\mathcal{I}^{dd}$ is positive definite thus, following Toro-Vizcarrondo and Wallace (1968), $\lambda'(\mathcal{I}^{dd} - R'\delta\delta'R)\lambda \geq 0$ if and only if

$$\Psi(\lambda) = \frac{(\lambda'R'\delta)^2}{\lambda'\mathcal{I}^{dd}\lambda} \leq 1. \tag{24}$$

(Since $\lambda'\mathcal{I}^{dd}\lambda$ is, by assumption, strictly positive.) But the results of Rao (1973) show that

$$\sup_{\lambda} \Psi(\lambda) = \delta'R\{\mathcal{I}^{dd}\}^{-1}R'\delta$$

and using this it can be shown that AMSE($\bar{\theta}$) - AMSE($\hat{\theta}$) is *nnd* if the non-centrality parameter $\upsilon = \delta'R\{\mathcal{I}^{dd}\}^{-1}R'\delta \leq 1$, as follows:

(i) suppose $\lambda'(\mathcal{I}^{dd} - R'\delta\delta'R)\lambda \geq 0$ for all $\lambda \neq 0$ then $\Psi(\lambda) \leq 1$ by (24). Thus in particular $\sup \Psi(\lambda) = \delta'R\{\mathcal{I}^{dd}\}^{-1}R'\delta \leq 1$;

(ii) now suppose that $\delta'R\{\mathcal{I}^{dd}\}^{-1}R'\delta \leq 1$, then $\sup \Psi(\lambda) \leq 1$, therefore $\Psi(\lambda) \leq 1$ for all $\lambda \neq 0$, thus $\lambda'(\mathcal{I}^{dd} - R'\delta\delta'R)\lambda \geq 0$ for all $\lambda \neq 0$.

Note that AMSE($\bar{\theta}$) - AMSE($\hat{\theta}$) is *nnd* if the matrix $\mathcal{I}^{dd} - R'\delta\delta'R$ is *nnd* which, by the above arguments, is true if and only if *ncp* of the test statistic T_n (given by $\upsilon = \delta'R\{\mathcal{I}^{dd}\}^{-1}R'\delta$) is less than unity. Thus if $\upsilon \leq 1$ it would appear that the reduction in asymptotic bias achieved by $\bar{\theta}$ is not enough to offset its asymptotic inefficiency. Thus, by the strong AMSE criterion, the usual MLE $\hat{\theta}$ will be superior, even in a misspecified model. Definite conclusions cannot be drawn if $\upsilon > 1$ since then $\mathcal{I}^{dd}$ - $R'\delta\delta'R$ may be indefinite.

A test of the hypothesis H_0: $\upsilon \leq 1$ can be carried out, using the result that $T_n \xrightarrow{d} \chi^2(q,\upsilon)$, as follows:

(i) define a critical value c_ε, such that $\Pr[\chi^2(q,\upsilon) > c_\varepsilon] = \varepsilon$, from non-central ξ^2 tables;

(ii) compute T_n and use the decision rule to accept H_0: $\upsilon \leq 1$ if $\mathcal{T}_n \leq c_\varepsilon$, suggesting no need for corrections.

It is important to note that only a sufficient condition has been established to ensure the superiority of $\hat{\theta}$ over $\bar{\theta}$ using the AMSE criterion. This result does not necessarily imply that $\bar{\theta}$ is to be preferred to $\hat{\theta}$ if $\upsilon > 1$; it merely suggests that, in certain situations, corrections may be useful if $\upsilon > 1$, be it only because they remove the asymptotic bias from $\hat{\theta}$. Indeed, on these grounds $\bar{\theta}$ may be preferred to $\hat{\theta}$ without qualification.

The fact that corrections may only be preferred (on AMSE grounds) when $\upsilon > 1$ creates a tricky problem. This is because the asymptotic results relating to corrections are only valid if the true DGP is sufficiently close to the null; that is, in a neighbourhood of the maintained model where the specification test statistic, T_n, (with *ncp* υ) has low power. If the misspecification is not small, so that the specification error cannot be regarded as $O(n^{-1/2})$, the asymptotic power of T_n will be unity (provided that it is consistent): the power increases with the sample size and, typically, in large samples large values of T_n will be realised. On the other hand if local specification error is of $O(n^{-1/2})$ the power of T_n will increase with υ, so that larger υ will create higher power. The problem is, then, how large can υ become before the implied large asymptotic power suggests that the true DGP is not in the neighbourhood of the maintained model and that, thus, the use of corrections is invalid? Since inferences about υ are based on T_n, and also that high power typically realises large values of T_n, an equivalent question concerns the magnitude of T_n. This question is difficult to answer. However, after limited experimentation, Chesher, Lancaster and Irish (1985) had some encouraging results and concluded the procedure worked well in correcting tobit MLEs for "quite substantial" neglected heteroskedasticity. Some Monte Carlo results, presented below, appear to confirm this conclusion.

Given the above interpretation that local specification error implies a neighbourhood of the maintained specification where T_n has low power one could define an upper bound on power, π^*, which would provide the boundary to the neighbourhood of the maintained specification.[21] If c is the asymptotic critical value for the specification test statistic, then an acceptable upper bound on υ could be achieved by solving $\Pr[\chi^2(q,\upsilon) > c] \leq \pi^*$. A suitable value for π^* would ensure that the upper bound on υ is greater than 1; $\pi^* = 60\%$ say, although it is completely arbitrary, as is the choice of c. This is admittedly not a very satisfactory solution to the problem.

A Monte Carlo based investigation would perhaps provide more insight to some of the above questions. The following section reports some findings of such an investigation.

5. Testing for parameter consistency

If not employed to adjust maximum likelihood estimates for neglected specification error, the local corrections can be used to form a contrast vector (in the sense of Hausman (1978)) designed to test the consistency of the obtained parameter estimates. This has recently been suggested by Davidson and MacKinnon (1989). The basis of the test is to examine the statistical significance of the contrast vector $\bar{\theta} - \hat{\theta}$ using the Hausman (1978) principle.

Given the asymptotic distribution results derived in Section 2 and how artificial regressions are used in the construction of

---

[21] Of course the test must be consistent; that is it must not exhibit low power whatever the extent of the misspecification.

maximum likelihood corrections, as outlined in Sections 3 and 4, the calculation of such a Hausman-type test statistic is relatively straightforward. The principle of the test is motivated by the asymptotic distribution results in equation (9), for the MLE, and equation (15) for the corrected estimator. These, together with Hausman's arguments, imply that under a sequence of local alternatives

$$\sqrt{n}(\bar{\theta} - \hat{\theta}) \xrightarrow{d} N\left(-\mathcal{I}_{\theta\theta}^{-1}\mathcal{I}_{\theta d}R'\delta,\ \mathcal{I}^{\theta\theta} - \mathcal{I}_{\theta\theta}^{-1} \right) \tag{25}$$

assuming that (11) holds. The variance in the above limiting distribution is positive semi-definite, as discussed previously. The test for parameter consistency is based on the following statistic

$$n(\bar{\theta} - \hat{\theta})'\left(\mathcal{I}^{\theta\theta} - \mathcal{I}_{\theta\theta}^{-1} \right)^{-1}(\bar{\theta} - \hat{\theta}) \tag{26}$$

which has a limiting χ^2 distribution on k degrees of freedom, under the null hypothesis, assuming that the variance matrix has full rank. A necessary condition for this is that $q > k$. Corrections based on an information matrix test procedure, as analysed in the context of the probit and tobit models in Section 3, satisfy this condition and we shall further assume that $\text{rank}(\mathcal{I}_{\theta d}) = k < q$. To make the test operational a consistent estimator is needed for this variance matrix and the procedure, using artificial regressions, described below suggests how this can be achieved.

The artificial regressions detailed in Sections 3 and 4 above have the following characteristics

$$\hat{g} = \hat{A}_1 b_1 + \hat{A}_2 b_2 + \text{errors}, \tag{27}$$

which can be of single or double length form, depending on the context, and where

(a) $\hat{A}_1'\hat{g} = 0$, by the likelihood equations of the null model;

(b) $\hat{A}_2'\hat{g} = nD_n(\mathbf{u};\hat{\theta}')$;

(c) the conditional moment test statistic, based on $D_n(\mathbf{u};\hat{\theta})$, is the difference between the residual sums of squares obtained by imposing $b_2 = 0$ and unrestricted ordinary least squares estimation;

(d) the contrast vector in the Hausman test is $\bar{b}_1 = \bar{\theta} - \hat{\theta}$, where $\bar{b}_1$ is the unrestricted ordinary least squares estimate of b_1 in (27);

(e) a consistent estimator of $\mathcal{I}^{\theta\theta} - \mathcal{I}_{\theta\theta}^{-1}$ is obtained as $n[(\hat{A}'\hat{A})^{11} - (\hat{A}_1'\hat{A}_1)^{-1}]$, where $(\hat{A}'\hat{A})^{11}$ is the top left $(k \times k)$ partition of $(\hat{A}'\hat{A})^{-1}$ and $\hat{A} = (\hat{A}_1, \hat{A}_2)$. Both these matrices are readily obtained as by-products of unrestricted and restricted least squares estimation of (27).

The test for parameter consistency is then constructed as

$$\bar{b}_1'\left((\hat{A}'\hat{A})^{11} - (\hat{A}_1'\hat{A}_1)^{-1} \right)^{-1}\bar{b}_1. \tag{28}$$

The large sample distribution of this statistic, under the null hypothesis of correct model specification, is χ^2_k and the test procedure looks for significantly large values of the statistic in order to reject the null. This test procedure is designed to detect possible inconsistency in maximum likelihood parameter estimates in the face of model misspecification; it may lack power in situations where the specification error does not cause parameter estimator inconsistency. To see why this is so, the analysis of Holly (1982) is directly applicable.

If, as assumed, $\text{rank}(\mathcal{I}_{\theta d}) = k < q$, then the Hausman test, based on (28), can have local asymptotic power equal to size for some forms of misspecification ($\delta \neq 0$). This is because the resultant non-centrality parameter of the Hausman test can be zero for some $\delta \neq 0$. (Observe that $R'\delta = 0$ if and only if $\delta = 0$, assuming R has full row rank.) This indicates that the test can lack power for some departures from the null specification.

On the other hand, again assuming $\text{rank}(\mathcal{I}_{\theta d}) = k < q$, the non-centrality parameter of the Hausman test procedure can never be larger than that of the conditional moment test, for any sequence of local alternatives (see Holly (1982)). This implies that for some $\delta \neq 0$ the two non-centrality parameters may be sufficiently close for the Hausman test to exhibit relatively greater local asymptotic power since the degrees of freedom for this test are smaller.

6. Some Monte Carlo results

The purpose of the experimentation is primarily to see, (i) whether the asymptotic distribution theory established for a sequence of local alternatives, in Section 2 of this chapter, can usefully be applied in the context of fixed, but small, specification error, and (ii) whether a prudent choice of indicators for the IM test statistic could effectively reduce the inefficiency of corrections, over MLEs.

Consider again the latent regression model $y^* = x'\theta + \varepsilon$. The maintained specification is that the ε are independently and identically distributed N(0,1). The regression function is defined as $x'\theta = \theta_1 + \theta_2 x_2 + \theta_3 x_3 = -1.0 + 0.5x_2 + 0.5x_3$, containing an intercept term (-1.0) and two non-constant regressors (x_2, x_3) which are both independently and identically distributed uniformly over (0,4), so that $u = (y^*, x')'$ form a sequence of *iid* random variables.

The following Monte Carlo results relate to ML corrections, based on the IM test statistic, for three sources of specification error denoted (I), (II) and (III), where

(I) θ_3 is a random variable drawn from an N(0.5,0.1) distribution independently of ε and the x's;

(II) the disturbances are heteroskedastic with distribution, conditional on x_3, being $N(0, (1.0 + 0.2x_3)^2)$;

(III) θ_2 and θ_3 are random variables: $\theta_2 \sim N(0.5,0.1)$ independently of $\theta_3 = 0.5 + (v - 1)$, where v is a gamma random variable with a mean of 1 and a variance of 0.25.

The above (I)-(III) represent quite substantial specification error; for example, for case (II) as x_3 moves from its minimum

TABLE I: θ_3 random, sample size = 200

Estimator	*True*	*Mean*	*St. Dev.*	*MSE*
1	-1.0	-0.76662	0.29299	0.14031
MLE 2	0.5	0.43774	0.10328	0.01454
3	0.5	0.34401	0.10296	0.03493
1	-1.0	-1.03141	0.85965	0.73998
COR 2	0.5	0.53965	0.57419	0.33127
3	0.5	0.49587	0.45390	0.20604
1	-1.0	-0.96726	0.39352	0.15593
COR1 2	0.5	0.49587	0.13552	0.01838
3	0.5	0.46787	0.19112	0.03756

TABLE II: $\sigma = 1.0 + 0.2x_3$, sample size = 200

Estimator	*True*	*Mean*	*St. Dev.*	*MSE*
1	-1.0	-0.66778	0.28437	0.19124
MLE 2	0.5	0.38633	0.09401	0.02176
3	0.5	0.29301	0.09899	0.05264
1	-1.0	-1.02647	0.82283	0.67775
COR 2	0.5	0.52641	0.55286	0.30635
3	0.5	0.49437	0.44822	0.20093
1	-1.0	-1.02337	0.56214	0.31655
COR1 2	0.5	0.52097	0.24165	0.05883
3	0.5	0.49753	0.30325	0.09197

TABLE III: θ_2 and θ_3 random, sample size = 200

Estimator	*True*	*Mean*	*St. Dev.*	*MSE*
1	-1.0	-0.57613	0.24754	0.24094
MLE 2	0.5	0.31808	0.08945	0.04110
3	0.5	0.23252	0.08565	0.07888
1	-1.0	-0.98706	0.74951	0.56193
COR 2	0.5	0.49927	0.47942	0.22984
3	0.5	0.44582	0.39919	0.16229
1	-1.0	-0.99128	0.48612	0.23639
COR1 2	0.5	0.49699	0.19212	0.03692
3	0.5	0.44582	0.23892	0.06002

value of 0 to its maximum value of 4, the variance of the disturbance increases from 1.0 to 3.24: an increase of over 200%.

Initially, as a check on the large sample distribution theory as applied to only a moderately sized sample, a sample of 200 observations were drawn on y^*, in each of the above cases (I)-(III), with the probit indicator vector, y, taking the value 1 if $y^* > 0$ and zero otherwise. Each experiment was replicated 500 times so that empirical sampling distributions of the corrections could be constructed. In all three cases, estimators (whether MLE or corrections) were examined on their ability to estimate $\theta^\circ = (-1.0, 0.5, 0.5)'$.

Firstly, corrections were calculated using the full IM test. The results are tabulated in Tables I to III. The summary statistics report the mean, standard deviation (St. Dev.) and mean square error (MSE) of MLEs and corrections (COR).

From these figures the inefficiency of the corrections is apparent. The standard deviation of the corrected estimate, based on the full set of IM test indicators (COR), is very much larger than that of the corresponding MLE. Thus even though the means of these corrections are far superior, their mean square errors are still much larger than the MLEs.

The efficiency of corrections can be vastly improved, however, if some information can be utilised about the nature of the specification error. For example, for Case I if it is known that only θ_3 is subject to random variation of unknown form then the inessential indicators can be dropped from the IM test indicator vector. This means that corrections can be based solely on the indicator which corresponds to the (3,3) element in the information matrix equality. These corrections are reported as COR1 in Table I.

Similar modifications can be carried out for Case III if the independence of the (unknown) variation between θ_2 and θ_3 is utilised. The appropriately modified corrections are also labelled COR1 in Table III and are based on the IM test using only two indicators, corresponding to the (2,2) and (3,3) elements in the information matrix equality. The results show that in this case the corrections are superior to the MLEs in mean square error.

The use of such (modified) correction procedures, where only a subset of indicators, or even only one indicator, is employed to correct MLEs for neglected parameter variation of unknown form was hinted at by Lancaster (1985) in the context of duration models. The above results indicate that this could provide useful updates even when the misspecification is quite substantial, despite the fact that the theoretical results are only valid under a sequence of local alternatives.

However, another point worth stressing is that the correction procedure is designed to work well asymptotically. In particular, when the misspecification is small (i.e., $\alpha - \alpha^\circ = c$, small) the approximate large sample distribution of the MLE, $\hat{\theta}$, is

$$\hat{\theta} \sim N\left(\theta^\circ + \mathcal{I}_{\theta\theta}^{-1}\mathcal{I}_{\theta\alpha}c, \ \frac{1}{n}\mathcal{I}_{\theta\theta}^{-1} \right)$$

whilst that of the correction, $\bar{\theta}$, is

$$\bar{\theta} \sim N\left(\theta^\circ, \ \frac{1}{n}\mathcal{I}^{\theta\theta}\right).$$

Thus, the standard errors of both are approximately $O(n^{-1/2})$; but the approximate bias of $\hat{\theta}$ is O(c), c small, whereas that of the correction is zero. Therefore, the mean square error of the correction should fall more rapidly than that of the MLE. Table IV extends the results of Table II to sample sizes of 500 and 1000. As conjectured, the bias of the MLE remains approximately constant as the sample size increases (but with fixed specification error). Also, the standard deviation, of the empirical sampling distribution, of all estimators falls with increased sample size. This implies, as confirmed by the entries in the last column of the table, that the mean square error will fall more rapidly for the corrections than for the MLE. For example, the correction based on the full set of IM indicators, COR, the mean square error falls by (approximately) two-thirds as the sample size increases from 200 to 500; and falls by more than 50% as the sample increases to 1000, at which point it compares very well with that

TABLE IV: $\sigma = 1.0 + 0.2x_3$

Estimator		True	Mean	St. Dev.	MSE
			sample size = 500		
MLE	1	-1.0	-0.64824	0.17044	0.15278
	2	0.5	0.37777	0.05770	0.01827
	3	0.5	0.28477	0.05705	0.04958
COR	1	-1.0	-0.96334	0.46837	0.22071
	2	0.5	0.49508	0.30983	0.09602
	3	0.5	0.46653	0.24494	0.06112
COR1	1	-1.0	-0.96857	0.33402	0.11256
	2	0.5	0.49606	0.14159	0.02006
	3	0.5	0.46793	0.17150	0.03044
			sample size = 1000		
MLE	1	-1.0	-0.65221	0.11619	0.13446
	2	0.5	0.37719	0.04012	0.01669
	3	0.5	0.28680	0.03955	0.04702
COR	1	-1.0	-0.98087	0.32364	0.10511
	2	0.5	0.49903	0.20917	0.04375
	3	0.5	0.47653	0.16908	0.02914

of the MLE. (The corrections which use only the correct contributions for the appropriate alternative of heteroskedasticity (COR1) can be seen to be out-performing the MLE at a sample size of 500.) Thus, regarding the use of corrections as an asymptotically valid procedure, the Monte Carlo results reported here are quite encouraging in that (on mean square error grounds) the performance of the corrections improves much more rapidly than that of the maximum likelihood estimators.

Turning now to the corresponding tobit model, the correction procedure (discussed at the end of Section 3 and using the first set of 5 indicators in the IM test statistic, corresponding to the $\beta\beta'$ partition of the information matrix equality) is considered in the context of the heteroskedastic alternative defined by Case II. We shall focus on the corrections as applied to the tobit regression parameter estimates. That is, as in the probit case, we shall consider the ability of maximum likelihood estimates and associated corrections to estimate $\theta^{\circ\prime} = (-1.0,\ 0.5,\ 0.5)$. For sample sizes of 200, 500 and 1000, based on 500 replications in each case, the results are presented in Table V.

TABLE V: $\sigma = 1.0 + 0.2x_3$, sample size = 200

Estimator		*True*	*Mean*	*St. Dev.*	*MSE*
			sample size 200		
	1	-1.0	-1.20203	0.31279	0.13865
MLE	2	0.5	0.51862	0.09426	0.00923
	3	0.5	0.56536	0.10206	0.01469
	1	-1.0	-0.91621	0.32430	0.11219
COR	2	0.5	0.49812	0.10143	0.01029
	3	0.5	0.47659	0.10328	0.01121
			sample size 500		
	1	-1.0	-1.18551	0.18802	0.06977
MLE	2	0.5	0.51520	0.05882	0.00369
	3	0.5	0.55862	0.05907	0.00693
	1	-1.0	-0.90884	0.20205	0.04913
COR	2	0.5	0.49551	0.06366	0.00407
	3	0.5	0.47238	0.06017	0.00438
			sample size 1000		
	1	-1.0	-1.19042	0.13512	0.05452
MLE	2	0.5	0.51575	0.04266	0.00268
	3	0.5	0.56090	0.04107	0.00540
	1	-1.0	-0.91568	0.14635	0.02803
COR	2	0.5	0.49660	0.04586	0.00211
	3	0.5	0.47488	0.04277	0.00246

The first thing to observe is that in the presence of neglected heteroskedasticity the maximum likelihood estimates tend to be expanded (in absolute size), which has been noted before (see, for example, Chesher, Lancaster and Irish (1985) for an explanation of this phenomenon). Secondly, the corrections do go in the right direction but, if anything, tend to over-correct. The most encouraging observation to come from the table is that the corrections in the tobit model are not as inefficient (compared to the maximum likelihood estimates) as the corresponding probit corrections were.[22] This results in a mean square error performance for the corrections which is on a par with (if not better than) that of the maximum likelihood estimates, in most cases.

Again, as observed in the probit simulations and as the general analysis suggests, for this fixed specification error the bias of the maximum likelihood estimates remains approximately constant, for increasing sample size, and the mean square error of the corrections falls more rapidly than that of the maximum likelihood estimates. For a sample size of 1000, all the corrected estimates are lower in mean square error than the corresponding maximum likelihood estimates. Thus, even though the corrections are based on a test statistic which is vague about the exact nature of the heteroskedastic alternative, they nevertheless perform remarkably well. Other experimentation (not reported here) using a different heteroskedastic specification (but still a function of the included regressors) provided similar results.

A final point worth mentioning is the apparent severity of the specification error, as indicated by the empirical power of the IM test statistic. For the heteroskedastic tobit model the observed rejection rates (at the nominal 5% level) were 86%, 99% and 100% for sample sizes 200, 500 and 1000 respectively. The test is consistent, as expected, but such large powers (for finite sample sizes) would seem to suggest that the (local) linear approximations upon which the corrections are based are untenable. What is found, however, is that even in this situation the correction procedure is still worth consideration.

Although more work is needed in this area, the results

---

[22] This is possibly due to the fact that in the probit model it was the estimation of β/σ that was of concern, not simply β as in the tobit model.

presented here indicate that maximum likelihood corrections can be useful in correcting tobit regression parameter estimates for heteroskedasticity of unknown form and, in particular, such corrections need not be tremendously inefficient, even when the specification error is significant.

7. Concluding remarks

The results presented in this chapter provide the limiting distribution of ML corrections in the face of local misspecification of unknown form and thus extend and complement the previous work of Chesher, Lancaster and Irish (1985) and Orme (1986). In particular the properties of estimators constructed to correct for local parameter variation of unknown form have now been provided. In cross-section data, allowing for certain parameters to vary randomly across individuals is highly plausible but causes problems in the context of non-linear budget sets (Hausman, 1985), e.g., discrete choice models where the budget set is just a collection of distinct points each of which defines the attributes of a particular choice that could be made. The probit/tobit models considered in this chapter are special cases which can be extended to cover multiple choice problems (see, for example, Hausman and Wise (1978)). However, acknowledging parameter variation requires a choice of distributional form for the varying parameters which, in particular, should be consistent with the utility maximisation theory on which the model is founded. As Hausman (1985, p.1266) points out, "distributional specification remains a difficult problem" but, moreover, as more parameters are allowed to vary, higher order integrals enter the likelihood specification and Hausman concludes that "this level of required integration would be intractable". Thus correcting for small amounts of parameter variation in these sorts of models (using the IM test statistic) may prove useful and is also relatively easy. For example, in censored normal linear models corrections can be obtained, and inferences drawn, from the artificial regression used to construct the asymptotically efficient IM test statistic. Further, these corrections based on the IM test statistic in the context of a maintained probit model could provide useful updates in the presence of a variety of unknown local specification errors.

The Monte Carlo results presented in Section 6 show that the asymptotic theory established in Section 2, but based on a sequence of local alternatives, provides an adequate approximation to the empirical sampling distribution of ML corrections in the face of quite substantial, fixed, specification error. However, in the case of the probit model, corrections based on the full set of IM test indicators are very inefficient (in that they can have very large sampling variances) compared to the MLEs. On the one hand, this inefficiency can be greatly reduced by excluding inessential indicators from the IM test statistic, provided that information exists regarding what indicators should be omitted. On the other hand, it should be remembered that the correction procedure is only asymptotically valid and is designed to remove the asymptotic bias from the MLE, for small amounts of misspecification. In particular, the bias of the correction will be approximately zero, whilst that of the MLE remains O(1); however, the standard error of corrections (as of MLEs) will be approximately of the order $n^{-1/2}$. This suggests that as the sample size grows, and since the bias of the MLE remains approximately O(1), the mean square error performance of the corrections should improve quite considerably relative to that of the MLE. The Monte Carlo evidence suggests that this does, in fact, appear to be the case.

Some encouraging results were obtained for the tobit model where it was observed that corrections based on the full set of indicators for the $\beta\beta'$ component of the IM test performed remarkably well. Not only did they correct the tobit regression parameter estimates back towards the "true" parameter values, but they did not suffer from the huge sampling variances encountered in the probit model. Indeed, the sampling standard deviations of these tobit corrections were only marginally higher than that of the tobit maximum likelihood estimates.

An analytic comparison between the corrections and MLEs (see

Section 4) reveals that, on the basis of an asymptotic mean square error criterion, the latter are still superior (even in the face of local specification error) if the non-centrality parameter, υ, of the specification test statistic is less than unity. This can easily be tested as outlined in Section 4. Although this suggests that corrections may be useful when $\upsilon > 1$, questions about the validity of corrections arise when (implied) values of υ are too large, since this may indicate that the specification error is not local. More work is needed to answer this problem.

If the correction procedure discussed in this chapter does not appear desirable as a means of adjusting maximum likelihood estimates, then it might still be considered to be of some value for the purposes of constructing a Hausman-type test statistic. The basis of this test was detailed in Section 5 and has been previously noted by Davidson and MacKinnon (1989). Within a Hausman (1978) framework, the test contrasts the correction with the maximum likelihood estimator, in order to check for the possible inconsistency of the latter. The test statistic can be obtained within the context of the same artificial regression that is employed to obtain the original misspecification test statistic upon which the correction is based. If the (full) information matrix test is the basis of the misspecification test procedure, then the corresponding test for parameter consistency can have higher asymptotic local power in some directions.

Appendix A

As given in the text, Section 2, the *pdf* of Γ is of the form

$$s(\gamma) = \eta^{-k/2}\, h\{(\gamma-\gamma_0)'(\gamma-\gamma_0)/\eta\}$$

giving the conditional *pdf* of y, marginal with respect to Θ, as

$$p_1(y|x;\theta,\eta) = \int_{\mathbb{R}^k} f_1(y|x;\theta(\gamma)s(\gamma)d\gamma$$

which can be written

$$p_1(y|x;\theta,\eta) = \int_{\mathbb{R}^k} f_1(y|x;\theta(\gamma_0+w\surd\eta))h(w'w)dw,$$

by making the transformation $w = (\gamma - \gamma_0)/\surd\eta$. Recall that $p_1(y|x;\theta_0,0) = f_1(y|x;\theta_0)$. The log-likelihood is then $\mathcal{L} = \Sigma \log\{p_1(y_i|x_i;\theta_0,\eta)\}$. To evaluate the score for η we need to calculate $\partial\log\{p_1(y_i|x_i;\theta,\eta)]/\partial\eta = \{\partial p_1/\partial\eta\}/p_1$, and in particular $\partial p_1/\partial\eta$. This latter term is (suppressing the arguments of p_1 and f_1)

$$\frac{\partial p_1}{\partial\eta} = \int_{\mathbb{R}^k} \frac{1}{2\surd\eta}\,\frac{\partial f_1}{\partial\theta'}\,\frac{\partial\theta}{\partial\gamma'}w\ h(w'w)\ dw \tag{A1}$$

As in the text we are only considering local unknown specification error, so without loss of generality we have that $\partial\theta/\partial\gamma' = Q$, a finite $(k \times k)$ matrix with elements not all equal to zero, but generally unknown. A score test of $\eta = 0$ is based on the mean score evaluated at $\eta = 0$ with other unknown parameters (θ_0) replaced by their MLEs $(\hat{\theta})$. However, at $\eta = 0$, $\partial p_1/\partial\eta$ is indeterminate, taking the form "0/0", since $f_1(.)$ at $\eta = 0$ does not involve w, and $E_h\{w\} = 0$.

Following Chesher (1984), re-write (A1)

$$\frac{\partial p_1}{\partial \eta} = \int_{\mathbb{R}^k} \frac{\sqrt{\eta}}{2} \frac{\partial f_1}{\partial \theta'} \frac{\partial \theta}{\partial \gamma'} w \ h(w'w) \ dw/\eta \tag{A2}$$

and take the limit, using l'Hôpital's rule, as η approaches zero through positive values. This gives

$$\zeta = \lim_{\eta \to 0_+} \frac{\partial p_1}{\partial \eta} = \frac{1}{2}\left\{ \zeta + \lim_{\eta \to 0_+} \sqrt{\eta} \frac{\partial}{\partial \eta} \int_{\mathbb{R}^k} \frac{\partial f_1}{\partial \theta'} Q \ w \ h(w'w) \ dw \right\},$$

$$= \frac{1}{2}\left\{ \zeta + \lim_{\eta \to 0_+} \Psi(\eta) \right\}.$$

Collecting terms gives $\zeta = \mathit{lim}\ \Psi(\eta)$, which thus needs to be calculated.

Now,

$$\Psi(\eta) = \sqrt{\eta} \int w'Q' \left\{ \frac{\partial}{\partial \eta} \left(\frac{\partial f_1}{\partial \theta} \right) \right\} h(w'w) \ dw \tag{A3}$$

since by assumption Q is not a function of η. Noting that $\gamma = \sqrt{\eta} w + \gamma_0$ and using the chain rule yields

$$\frac{\partial}{\partial \eta}\left(\frac{\partial f_1}{\partial \theta} \right) = \frac{\partial^2 f_1}{\partial \theta \partial \theta'} \frac{\partial \theta}{\partial \eta} = \frac{1}{2\sqrt{\eta}} \frac{\partial^2 f_1}{\partial \theta \partial \theta'} \frac{\partial \theta}{\partial \gamma} w,$$

$$= \frac{1}{2\sqrt{\eta}} \frac{\partial^2 f_1}{\partial \theta \partial \theta'} Qw. \tag{A4}$$

Substituting (A4) into (A3) gives

$$\Psi(\eta) = \frac{1}{2} \int w'Q' \frac{\partial^2 f_1}{\partial \theta \partial \theta'} Qw \ h(w'w) \ dw. \tag{A5}$$

Observing that $E\{ww'\} = \lambda^2 I_k$ and letting $\eta \to 0_+$ gives

$$\lim_{\eta \to 0_+} \Psi(\eta) = \frac{\lambda^2}{2} \text{trace} \left\{ Q' \frac{\partial^2 f_1(y|x;\theta_0)}{\partial \theta \partial \theta'} Q \right\}$$

$$= \frac{\lambda^2}{2} \text{trace} \left\{ QQ' \frac{\partial^2 f_1(y|x;\theta_0)}{\partial\theta\partial\theta'} \right\}. \tag{A6}$$

Finally, using the result in (A6) we obtain the mean score for η at $\eta = 0$, $\theta_0 = \hat{\theta}$ as

$$\frac{1}{n} \frac{\partial \mathcal{L}}{\partial \eta} \bigg|_{\eta=0,\theta_0=\hat{\theta}} = \frac{\lambda^2}{2} \frac{1}{n} \sum \text{trace} \left\{ QQ' \frac{1}{f_1(y|x;\hat{\theta})} \frac{\partial^2 f_1(y|x;\hat{\theta})}{\partial\theta\partial\theta'} \right\},$$

which is the result given in the text, equation (3), recalling that (suppressing arguments)

$$\{f_1\}^{-1}\{\partial^2 f_1/\partial\theta\partial\theta'\} = \partial^2 \log f_1/\partial\theta\partial\theta' + \{\partial \log f_1/\partial\theta\}\{\partial \log f_1/\partial\theta'\}.$$

Appendix B

Expanding $p_1(y|x;\theta_0,\eta)$ about $\eta = 0$ gives the following approximation to the true *pdf* of y, marginal with respect to Θ as

$$p_1(y|x;\theta_0,\eta) = p_1(y|x;\theta_0,0) + \frac{\partial p_1(y|x;\theta_0,\eta)}{\partial \eta}\eta.$$

Recalling the results that were established in Appendix A, in particular (A6), noting that $p_1(y|x;\theta_0,0) = f_1(y|x;\theta_0)$ and substituting $\Omega = \eta\lambda^2 QQ'$, we can write the above expression (in terms of θ_0 and Ω) as

$$p_1(y|x;\theta_0,\Omega) = f_1(y|x;\theta_0)\left\{1 + \frac{1}{2}\,\text{trace}\{\Omega D(u;\theta_0)\}\right\},$$

where

$$D(u;\theta_0) = \frac{\partial \log f_1}{\partial \theta}\frac{\partial \log f_1}{\partial \theta'} + \frac{\partial^2 \log f_1}{\partial \theta \partial \theta'}$$

which is exactly the approximation obtained by Chesher (1983); see also Cox (1983).

References

Arabmazar, A. and P. Schmidt (1981), 'Further Evidence on the Robustness of the Tobit Estimator to Heteroskedasticity', *Journal of Econometrics,* 17, 253-8. -8.

Arabmazar, A. and P. Schmidt (1982), 'An Investigation of the Robustness of the Tobit Estimator to Non-Normality', *Econometrica,* 50, 1055-63.

Bera, A.K., C.M. Jarque and L - F. Lee (1984), 'Testing for the Normality Assumption in Limited Dependent Variable Models', *International Economic Review,* 25, 563-78.

Breusch, T.S. and A.R. Pagan (1980), 'The Lagrange Multiplier Test and its Application to Model Misspecification in Econometrics', *Review of Economic Studies,* 47, 239-54.

Chesher, A.D. (1983), 'The Information Matrix Test: A Simplified Calculation via a Score Test Interpretation', *Economics Letters,* 13, 45-8.

Chesher, A.D. (1984), 'Testing for Neglected Heterogeneity', *Econometrica,* 52, 865-72.

Chesher, A.D. and M. Irish (1987), 'Residual Analysis in Grouped and Censored Normal Linear Models', *Journal of Econometrics,* 34, 33-61.

Chesher, A.D, T. Lancaster and M. Irish (1985), 'On Detecting the Failure of Distributional Assumptions', *Annals de l'Insee,* 59/60, 7-44.

Cox, D.R. (1983), 'Some Remarks on Overdispersion', *Biometrika,* 70, 269-74.

Davidson, R. and J.G. MacKinnon (1989), 'Testing for Consistency Using Artificial Regressions', *Econometric Theory,* 5, 363-84.

Godfrey, L.G. (1988), *Misspecification Tests in Econometrics: The Lagrange Multiplier Principle and Other Approaches,* Cambridge University Press, Cambridge.

Godfrey, L.G. and M.R. Wickens (1982), 'Tests of Misspecification Using Locally Equivalent Alternative Models', in *Evaluating the Reliability of Macroeconomic Models,* eds., G.C.Chow and P.Corsi, Wiley, London.

Gourieroux, C., A. Monfort and A. Trognon, (1983), 'Testing Nested or Non-Nested Hypotheses', *Journal of Econometrics,* 21, 83-115.

Hall, A. (1987), 'The Information Matrix Test for the Linear Model', *Review of Economic Studies,* 54, 257-63.

Hausman, J.A. (1978), 'Specification Tests in Econometrics', *Econometrica,* 46, 1250-71.

Hausman, J.A. (1985), 'The Econometrics of Non-Linear Budget Sets', *Econometrica,* 53, 1255-82.

Hausman, J.A. and D. Wise (1978), 'A Conditional Probit Model for Qualitative Choice', *Econometrica,* 46, 403-26.

Holly, A. (1982), 'A Remark on Hausman's Specification Test', *Econometrica,* 50, 749-59.

Johnston, J. (1984), *Econometric Methods,* 3rd Edition, McGraw-Hill, New York.

Kiefer, N.M. and G.R. Skoog (1984), 'Local Asymptotic Specification Error Analysis', *Econometrica,* 52, 873-85.

Lancaster, T. (1984), 'The Covariance Matrix of the Information Matrix Test', *Econometrica,* 52, 1051-3.

Lancaster, T. (1985), 'Generalised Residuals and Heterogeneous Duration Models', *Journal of Econometrics,* 28, 155-69.

Lancaster, T. and A.D. Chesher (1985), 'Residuals, Tests and Plots with a Job Matching Application', *Annals de l'Insee,* 59/69, 47-69.

Lee, L - F. and G.S. Maddala (1985), 'The Common Strucure of Tests for Selectivity Bias, Serial Correlation, Heteroskedasticity and Non-Normality in the Tobit Model', *International Economic Review,* 26, 1-20.

Magnus, J.R. and H. Neudecker (1980), 'The Elimination Matrix: Some Lemmas and Applications', *SIAM Journal on Algebraic and Discrete Methods,* 1, 422-49.

Newey, W.K. (1985), 'Maximum Likelihood Specification Testing and Conditional Moment Tests', *Econometrica,* 53, 1047-70.

Orme, Chris (1986), 'A Simple Correction for Local Misspecification', *Bulletin of Economic Research,* 38, 177-81.

Orme, Chris (1988), 'The Calculation of the Information Matrix Test for Binary Data Models', *The Manchester School,* 55,

370-6.

Orme, Chris (1989a), 'Evaluating the Performance of Maximum Likelihood Corrections in the Face of Local Misspecification', *Bulletin of Economic Research,* 41, 29-44.

Orme, C.D. (1989b), *Misspecification and Inference in Micro-Econometrics,* Unpublished D.Phil. Thesis, University of York.

Orme, Chris (1990a), 'The Small Sample Performance of the Information Matrix Test', *Journal of Econometrics,* 46, 309-31.

Orme, Chris (1992), 'Efficient Score Tests for Heteroskedasticity in Microeconometrics', (forthcoming) *Econometric Reviews.*

Rao, C.R. (1973), *Linear Statistical Inference and its Applications,* Wiley, New York.

Rothenberg, T.J. and C.T. Leenders (1964), 'Efficient Estimation of Simultaneous Equation Systems', *Econometrica,* 32, 57-76.

Ruud, P.A. (1981), *Misspecification in Limited Dependent Variable Models,* Unpublished Ph.D. Thesis, M.I.T.

Ruud, P.A. (1983), 'Sufficient Conditions for the Consistency of Maximum Likelihood Estimation despite Misspecification of the Distribution in Multinomial Discrete Choice Models', *Econometrica,* 51, 225-8.

Toro-Vizcarrondo, C. and T.D. Wallace (1968), 'A Test of Mean Square Error Criterion for Restrictions in Linear Regression', *Journal of the American Statistical Association,* 63, 558-72.

White, H. (1982), 'Maximum Likelihood Estimation of Misspecified Models', *Econometrica,* 50, 1-25.

Yatchew, A. and Z. Griliches (1985), 'Specification Error in Probit Models', *Review of Economics and Statistics,* 67, 134-9.

C. D. ORME

7. THE SAMPLING PERFORMANCE OF THE INFORMATION MATRIX TEST [1]

1. Introduction

The information matrix (IM) test was originally proposed by White (1982) as a pure specification test and is based on the information matrix equality which obtains when the model specification is correct. This equality, or moment condition, implies the asymptotic equivalence of the hessian and outer product forms of Fisher's information matrix and the test is designed to detect the failure of this equality. As White (1982, page 9) points out, the failure of this equality implies model misspecification which "can have serious consequences when standard inferential techniques are applied". The test procedure proposed by White is a test for model misspecification that invalidates standard (statistical) inferences; a rejection of the null hypothesis of correct model specification implies "the inconsistency of the usual maximum likelihood covariance matrix estimators ... at the very least, as well as possible inconsistency of the QMLE (Quasi-Maximum Likelihood Estimator) for parameters of interest", White (1982, page 11).[2] This conjecture

---

[1] A large part of the material in this chapter is drawn from Orme (1990a), an earlier version of which was presented at the European Meeting of the Econometric Society, Copenhagen, August 1987.

[2] The IM test procedure, by itself, is not a test of parameter consistency alone. However, in conjunction with other test

is not true, however, for certain sources of model misspecification. Hall (1987) reviews the IM test statistic for the classical normal linear regression model and shows that it will be sensitive to non-normality of the regression disturbances. In this case, however, the usual parameter estimators remain consistent and all standard inferential techniques are asymptotically valid (although not optimal) for many types of non-normality. Thus although the IM test procedure, in this case, may indicate model misspecification, such specification error (i.e., non-normality of the disturbances) is essentially inconsequential to the asymptotic validity of procedures.

The calculation of the IM test in the context of binary response models has been detailed by Orme (1988, 1989b). Indeed, as was shown by Orme (1988), the calculation of the test statistic is particularly simple in binary response models (including probit and logit), being the "nR^2" statistic obtained from running an artificial regression after model estimation. In particular, for the probit model, Newey (1985) has pointed out that the IM test will be sensitive to heteroskedasticity and non-normality, as in a normal linear model. However, unlike in the normal linear model, such specification errors give rise to inconsistent maximum likelihood estimators in the context of probit (and tobit) models; see Arabmazar and Schmidt (1981, 1982) and Yatchew and Griliches (1985). The method of calculation, described by Orme (1988) is specific to those sorts of models and thus is not generally applicable. In this chapter we shall review the IM test procedure in more general settings and then examine various methods of calculation which are widely applicable.

One variant of the test that is generally applicable is White's original test statistic. This, however, requires analytical third derivatives of the log-density in the construction of its covariance matrix and, therefore, can be

---

procedures White's IM test is used to develop a program for testing parameter estimator consistency (see White, 1982).

burdensome to compute. Chesher (1983) and Lancaster (1984) derived a simplified version of the IM test by exploiting the null hypothesis in order to construct a different covariance matrix estimator. The attractiveness of this simplified version is that it does not require analytical third derivatives of the log-density and, further, can be easily computed as the sample size (n) times the uncentred R^2 from a suitable artificial regression. Chesher (1983) introduced this simplified version of the IM test statistic by interpreting the test procedure as a score/Lagrange multiplier test for local parameter variation. This score test procedure was detailed in the two appendices to Chaper 6 of this volume. In point of fact, using this interpretation, he utilised the Godfrey and Wickens (1982) "Outer Product of the Gradient" (OPG) method of calculating score test statistics. Davidson and MacKinnon (1983) argue that the OPG method yields a test statistic having a small sample distribution which may not be well approximated by its asymptotic distribution. This, it is argued, is because the OPG method uses a highly inefficient covariance matrix estimator which employs high sample moments to estimate high population moments. If the conjecture of Davidson and MacKinnon is true, we might expect poor small sample performance of the Lancaster/Chesher (OPG) variant of the IM test procedure. Indeed, recent work by Taylor (1987) indicates that this is in fact the case, and this matter obviously merits investigation.[3] This could have similar implications for the sampling performance of ML corrections constructed using this OPG variant of the IM test statistic. Moreover, the Lancaster/Chesher IM test statistic is just an example of a conditional moment test statistic, as discussed by Newey (1985).[4] Thus the same criticisms

[3] See also Kennan and Neumann (1987).

[4] In particular, the equality derived by Lancaster (1984) is just a special case of the "generalised information matrix equality" given by Newey (1985).

should apply to Newey's test statistics as well.[5]

In the next section the basis of the IM test procedure is reviewed. In Section 3 both White's IM test statistic and the Lancaster/Chesher simplified version are shown to be members of a set of IM test statistics where each member, although employing different covariance matrix estimators, still retains a nR^2 interpretation. Such nR^2 formulations offer computational simplicity and are, therefore, attractive to applied workers. These formulations have previously been proposed in many different testing situations (for example, Godfrey and Wickens (1982) as cited previously, Godfrey (1978 a,b), Hausman (1984) and more recently Newey (1985)). Interestingly, amongst this set of IM test statistics, the Lancaster/Chesher variant is shown to be numerically the largest.[6]

In Section 4, some Monte Carlo evidence is offered on the relative merits of these alternative nR^2 IM test statistics over different econometric models. It appears that substantial improvements can be made over both the Lancaster/Chesher variant and White's original test statistic by incorporating expected values of third derivatives of the log-density into the nR^2 formulation. Moreover the results indicate that the performance of the Lancaster/Chesher variant deteriorates as the degrees of freedom increase whereas other variants do not necessarily exhibit this feature. Although all these test statistics are asymptotically equivalent, their sampling performances under the null hypothesis can vary widely and thus there is a clear need to guard against incorrect inferences.

In Section 5, alternative "explained sum of squares"

---

[5] Following this line of argument, one could also criticise the so-called Berndt, Hall, Hall, Hausman (1974) (BHHH) algorithm which uses the OPG approximation to the hessian.

[6] A similar result is established by Kennan and Neumann (1987), although they only derive the numerical inequality between White's and Lancaster/Chesher's IM test statistics.

procedures are provided for a certain class of models, which include the probit model, Poisson regression model and exponential duration model. These procedures employ a double length regression in order to compute the "asymptotically efficient" variant of any conditional moment test and thus do not attract the criticism of Davidson and MacKinnon (1983). Indeed, the sampling performance of the asymptotically efficient variant of the IM test in probit models (as reported in Section 4 of this chapter) indicates that the use of this variant is to be encouraged in order to obtain a test statistic which exhibits reasonable size properties. Evidence suggests that this conclusion holds in other models as well; see Orme (1990b) and Chesher and Spady (1991). It is hoped that the general procedure developed in Section 5 will help facilitate the construction of such test statistics, at least in simple models. The analysis is extended to the Weibull duration model where it is shown that an asymptotically efficient information matrix test statistic can be obtained as the explained sum of squares test statistic from a triple length regression. To conclude this section, we consider again the truncated regression model and update the work of Orme (1990a) by providing a double length regression which can be used to calculate the asymptotically efficient variant of the IM test statistic for the truncated regression model. Further Monte Carlo experimentation supports the conjecture that, for this model too, the efficient variant is vastly superior to the OPG Chesher/Lancaster variant in terms of its size properties. The double length regression needed to construct the efficient version of the IM test for the tobit model was detailed in Chapter 6 of this volume, as developed by Orme (1991a). As with the tobit model, the artificial regressions used to calculate the efficient version of the IM test statistic can also be used to provide the maximum likelihood corrections, which were discussed in Chapter 6.

Finally, some concluding remarks are presented in Section 6.

2. The IM test procedure

Let $\ell(u;\theta) = \log\{f(u;\theta)\}$ be the logarithm of a density function, for a variate U, depending on a (k × 1) parameter vector θ.[7] Following a notation similar to that of White (1982), the following (k × k) matrices, when they exist, are defined by

$$A_n(\mathbf{U};\theta) = \frac{1}{n}\sum_{=1}^{n} \frac{\partial^2 \ell(U_i;\theta)}{\partial\theta\,\partial\theta'}, \quad B_n(\mathbf{U};\theta) = \frac{1}{n}\sum_{=1}^{n} \frac{\partial \ell(U_i;\theta)}{\partial\theta}\frac{\partial \ell(U_i;\theta)}{\partial\theta'},$$

$$A(\theta) = E\left\{\frac{\partial^2 \ell(U_i;\theta)}{\partial\theta\,\partial\theta'}\right\}, \quad B(\theta) = E\left\{\frac{\partial \ell(U_i;\theta)}{\partial\theta}\frac{\partial \ell(U_i;\theta)}{\partial\theta'}\right\}$$

where expectations (in the above expressions and in what follows) are taken with respect to the true density function, and $\mathbf{U}' = (U_1,...,U_n)$. When the model is correctly specified, the true density is $\exp\{\ell(u;\theta°)\}$ where θ° is the true parameter point. It is important, for what follows, that the reader recognises the difference between $A_n(\mathbf{U};\theta)$ and $A(\theta)$ (and between $B_n(\mathbf{U};\theta)$ and $B(\theta)$): the latter does not in general depend upon **U**, the former does. (For example, if U is a Poisson random variable with true mean equal to the scalar θ°, then $A_n(\mathbf{U};\theta°) = n^{-1}(-\{\theta°\}^{-2}\Sigma\, U_i)$, whilst $A(\theta°) = -\{\theta°\}^{-1}$.)

The IM test procedure is based on the information matrix equality which states that, when the model is correctly specified, $A(\theta°) = -B(\theta°)$ or equivalently $A(\theta°) + B(\theta°) = 0$, which is a standard result from likelihood theory, provided that $\ell(U;\theta)$ is continuously twice differentiable with respect to θ and that the range of possible values of U does not depend on θ. On the basis of n independent realisations of U, $u_1,u_2,...,u_n$, the test procedure investigates the statistical significance of $A_n(\mathbf{u};\hat{\theta}) + B_n(\mathbf{u};\hat{\theta})$, where $\mathbf{u}'=(u_1,...,u_n)$ and $\hat{\theta}$ is the maximum likelihood estimate of θ° which solves the following optimisation problem

---

[7] Upper case letters will usually be used to denote random variables and lower case letters their realisations.

$$\max_{\theta} \sum_{i=1}^{n} \ell(u_i;\theta).$$

The solution to the above problem in which realisations u_i are replaced by random variables U_i defines the maximum likelihood estimator which is also commonly denoted by $\hat{\theta}$. To economise on notation, $\hat{\theta}$ is used for both estimate and estimator.

Since $A_n(\mathbf{u};\theta) + B_n(\mathbf{u};\theta)$ is a symmetric matrix some of its elements will be redundant and so the $q \le k(k+1)/2$ non-redundant elements are placed in a $(q \times 1)$ column vector of indicators defined as

$$D_n(\mathbf{u};\theta) = \frac{1}{n}\sum_{i=1}^{n} d(u_i;\theta)$$

where $d(u_i;\theta)$ is a $(q \times 1)$ vector of contributions having typical element

$$d_r(u_i;\theta) = \partial^2\ell(u_i;\theta)/\partial\theta_j\partial\theta_h + \partial\ell(u_i;\theta)/\partial\theta_j \; \partial\ell(u_i;\theta)/\partial\theta_h \,,$$
$$(r=1,...q;\ j=1,...k;\ h=j,...k).$$[8]

White (1982) showed that if the model is correctly specified, then the asymptotic sampling distribution of $\sqrt{n}D_n(\mathbf{U};\hat{\theta})$ is multivariate normal with mean zero and covariance matrix $V(\theta^\circ)$, defined by

$$V(\theta^\circ) = E[\{w(U_i;\theta^\circ)\}\{w(U_i;\theta^\circ)\}']$$

where

$$w(U_i;\theta) = d(U_i;\theta) - \nabla D(\theta)\{A(\theta)\}^{-1}\nabla\ell(U_i;\theta)'$$

---

[8] Redundant indicators have been assumed away so that $q \le k(k+1)/2$.

and

$$\nabla\ell(U_i;\theta) = \frac{\partial\ell(U_i;\theta)}{\partial\theta'}, \qquad \nabla D(\theta) = E\left\{\frac{\partial d(U_i;\theta)}{\partial\theta'}\right\}.$$
$$(1 \times k) \qquad\qquad\qquad (q \times k)$$

Then the statistic defined as

$$\omega = nD_n(\mathbf{U};\hat{\theta})'\{\overline{V}(\hat{\theta})\}^{-1}D_n(\mathbf{U};\hat{\theta}), \tag{1}$$

where $\overline{V}(\hat{\theta})$ is any consistent estimator for $V(\theta°)$, has an asymptotic sampling distribution which is χ^2_q when the model is correctly specified.[9] To carry out the test, the IM test statistic ω, given in (1), is computed on the basis of the n independent realisations of U and is compared to the critical value of the χ^2_q distribution for the given nominal size of the test. A computed test statistic exceeding this critical value is interpreted as indicating model misspecification.

White's original IM test used the following estimator for $V(\theta°)$

$$\frac{1}{n}\sum_{i=1}^{n}[d(U_i;\hat{\theta})-\nabla D_n(\mathbf{U};\hat{\theta})\{A_n(\mathbf{U};\hat{\theta})\}^{-1}\nabla\ell(U_i;\hat{\theta})'].$$
$$[d(U_i;\hat{\theta})-\nabla D_n(\mathbf{U};\hat{\theta})\{A_n(\mathbf{U};\hat{\theta})\}^{-1}\nabla\ell(U_i;\hat{\theta})']', \tag{2}$$

where

$$\nabla D_n(\mathbf{U};\theta) = \frac{1}{n}\sum_{i=1}^{n}\frac{\partial d(U_i;\theta)}{\partial\theta'} \qquad (q \times k).$$

---

[9] Indeed, this will be the correct sampling distribution even if the model is misspecified provided that the usual maximum likelihood techniques remain valid. This point is taken up again later.

An asymptotically efficient form of the IM test statistic is constructed from (1) with the covariance matrix $V(\hat{\theta})$, which is obtained by, firstly, calculating $V(\theta^\circ)$ under the null hypothesis, (thus eliminating U_i) and then replacing θ° by $\hat{\theta}$ in the resulting expression. The first of these computations might be slightly eased by appealing to Lancaster's (1984) result (see also Newey (1985)) which states that under the correct model specification

$$\nabla D(\theta^\circ) = - E\{d(U_i;\theta^\circ)\nabla \ell(U_i;\theta^\circ)'\}, \tag{3}$$

so that $V(\theta^\circ)$ under the null hypothesis, may be written as

$$V(\theta^\circ) = E[\{d(U_i;\theta^\circ)\}\{d(U_i;\theta^\circ)\}'] + \nabla D(\theta^\circ)\{A(\theta^\circ)\}^{-1}\nabla D(\theta^\circ)'. \tag{4}$$

Substituting $\hat{\theta}$ for θ°, after evaluating the right hand side in (4), gives $V(\hat{\theta})$, sometimes said to be "asymptotically efficient" since it depends on the data only implicitly through the asymptotically efficient maximum likelihood estimator, $\hat{\theta}$. (Indeed sometimes θ° may not appear in $V(\theta^\circ)$ at all; for example, in the simple model $U_i = \theta + \varepsilon_i$, where $\{\varepsilon_i\}$ are a sequence of *iid* standard normal variates, the IM test statistic has asymptotic variance equal to 2.) From a practical viewpoint this means that it estimates the high population moments that occur in $\{w(U_i;\theta)\}\{w(U_i;\theta)\}'$ by functions of low population moments. Davidson and MacKinnon (1983) would argue that this estimator should be used to compute IM test statistics in order for asymptotic distribution to be a good approximation to the true sampling distribution of (1), under the null hypothesis of correct model specification (see, however, Efron and Hinkley (1978) for a discussion of the relative merits of using observed information rather than expected information in statistical inference). However, (4) requires evaluation not only of $\nabla D(\theta)$ but also of $E[\{d(U_i;\theta)\}\{d(U_i;\theta)\}']$.

White's test does not require all these burdensome computations, but does require analytical third derivatives of the log-density. Chesher (1983) and Lancaster (1984) proposed an elegant way of avoiding even this calculation. By using the relationship given in (3) and the information matrix equality, Lancaster established the following consistent estimator for $V(\theta°)$ under the null hypothesis

$$\frac{1}{n}\sum_{i=1}^{n} [d(U_i;\hat{\theta})-\nabla G_n(\mathbf{U};\hat{\theta})\{B_n(\mathbf{U};\hat{\theta})\}^{-1}\nabla\ell(U_i;\hat{\theta})'] . \\ [d(U_i;\hat{\theta})-\nabla G_n(\mathbf{U};\hat{\theta})\{B_n(\mathbf{U};\hat{\theta})\}^{-1}\nabla\ell(U_i;\hat{\theta})']',$$

where

$$G_n(\mathbf{U};\theta) = \frac{1}{n}\sum_{i=1}^{n} \{d(U_i;\theta)\nabla\ell(U_i;\theta)\}.$$

Using this, he showed that an IM test statistic can be calculated as n times the uncentred R^2 from an artificial regression of a $(n \times 1)$ vector of ones on a $(n \times q+k)$ matrix having rows $\{d(u_i;\hat{\theta})',\nabla\ell(u_i;\hat{\theta})\}$. This variant of the IM test statistic is also discussed in two articles by Chesher (1983, 1984), who provides a score/Lagrange multiplier test interpretation; this is the so-called Outer Product of the Gradient (OPG) version.

Although the OPG form of the IM test statistic is relatively simple to compute, the same arguments that advise using the asymptotically efficient covariance matrix estimator (Davidson and MacKinnon, 1983) would strongly advise against the use of this OPG variant. The criticism is that the OPG method of calculating test statistics employs a very inefficient covariance matrix estimator which estimates high population moments with high sample moments, so that the derived asymptotic sampling distribution of the test statistic does not provide an adequate approximation to the true sampling distribution, even in quite large samples. In particular it is argued that it will be "over-sized" in that the null hypothesis will be rejected far too often even when it is true. If this conjecture is true then we might expect poor small (and even moderately large) sample performance from the OPG variant of the IM test statistic. Indeed, recent work by Taylor (1987) and Kennan and Neumann (1987) indicates that this is in fact the case and, thus, it would seem that this matter merits further investigation.

On the other hand, apart from the computational burden of calculating the covariance matrix estimator, expressing the asymptotically efficient version of the IM test statistic in an easy to compute nR^2 or explained sum of squares formulation is case specific. Instances where this can be achieved are given by Hall (1987) for the normal linear regression model and for general binary data models as detailed by Orme (1988). Also, a slightly different "explained sum of squares statistic" form of the asymptotically efficient variant is presented in Section 5 for the Poisson regression model and the exponential duration model. Instances of when this cannot be achieved occur when $V(\theta^{\circ})$ is impossible to compute, even after exploiting the null hypothesis; e.g., duration models subject to random censoring. An example of where $V(\theta^{\circ})$ could be particularly tedious to compute, and which even then does not at first sight yield a nR^2 formulation, is the normal truncated regression model (see Section 4, below). Moreover, the work of Kennan and Neumann (1987) also appears to indicate that in certain econometric models, where a large number of test indicators are included, even the asymptotically efficient IM test statistic can be slightly over-sized, although it remains a substantial improvement over the OPG variant.

Given these observations, the route taken here is to define, in general, a set of IM test statistics, in which each member can be computed from an nR^2 formulation using the same artificial regression used to construct the Lancaster/Chesher OPG variant. This set is important because it can be shown that (i) White's original test is a member of this set and (ii) the Lancaster/Chesher variant is numerically the largest member of the set. Each member is, in general, easier to compute than the asymptotically efficient variant and, in particular, does not require evaluation of $E[d(U_i;\theta)d(U_i;\theta)']$. Computational convenience is not the only criterion by which a test should be judged, however, and finite sample behaviour is also important. The concern of this chapter is to investigate whether any member of this set exhibits an acceptable sampling performance, in terms of size, in absolute terms or, if not, performs equally as well as the, (perhaps) computationally more burdensome, asymptotically efficient variant.

Before proceeding, a point which, perhaps, should be made is that different variants of the IM test statistic (constructed by employing different covariance matrix estimators) will possess different finite sample power properties; although they will have the same asymptotic power against a local alternative which approaches the null at a rate of $n^{-1/2}$. This aspect is not considered here. A related point, however, is that White's original version of the IM test will have power equal to size in situations where both $\hat{\theta}$ is consistent for $\theta°$ and the usual maximum likelihood covariance matrix estimators (viz. $\{B_n(\mathbf{U};\hat{\theta})\}^{-1}$ or $-\{A_n(\mathbf{U};\hat{\theta})\}^{-1}$, see Efron and Hinkley (1978)) are consistent even if the model is misspecified. This situation arises when $E[\nabla\ell(U_i;\theta°)] = 0$ and $E[d(U_i;\theta°)] = 0$, so that White's covariance estimator remains consistent for $V(\theta°)$, where expectations are taken with respect to the true density which may not be $\exp\{\ell(U_i;\theta°)\}$. Exploiting the null hypothesis to construct different estimators for $V(\theta°)$ may thus yield IM test statistics which reject the null hypothesis (when the model is misspecified) even though the standard maximum likelihood techniques are valid, since the covariance matrix employed will be inconsistent, in general, when such a misspecification obtains.[10] It is not enough, as is sometimes thought, that $\hat{\theta}$ be consistent for $\theta°$ in order for White's test to have power equal to size; e.g., in the familiar normal linear regression model, $y = X\beta° + \varepsilon$, where the assumed specification is $e \sim N(0,I)$ when in truth the disturbances are heteroskedastic, $e \sim N(0,\Omega)$ with $\Omega = \mathrm{diag}(\omega_{ii})$, the ordinary least squares estimator of $\beta°$ is consistent but the power of White's test will not equal its size because (amongst other things) the

---

[10] This may not, however, be true if the covariance matrix estimator employed is $-A(\hat{\theta})^{-1}$ rather than $-A_n(\mathbf{u};\hat{\theta})^{-1}$ or $B_n(\mathbf{u};\hat{\theta})^{-1}$.

probability limit of the IM test indicator vector will not be zero. A slightly different situation arises if the disturbances are homoskedastic but serially correlated. The IM test indicator vector is not sensitive to serial correlation (Hall, 1987) so the test will not be consistent (i.e., the test statistic is $O_p(1)$ when the errors are serially correlated). However, the test will not have power equal to size because it will no longer possess the null distribution, due to the construction of the variance matrix. This is contrary to White's conjecture that the IM test is expected to be consistent (i.e., have unit power asymptotically) "against any alternative which renders the usual maximum likelihood inference techniques invalid" (White, 1982, page 12).

3. A set of IM tests

A set of IM test statistics is defined by considering the following, non-negative definite, estimator for $V(\theta^\circ)$

$$\left.\begin{aligned} \bar{V}(\mathbf{U};\hat{\theta}) &= \frac{1}{n}\sum_{i=1}^{n} \{\bar{w}(U_i;\hat{\theta})\}\{\bar{w}(U_i;\hat{\theta})\}', \\ \bar{w}(U_i;\hat{\theta}) &= d(U_i;\hat{\theta}) - \nabla\bar{D}(\hat{\theta})\{\bar{A}(\hat{\theta})\}^{-1}\nabla\ell(U_i;\hat{\theta})' \end{aligned}\right\} \qquad (5)$$

where $\bar{A}(\hat{\theta})$ is any consistent estimator of $A(\theta^\circ)$, under the assumption that $\ell(u;\theta)$ is correctly specified (e.g., $A(\hat{\theta})$, $A_n(\mathbf{U};\hat{\theta})$ or $-B_n(\mathbf{U};\hat{\theta})$ [11]) and $\nabla\bar{D}(\hat{\theta})$ is any consistent estimator of $\nabla D(\theta^\circ)$ (e.g., $\nabla D(\hat{\theta})$, $\nabla D_n(\mathbf{U};\hat{\theta})$ or, using Lancaster's result, $-G_n(\mathbf{U};\hat{\theta})$). If we define, in a somewhat shorthand notation, the $(n \times q)$ matrix $\bar{W}$ to have rows $\bar{w}(U_i;\hat{\theta})'$ $(i=1,\ldots,n)$, then the covariance matrix estimators, (5), can be written $\bar{V}(U;\hat{\theta}) = n^{-1}\bar{W}'\bar{W}$. The corresponding estimate is obtained by substituting realisations u_i for random variables U_i in W. Using any one of these estimators yields an nR^2 formulation for calculating an IM test statistic (1) as follows.

---

[11] In certain circumstances $A(\hat{\theta})$ may be identical to $A_n(\mathbf{U};\hat{\theta})$; for example, where the U_i are simple Poisson random variables.

The likelihood equations, which are solved to give the maximum likelihood estimate $\hat{\theta}$, are $\nabla\mathcal{L}(\mathbf{u};\hat{\theta})' = \Sigma\nabla\ell(u_i;\hat{\theta})' = 0$, where the summation is over $i = 1,...,n$. Therefore the vector of indicators for the IM test can be expressed

$$D_n(\mathbf{u};\hat{\theta}) = \frac{1}{n}\sum_{i=1}^{n} d(u_i;\hat{\theta}) = \frac{1}{n}\sum_{i=1}^{n} [d(U_i;\hat{\theta}) - \nabla\overline{D}(\hat{\theta})\{\overline{A}(\hat{\theta})\}^{-1}\nabla\ell(U_i;\hat{\theta})'].$$

Substituting this and (5) into (1) gives

$$\omega = nD_n(\mathbf{u};\hat{\theta})'\{\overline{V}(\mathbf{u};\hat{\theta})\}^{-1}D_n(\mathbf{u};\hat{\theta}) = s'\overline{W}(\overline{W}'\overline{W})^{-1}\overline{W}'s$$

where s is a $(n \times 1)$ vector of ones (sometimes referred to as the "sum vector"). From this it follows, then, that employing a covariance matrix estimator of the form (5) defines a class of IM test statistics as

$$\Omega = \{\omega : \omega = s'\overline{W}(\overline{W}'\overline{W})^{-1}\overline{W}'s = nR^2_{s.\overline{W}} = n - RSS_{s.\overline{W}}\} \qquad (6)$$

where the R^2 statistic is the uncentred R^2 from an artificial regression of s on $\overline{W}$, and RSS is the associated residual sum of squares. The members of Ω will be numerically different according to the choice of $\overline{W}$. White's original IM test statistic is a member of Ω corresponding to the matrix $\overline{W}$ having rows

$$\overline{w}(u_i;\hat{\theta})' = \{d(u_i;\hat{\theta}) - \nabla D_n(\mathbf{u};\hat{\theta})[A_n(\mathbf{u};\hat{\theta})]^{-1}\nabla\ell(u_i;\hat{\theta})'\}'.$$

The Lancaster/Chesher (OPG) variant is also a member, defined by

$$\overline{w}(u_i;\hat{\theta})' = \{d(u_i;\hat{\theta}) - \nabla G_n(\mathbf{u};\hat{\theta})[B_n(\mathbf{u};\hat{\theta})]^{-1}\nabla\ell(u_i;\hat{\theta})'\}'$$

which can be interpreted as n times the uncentred R^2 from regressing s on $\hat{Y} = (\hat{Y}_1,\hat{Y}_2)$, where $\hat{Y}_1$ has rows $d(u_i;\hat{\theta})'$ and $\hat{Y}_2$ has rows $\nabla\ell(u_i;\hat{\theta})$. It will be useful to call this particular member ω_1, say. (The asymptotically efficient variant is not, however, a member since this requires evaluation of $E[d(u_i;\theta)d(u_i;\theta)']$.) In fact all members of Ω can be constructed from the regression of s on $\hat{Y}$ by imposing certain restrictions on the "coefficients"; the OPG variant corresponds to "unrestricted" estimation. This is a convenient interpretation since it readily establishes that the OPG variant will be numerically the largest member of the class.

Consider the artificial regression employed to calculate the OPG variant of the IM test statistic:

$$s = \hat{Y}b + e \tag{7}$$

where b is partitioned as $b' = (b_1', b_2')$ conformably with $\hat{Y} = (\hat{Y}_1, \hat{Y}_2)$. Then

$$\omega_1 = n - e'e = n - RSS_{s.\hat{Y}} \; .$$

The artificial regression employed to compute any $\omega \in \Omega$ is

$$s = \overline{W}b^* + e^* \tag{8}$$

with

$$\omega = n - e^{*\prime}e^* = n - RSS_{s.\overline{W}} \; .$$

However, note that $\overline{W}$ can be written as

$$\overline{W} = [\hat{Y}_1, \hat{Y}_2] \begin{bmatrix} I_q \\ -\{\overline{A}(\hat{\theta})\}^{-1}\nabla\overline{D}(\hat{\theta})' \end{bmatrix} = [\hat{Y}_1 - \hat{Y}_2\overline{M}(\hat{\theta})]$$

where $\overline{M}(\hat{\theta}) = \{\overline{A}(\hat{\theta})\}^{-1}\nabla\overline{D}(\hat{\theta})'$ and I_q is the $(q \times q)$ identity matrix. Thus (8) may be written as

$$s = [\hat{Y}_1 - \hat{Y}_2\overline{M}(\hat{\theta})]b^* + e^* \; .$$

From this it is clear that estimating (8) is equivalent to estimating (7) but subject to the constraints $b_2 = -\overline{M}(\hat{\theta})b_1$ (i.e., $\overline{H}b=0$, $\overline{H} = [\overline{M}(\hat{\theta}), I_k]$ $(k \times \{q+k\}$, rank$(\overline{H})=k)$. Assuming, then, that

$\hat{Y}'\hat{Y}$ is positive definite, standard econometric methods show that

$$e^{*\prime}e^{*} = e'e + s'\hat{Y}'(\hat{Y}'\hat{Y})^{-1}\overline{H}'\{\overline{H}(\hat{Y}'\hat{Y})^{-1}\overline{H}'\}\overline{H}(\hat{Y}'\hat{Y})^{-1}\hat{Y}'s > e'e\ .$$

Therefore $\omega = n - e^{*\prime}e^{*} < n - e'e = \omega_1$.[12] Equality obtains when $\overline{M}(\hat{\theta}) = [B_n(\mathbf{u};\hat{\theta})]^{-1}G_n(\mathbf{u};\hat{\theta})'$, since the unrestricted coefficient estimators from (7) satisfy $b_2 = - [B_n(\mathbf{u};\hat{\theta})]^{-1}G_n(\mathbf{u};\hat{\theta})'b_1$. Thus any member of Ω can be obtained from a common artificial regression (7) but subject to certain "restrictions" being placed on the coefficients.

The inequality obtained above may be of some use if the conjecture that the OPG variant will, in general, be over-sized is true. In that case, any over-rejection exhibited by the OPG variant (ω_1), when $\ell(u;\theta)$ is correctly specified, is guaranteed to be reduced by employing another member of Ω. This naturally begs the question as to which member of Ω will exhibit the best performance, in terms of size. Equivalently this question addresses the choice of $\overline{V}(\mathbf{u};\hat{\theta})$, defined in (5), which is an estimate of $V(\theta°)$. In estimating $V(\theta°)$ essentially we require the conditional variance of $d(U_i;\theta°)'$ given $\nabla\ell(U_i;\theta°)$. $\overline{V}(\mathbf{u};\hat{\theta})$ estimates this by the residual variance following some sort of "regression" of $d(u_i;\hat{\theta})'$ on $\nabla\ell(u_i;\hat{\theta})$. The OPG variant estimates the conditional mean of $d(U_i;\theta°)'$ given $\nabla\ell(U_i;\theta°)$ by a least squares regression of $d(u_i;\hat{\theta})'$ on $\nabla\ell(u_i;\hat{\theta})$. However, the true conditional mean of $d(U_i;\theta°)'$ is known to be $\{\nabla D(\theta°)[A(\theta°)]^{-1}\nabla\ell(U_i;\theta°)'\}'$; so that, utilising the asymptotic efficiency of $\hat{\theta}$, a natural way of estimating the "population"

---

[12] The covariance matrix estimate employed by Lancaster is, effectively, the OLS residual variance from regressing $\hat{Y}_1$ on $\hat{Y}_2$. The alternative estimates employed by other members of Ω are the residual variances when estimates other than OLS are employed. By the nature of least squares, this leads to a higher residual variance; the inequality then follows immediately. Many thanks to Tony Lancaster for pointing this out in a private correspondence.

coefficients $\nabla D(\theta°)[A(\theta°)]^{-1}$, under the null hypothesis of correct model specification, is to employ the corresponding maximum likelihood estimate, $\nabla D(\hat{\theta})[A(\hat{\theta})]^{-1}$. This generates a particular member of Ω which, because of its construction (although requiring expected values of third derivatives of the log-density and expected values of second derivatives of the log-density), might be expected to be superior to the OPG variant, in terms of size, whilst still being available in a nR^2 formulation. Other members of Ω can be thought of as being generated in a similar way; including, of course, White's original version of the IM test statistic which does not require expected values of third derivatives or expected values of second derivatives of the log-density.

In the next section we offer some Monte Carlo evidence on the relative performances of different members of Ω in two non-linear models: truncated normal regression model and probit model. In the latter case the asymptotically efficient version of the IM test statistic (and also a slight variant of this) is readily available which, moreover, can be calculated from explained sum of squares and nR^2 formulations (see Orme (1988)). This is included in the study as a bench-mark against which to compare the different members of Ω.

4. Monte Carlo evidence

(a) The truncated normal regression model

Consider the latent regression model

$$y_i^* = x_i'\beta + \varepsilon_i \ , \tag{9}$$

where x_i is a $(k \times 1)$ vector of explanatory variables and the ε_i represent a sequence of $NID(0,\sigma^2)$ disturbances. In the truncated regression model, sample values $u_i' = (y_i^*, x_i')$ are only observed when y_i^* is positive. Then the corresponding observed limited dependent variable is

$$y_i = y_i^* \ , \qquad x_i'\beta + \varepsilon_i > 0, \tag{10}$$

The algebra is considerably simplified if (9) is re-parameterised as

$$\tau\, y_i^* = x_i'\delta + \upsilon_i; \qquad \tau = 1/\sigma,\ \delta = \beta/\sigma. \tag{11}$$

The density function of y_i, conditional on x_i, is the truncated normal defined by

$$f(y_i;\delta',\tau) = \begin{cases} \tau\, \dfrac{\phi(\tau y_i - x_i'\delta)}{\Phi(x_i'\delta)}, & y_i > 0, \\ 0, & \text{otherwise,} \end{cases}$$

where $\phi(.)$ is the standard normal *pdf* and $\Phi(.)$ the standard normal *cdf*. It follows that the log-likelihood (ignoring constants) for this model, based on a sample of n observations $y'=(y_1,\ldots,y_n)$, is

$$\mathcal{L}(\delta',\tau) = n\log(\tau) - \frac{1}{2}\sum_{i=1}^{n}\{\tau y_i - x_i'\delta\}^2 - \sum_{i=1}^{n}\Phi(x_i'\delta)$$

giving the following likelihood equations

$$\hat{\tau}\, X'y - (X'X)\hat{\delta} - X'\hat{h} = 0, \tag{12a}$$

$$n\hat{\tau}^{-1} - \hat{\tau}\, y'y + \hat{\delta}'X'y = 0, \tag{12b}$$

where h is the $(n \times 1)$ vector having elements $h_i = \phi(x_i'\delta)/\Phi(x_i'\delta)$ and X is a $(n \times k)$ matrix of rank $k < n$ with rows x_i'.[13]

Turning now to second derivatives it is readily established that (in the notation of Section 2)

---

[13] Observe that here there are k+1 parameters.

$$A_n(\mathbf{u};\theta) = \frac{1}{n}\begin{bmatrix} -X'(I-\Lambda)X & X'y \\ y'X & -n\tau^{-2} - y'y \end{bmatrix} = \frac{1}{n}\sum_{i=1}^{n} a(u_i;\theta), \tag{13}$$

with

$$a(u_i;\theta) = \begin{bmatrix} -(1-\lambda_i)x_i x_i' & x_i y_i \\ y_i x_i' & -\tau^{-2} - y_i^2 \end{bmatrix} \tag{14}$$

where Λ is a diagonal matrix with non-zero elements equal to $\lambda_i = h_i(h_i + x_i'\delta)$ and $0 < \lambda_i < 1$. Since the model incorporates regressors (x_i'), $A(\theta°)$ is defined as *lim* $n^{-1}\Sigma\ E[a(u_i;\theta°)|x_i]$, and is estimated by taking the expectation of (14), conditional on x_i (thus eliminating y_i), and substituting $\hat{\theta}$ for any unknown parameters, then summing over observations and dividing by n; this will be denoted $A(\hat{\theta})$. Equations (12) and a knowledge of the first two moments of a truncated normal variate show that the maximum likelihood estimators, $\hat{\theta} = (\hat{\delta}', \hat{\tau})$, are "method of moments" (MOM) estimators. The fact that MLEs are MOM is important here because (13) evaluated at $\hat{\theta}$ can be obtained by substituting (12) into (13) and gives $A_n(\mathbf{u};\hat{\theta})$ which, in this case, is identical to $A(\hat{\theta})$.

Contributions to the IM test indicator vector are

$$d(u_i;\theta) = \begin{bmatrix} \{(\upsilon_i - h_i)^2 - (1-\lambda_i)\}z_i \\ \{(\upsilon_i - h_i)(\tau^{-1} - \upsilon_i y_i) + y_i\}x_i \\ (\tau^{-1} - \upsilon_i y_i)^2 - \tau^{-2} - y_i^2 \end{bmatrix} \tag{15}$$

evaluated at $\hat{\theta} = (\hat{\delta}', \hat{\tau})$, where z_i is a column vector containing distinct elements of the symmetric matrix $x_i x_i'$ (i.e., containing regressors and cross products of regressors, but omitting redundant terms, e.g., squares of dummy variables and any cross products of dummy variables).

In the sampling experiment the simple truncated regression model, described by (10), was employed, initially with a single standard normal regressor plus intercept and then with two independent standard normal regressors plus intercept. The reason for this was to provide some indication of the effects of introducing more regressors on the performance of different versions of IM test statistics, detailed below. Regressors and disturbances were generated by NAG routine G05DDF with simulated values of y being discarded if non-positive, in which case new values of both the disturbance and regressors were generated.[14]

Several IM test statistics belonging to the class Ω defined in Section 3 were computed as the sample size (n) times the uncentred R^2 from the following artificial regression

$$1 = d(u_i;\hat{\theta})'b_1 + \nabla\ell(u_i;\hat{\theta})b_2 + e_i,\ i = 1,\dots,n, \tag{16}$$

but subject to different "restrictions" being placed on the coefficients. Firstly ω_1, the OPG variant, was calculated with no restrictions imposed. Five other variants, $\omega_2,\dots,\omega_6$, were calculated by imposing, respectively, the following restrictions on the coefficients:

$$b_2 = -\,\overline{M}_j(\hat{\theta})b_1,\ j = 2,\dots,6, \tag{17}$$

with

$$\overline{M}_2(\hat{\theta}) = -[B_n(\mathbf{u};\hat{\theta})]^{-1}\nabla D_n(\mathbf{u};\hat{\theta})',\ \overline{M}_3(\hat{\theta}) = -[B_n(\mathbf{u};\hat{\theta})]^{-1}\nabla D(\hat{\theta})',$$
$$\overline{M}_4(\hat{\theta}) = -[A_n(\mathbf{u};\hat{\theta})]^{-1}G_n(\mathbf{u};\hat{\theta})',\ \overline{M}_5(\hat{\theta}) = [A_n(\mathbf{u};\hat{\theta})]^{-1}\nabla D_n(\mathbf{u};\hat{\theta})'$$

and

$$\overline{M}_6(\hat{\theta}) = [A_n(\mathbf{u};\hat{\theta})]^{-1}\nabla D(\hat{\theta})'$$

---

[14] This situation is slightly unusual for this sort of study in that the regressors will not be fixed in repeated sampling and, because of the inherent truncation, the distribution of the observed regressors will not be standard normal. It is felt, however, that this procedure is in the spirit of a truncated regression model.

and where

$$\nabla D(\theta) = \frac{1}{n}\sum_{i=1}^{n} E\{\nabla d(u_i;\theta)\,|\,x_i\}$$

where, in the above expressions, expectations are taken under the assumption of correct model specification.[15] Observe that ω_1 (the OPG formulation) and ω_4 are the only variants that do not employ third derivatives, or expected values thereof, of the log-density. Each test statistic was simulated 500 times and the results are presented in Table 1 for the "full" set of IM test indicators (six for the single regressor case and ten for the two regressor case) and also for the subset of indicators which are sensitive to heteroskedasticity (the first three in the single regressor case and the first six in the two regressor case).

The rejection rates in the tables give the percentage of the relevant test statistics exceeding the critical values indicated by the two nominal sizes (5% or 10%) as implied by the asymptotic χ^2 sampling distribution. The figures in the main body of the table under the column headings "mean" and "st.dev." give the mean and standard deviations of the particular test statistic calculated from the 500 replicated values and should be compared to the "asymptotic" values which are given at the head of the respective columns.

---

[15] The derivation of the matrix $\nabla D(\theta)$ is given in Appendix A.

TABLE 1

Truncated Regression Model: IM Test Statistics (all indicators)

No. of regressors:	k = 2				k = 3			
statistic	mean	st.dev.	rej. rates[a]		mean	st.dev.	rej. rates	
	6.00	3.46	5%	10%	10.00	4.47	5%	10%
sample size = 50								
ω1	15.85	8.10	59.4	73.6	23.75	7.57	75.6	83.8
ω2	10.80	6.53	23.0	43.8	12.31	5.20	13.0	20.4
ω3	7.63	5.05	13.6	22.2	12.26	5.00	12.4	21.0
ω4	14.12	7.73	50.0	61.8	21.48	7.05	65.0	76.4
ω5	9.05	4.14	18.2	33.0	12.08	4.94	10.0	19.6
ω6	5.62	2.53	1.8	3.2	12.81	4.91	14.0	23.0
sample size = 200								
ω1	12.00	8.29	36.8	46.8	20.12	11.02	48.6	57.4
ω2	9.18	6.39	23.2	31.8	11.46	5.93	11.2	20.2
ω3	7.28	4.82	13.2	20.4	11.27	5.59	10.2	19.4
ω4	11.07	8.00	33.0	41.8	19.21	10.67	45.4	54.2
ω5	8.89	5.00	21.0	31.6	11.43	5.76	11.2	19.4
ω6	6.52	2.98	4.2	10.8	11.63	5.77	11.6	20.8
sample size = 500								
ω1	9.11	7.58	25.2	35.0	16.40	8.97	32.4	42.4
ω2	8.19	5.96	16.2	25.2	10.88	5.58	10.0	16.8
ω3	7.02	4.67	10.4	14.8	10.71	5.66	9.6	16.6
ω4	9.29	7.28	22.0	30.6	15.92	8.81	30.4	40.0
ω5	8.44	5.49	19.4	28.2	10.89	5.58	10.0	15.4
ω6	7.03	3.93	8.4	16.6	10.90	5.88	10.4	26.8
sample size = 1000								
ω1	8.33	5.79	20.2	27.0	14.19	7.77	26.2	33.4
ω2	7.36	4.89	13.0	21.0	10.18	5.71	8.4	15.0
ω3	6.72	4.24	8.4	15.0	9.92	5.54	8.4	13.4
ω4	7.94	5.55	17.8	24.8	13.85	7.61	24.6	31.4
ω5	7.68	4.94	16.4	23.8	10.16	5.64	8.2	13.6
ω6	6.93	4.09	10.0	19.0	9.99	5.58	8.0	13.4

[a] Standard error of rejection rates are 0.975 and 1.342 resp.

TABLE 2

Truncated Regression Model: IM Test Statistics
(first set of k(k+1)/2 indicators)

No. of regressors:	k = 2				k = 3			
statistic	mean	st.dev.	rej. rates[a]		mean	st.dev.	rej. rates	
	3.00	2.45	5%	10%	6.00	3.46	5%	10%
sample size = 50								
ω1	4.67	3.23	16.2	28.6	9.41	4.13	21.8	36.2
ω2	2.64	2.21	3.0	8.8	6.03	2.99	3.4	8.4
ω3	2.40	1.86	1.2	3.4	6.16	2.76	1.4	5.6
ω4	3.53	2.91	8.8	16.8	8.03	2.76	11.4	24.0
ω5	3.53	2.45	5.8	14.6	6.46	3.23	4.6	10.2
ω6	3.10	1.90	1.4	5.8	6.59	3.05	3.0	10.6
sample size = 200								
ω1	4.15	3.30	14.8	22.2	8.25	4.91	17.2	26.2
ω2	2.96	2.58	6.2	12.0	6.16	3.76	6.8	11.6
ω3	2.79	2.39	4.6	9.4	6.24	3.81	8.4	13.4
ω4	3.32	2.94	9.4	14.2	7.45	4.55	13.2	21.0
ω5	3.79	2.99	11.2	17.8	6.35	3.96	8.2	14.4
ω6	3.43	2.51	6.2	15.4	6.39	3.99	8.8	14.0
sample size = 500								
ω1	3.84	3.54	12.6	17.8	7.73	4.52	14.0	21.4
ω2	3.11	2.94	6.8	11.6	6.20	3.90	6.6	13.0
ω3	2.94	2.66	5.8	10.0	6.21	3.91	6.6	11.6
ω4	3.29	3.15	8.4	13.0	7.21	4.30	11.2	18.2
ω5	3.69	3.37	11.6	16.2	6.27	3.98	7.6	12.8
ω6	3.47	3.01	9.2	15.0	6.27	3.99	7.0	12.2
sample size = 1000								
ω1	3.40	2.95	9.6	15.4	7.12	4.40	11.4	19.0
ω2	2.97	2.62	7.2	11.0	5.74	3.81	5.2	11.4
ω3	2.86	2.46	6.4	10.4	5.73	3.81	5.2	11.2
ω4	3.06	2.73	7.8	12.8	6.78	4.20	9.2	16.4
ω5	3.33	2.87	9.4	14.4	5.76	3.83	5.2	11.6
ω6	3.22	2.69	7.8	13.6	5.75	3.83	5.2	11.4

[a] Standard error of rejection rates are 0.975 and 1.342 resp.

From Table 1 it is clear that the OPG variant, ω_1, is rejecting far too often at all nominal levels considered and for all sample sizes, although its performance does improve as the sample size increases. More importantly, however, its performance deteriorates as the number of regressors is increased from k = 2 to k = 3 where it rejects the correctly specified model over 75% of the time at the nominal 5% level for a sample size of 50 and, correspondingly, over 26% of the time for a sample size as large as 1000. Its mean decreases gradually from a value of 23.75 (for a sample size of 50) to a value of 14.19 for a sample size of 1000, compared to the implied asymptotic value of 10. Its standard deviation behaves somewhat more strangely: increasing, initially, from a value of 7.57 to 11.02 (for a sample size of 200) and then falling back again to 7.77 (for a sample size of 1000), but always remaining much larger than 4.47. These results appear to confirm the pessimistic conclusions drawn by Taylor (1987) concerning the size bias of the Lancaster/Chesher (OPG) variant of the IM test. It is also interesting that variant ω_4 also exhibits substantial size bias, rejecting the correctly specified model 65% of the time at the nominal 5% level, for a sample size of 50, and nearly 25% of the time for a sample size of 1000. Variants ω_1 and ω_4 which do not employ third derivatives of the log-density in their construction, are consistently the worst performers in all experiments reported in Tables 1 and 2.

Turning now to Table 2, as conjectured, the performance of the OPG variant becomes better when only a subset of the full set of IM test indicators is employed. The test still over-rejects, but less than before, suggesting a reduction in the size bias as the number of indicators decreases relative to the sample size. For example, the performance of the test statistic with 6 indicators (k = 3), as reported in Table 2, for a sample size of 50 is roughly comparable with the performance using the full set of 10 indicators (Table 1, k = 3) for a sample size of 1000. The performance of variant ω_4 is similarly improved, but these two still exhibit the worst size bias within the set.

Other variants, as expected, do not exhibit as large an

upward size bias as the OPG variant; this is true even of variant ω_4. However, as observed above, very little appears to be gained from using variant ω_4 whose performance using the full set of indicators is not markedly different from that of ω_1 and in particular the size bias gets larger as the number of regressors increases. However, like the performance of OPG, the size bias might be expected to get worse, for each sample size, as the number of regressors is increased further. This characteristic of both ω_1 and ω_4 is certainly not evident in the performance of the remaining variants (which include White's test statistic, ω_5). As pointed out above, a distinguishing feature in the construction of these remaining variants is the use of third derivatives, or expected values of third derivatives, of the log-density. Using the full set of indicators with a model which has an intercept and just one regressor the performance of variants ω_2, ω_3 and ω_5 is very poor to begin with but gets better in larger samples. However, ω_3 being the best of these still rejects at a rate of 8.4% at the 5% level for a sample size of 1000. Interestingly ω_6 is initially under-sized but the rejection rates increase steadily so that it becomes over-sized, performing marginally worse than ω_3 for a sample size of 1000. The striking feature is that all these remaining 4 variants have a remarkably improved performance for sample size equal to 1000 when the number of regressors has been increased from k = 2 to k = 3, rejecting 8.4%, 8.4%, 8.2% and 8.0% of the time, respectively, at the nominal level of 5%. Indeed, the performance of variants ω_2, ω_3 and ω_5 is improved over all sample sizes, for the k = 3 case. For the reduced set of indicators, k = 3, the performances of all these remaining variants, again, show a distinct improvement over both ω_1 and ω_4.

In conclusion, then, the results of the sampling experiments presented above indicate that, for the truncated regression model, substantial reductions in size bias can be achieved over variants ω_1 and ω_4 by employing third derivatives of the log-density in the construction of the covariance matrix estimator. Whether this is true in other models is open to question. Since only marginal gains over OPG were apparent in the above model, variant ω_4 was dropped from further sampling experiments which investigated the relative advantages of using third derivatives of the log-density in constructing IM test statistics in the binary probit model.

(b) The binary probit model

Consider again the latent regression model (9) and (10). The binary probit model can be thought of as a transformation of this model where only the sign of y*, and thus τy*, is observed. This defines the binary variable y_i as

$$\left.\begin{aligned} y_i &= 1,\ x_i'\delta + \upsilon_i > 0, \\ y_i &= 0,\ x_i'\delta + \upsilon_i \leq 0. \end{aligned}\right\} \quad (18)$$

Notice that, since the sign of y* is indistinguishable from the sign of τy*, only $\delta = \beta/\sigma$ is identified. It then follows that, conditional on x_i, the probability that y_i takes the value 1 is $\Phi(x_i'\delta)$ and the log-likelihood based on a sample of n realisations of y is

$$\mathcal{L}(\delta) = \sum_{i=1}^{n} \{y_i \log\Phi(x_i'\delta) + (1-y_i)\log(1-\Phi(x_i'\delta))\}.$$

Differentiating this gives the likelihood equations

$$\sum_{i=1}^{n} \{y_i - \Phi(x_i'\hat{\delta})\} \frac{\Phi(x_i'\hat{\delta})}{\{\Phi(x_i'\hat{\delta})(1-\Phi(x_i'\hat{\delta}))\}} x_i = 0 \quad (19)$$

or $X'\hat{\Lambda}\hat{r} = 0$, where, in this case, $\hat{\Lambda}$ is a diagonal matrix with non-zero elements equal to

$$\hat{\lambda}_i = \phi(x_i'\hat{\delta})/\{\Phi(x_i'\hat{\delta})(1 - \Phi(x_i'\hat{\delta}))\}^{1/2}$$

and $\hat{r}$ is a residual vector with typical element equal to

$$\hat{r}_i = (y_i - \Phi(x_i'\hat{\delta}))/\{\Phi(x_i'\hat{\delta})(1 - \Phi(x_i'\hat{\delta}))\}^{1/2}.$$

Differentiating the log-likelihood twice gives the matrix of second partial derivatives

$$A_n(\mathbf{u};\delta) = \frac{1}{n}\,[-X'\Gamma X\,] = \frac{1}{n}\sum_{i=1}^{n} -a(u_i;\delta), \tag{20}$$

$$a(u_i;\delta) = \left\{ y_i \left(\frac{\phi(x_i{'}\delta)}{\Phi(x_i{'}\delta)} \left[\frac{\phi(x_i{'}\delta)}{\Phi(x_i{'}\delta)} + x_i{'}\delta \right] \right) \right.$$

$$\left. + \; (1-y_i) \left(\frac{\phi(x_i{'}\delta)}{1-\Phi(x_i{'}\delta)} \left[\frac{\Phi(x_i{'}\delta)}{1-\Phi(x_i{'}\delta)} - x_i{'}\delta \right] \right) \right\} x_i x_i{'}$$

$$= \gamma_i x_i x_i{'}, \text{ say,} \tag{21}$$

where Γ is a diagonal matrix with non-zero elements equal to γ_i. Taking expectations in (20) conditional on x_i and replacing unknown parameters by $\hat{\delta}$ gives $A(\hat{\delta}) = n^{-1}X'\hat{\Gamma}^E X$, where Γ^E is a diagonal matrix with non-zero elements equal to $E(\gamma_i \,|\, x_i) = \phi(x_i'\delta)/\{\Phi(x_i'\delta)(1 - \Phi(x_i'\delta))\}$ and $\hat{\Gamma}^E$ is simply Γ^E evaluated at $\delta = \hat{\delta}$; thus in the probit model $A_n(\mathbf{u};\hat{\delta})$ is not equal to $A(\hat{\delta})$.

Contributions to the IM test indicator vector are $d(u_i;\hat{\delta}) = -\,\hat{r}_i\hat{\lambda}_i(x_i'\hat{\delta})z_i$, where z_i is defined as before except that if the regression function, $x_i'\delta$, contains an intercept term the first element of $D_n(\mathbf{u};\hat{\delta})$ must be eliminated from the test statistic since it will always be identically zero. For this model Newey (1985) proposed the OPG variant of the IM test procedure. However, Orme (1988) showed that the asymptotically efficient version of the IM test statistic can easily be calculated as the uncentred explained sum of squares from an artificial regression where the observations on the left hand side variable are $\hat{r}_i$ and observations on the right hand side variables are $\hat{\lambda}_i x_i{'}$ and $-\hat{\lambda}_i(x_i'\hat{\delta})z_i{'}$. An asymptotically equivalent version (under the null) of this test statistic can be calculated from the same artificial regression as the sample size times the uncentred R^2. These two variants were included in the sampling experiment detailed below so that the asymptotically efficient variant could be compared with other (sub-optimal) variants.[16]

In the sampling experiment, the binary probit model (18) was employed with two standard normal regressors plus an intercept, the disturbances υ being realisations of a standard normal variate. Prior to experimentation 1000 values of the two regressors were generated and were held constant across all replications with the first n values being used in a simulation of sample size n (i.e., the regressors were fixed in repeated sampling). This is common practice in such simulation studies so that attention can be focused on the performance of a statistic for a particular sample size without allowing for the changing configuration of the regressors over the replications. To check the generality of the results obtained different seeds for the random number generator were used for a few cases which yielded similar results those reported below.

Apart from the two variants mentioned above (which will be denoted ω_A and ω_B, respectively), seven other IM test statistics were computed $(\omega_1,\omega_2,\omega_3,\omega_5,\omega_6,\omega_7,\omega_8)$ using the artificial regression defined in (16), but appropriate for this model (variant ω_4 was not considered here for reasons detailed previously). The OPG variant ω_1, as before, is the unrestricted uncentred R^2 from this regression and variants ω_j, j = 2,3,5,6, are defined as before. The two extra variants (ω_7 and ω_8) were included in the experiment due to the difference, in this model, between $A_n(\mathbf{u};\hat{\delta})$ and $A(\hat{\delta})$. These, again, were calculated as n times the uncentred R^2 from the artificial regression of the form (16) subject to the restrictions of the form (17) but where $\overline{M}_7(\hat{\delta}) = [A(\hat{\delta})]^{-1}\nabla D_n(\mathbf{u};\hat{\delta})'$ and $\overline{M}_8(\hat{\delta}) = [A(\hat{\delta})]^{-1}\nabla D(\hat{\delta})'$. Each test statistic was simulated 500 times and the results are presented in Table 3, where the reported means and standard deviations should be compared to the asymptotic values of 5.00 and 3.16 respectively. The standard errors of the reported rejection rates are as before.

---

[16] Sub-optimal in that the covariance matrix estimator is not the so-called asymptotically efficient one.

These results are interesting since variants ω_3, ω_6 and ω_8, which all employ expected values of third derivatives of the log-density, initially exhibit under-rejection (sample sizes 50 and 200) whilst the performances of the remaining members of Ω are appalling by comparison. The OPG variant for the probit model was proposed by Newey (1985) and, in the experiments, rejected the correct specification over 87% of the time at the nominal 5% level for a sample size of 50 and rejected it nearly 30% of the time for a sample size as large as 1000. These results, together with those in Tables 1 and 2 and those reported by Taylor (1987), clearly indicate that inferences based on this variant of the IM test can be, potentially, very misleading. White's test statistic also behaves very poorly, performing only marginally better than the OPG variant, in this case.

The performance of the variants which employ expected values of third derivatives of the log-density is striking by comparison and perhaps encouraging since, in particular, variant ω_3 also performed reasonably well in the truncated regression model (at least in the three regressor case). However, the performances of the asymptotically efficient variant, ω_B, and ω_A are markedly superior to all other variants considered with the only cause for concern being that the sampling distribution has a slightly higher variance than expected due to a few large realised values of the test statistic. (Indeed the performances showed little or no deterioration from that in the two regressor probit model, the results of which are not reported here but are available upon request.) The rejection rates at the nominal 5% level are all acceptable, except perhaps for ω_A which rejected the correct specification 7.4% of the time for a sample size of 50.

The results for the probit model indicate that substantial improvements (in terms of a reduction in size bias) can be made over both the OPG variant and White's test statistic by employing expected values of third derivatives of the log-density. In absolute terms, however, the asymptotically efficient variant performs very well indeed and since it, also, is available from an artificial regression (and so is easily constructed) its use in

TABLE 3

Binary Probit Model: IM Test Statistics
(k(k+1)/2 set of indicators)

Number of regressors: k = 3					
sample size	statistic	mean	st.dev.	rej. rates[a]	
		5.00	3.16	5%	10%
50	ωA	5.12	3.63	7.4	11.4
50	ωB	3.95	3.39	4.4	6.2
50	ω1	18.80	6.35	87.6	91.8
50	ω2	9.14	3.79	26.4	43.2
50	ω3	4.28	2.49	1.6·	5.4
50	ω5	12.31	5.55	56.0	67.8
50	ω6	4.42	1.95	0.8	2.8
50	ω7	12.37	5.62	56.6	67.4
50	ω8	4.67	2.01	1.2	3.4
200	ωA	5.04	3.80	6.0	10.4
200	ωB	4.71	5.70	6.6	8.4
200	ω1	17.69	14.30	58.6	64.2
200	ω2	13.08	9.51	49.0	56.0
200	ω3	4.11	2.28	2.0	4.0
200	ω5	13.65	11.48	47.4	55.0
200	ω6	4.96	2.44	3.2	6.2
200	ω7	13.61	11.55	46.6	54.8
200	ω8	5.15	2.55	3.4	7.0
500	ωA	4.99	3.36	6.6	10.2
500	ωB	4.52	3.37	4.6	8.8
500	ω1	13.82	12.42	44.6	53.2
500	ω2	12.09	11.68	39.2	44.6
500	ω3	4.60	2.57	2.4	5.2
500	ω5	11.70	11.52	37.0	43.2
500	ω6	5.70	3.11	6.0	14.4
500	ω7	11.64	11.44	36.8	43.0
500	ω8	5.89	3.27	6.8	16.8
1000	ωA	5.00	3.57	6.6	10.2
1000	ωB	4.78	3.86	6.6	9.8
1000	ω1	10.10	9.31	29.6	37.6
1000	ω2	8.67	8.65	24.8	30.0
1000	ω3	5.08	2.83	3.0	8.6
1000	ω5	8.40	8.27	24.2	29.6
1000	ω6	5.91	3.49	10.0	17.2
1000	ω7	8.37	8.22	23.8	29.6
1000	ω8	6.03	3.63	11.0	18.4

[a] Standard error of rejection rates are 0.975 and 1.342 resp.

this model, at least, seems reasonable in order to achieve a test procedure which has good size properties. Whether this is true in other models is open to question. Further, since the asymptotically efficient variant may not be attainable as an explained sum of squares from some artificial regression in other models a researcher may be loath to construct it (if indeed that is possible). Does then the remarkable reduction in size bias achieved by ω_3, say, in the above models hold true in other models ? If it does, then the use of this variant may be a tempting alternative to the use of the OPG variant, whose performance appears to deteriorate as the number of regressors increases whereas the results presented so far suggest that the performance of ω_3 does not. An interesting area for future research would be to investigate this matter and also to consider the power properties of such a variant of the IM test statistic.

It does appear, however, that the use of the asymptotically efficient variant does produce the best results in terms of size. With this in mind, in the next section an explicit derivation of this form of the IM test statistic is given for two other commonly used non-linear econometric models: the Poisson regression model, employed when modelling count data, and the exponential model, employed when modelling duration data. The convenient result in these two cases, as well, is that the test statistic is obtainable as an explained sum of squares statistic. Moreover, if we define a class of models of which the Poisson, exponential and probit are members then the asymptotically efficient variant of any conditional moment test statistic is obtainable as an explained sum of squares statistic. This fact does not appear to have been observed previously.[17] This interpretation provides a simple calculation procedure for applied workers which does not require extensive or sophisticated programming.

---

[17] Kennan and Neumann (1987) discuss the IM test statistic for the exponential duration model. They do not, however, recognise this simple interpretation of the asymptotically efficient variant.

5. Alternative model specifications

(a) The Poisson regression model

In a Poisson regression model a Poisson random variable, y_i, has probability function

$$f(y_i;\beta) = \frac{\{\lambda_i\}^{y_i}}{y_i!}\exp(-\lambda_i), \qquad y_i = 0,1,2,\ldots$$

where $\log(\lambda_i) = x_i'\beta$, which contains an intercept term.

Poisson models, such as this, are of interest in applied work when modelling count data since they are relatively simple to estimate (see Maddala, 1983; Cameron and Trivedi, 1986; Lee, 1986; Mullahy, 1986) and have been used as a starting point in recent studies (see, for example, Hausman et al, 1984). However, because of the "mean = variance" restriction that is imposed, the practical usefulness of such a model might be open to question and the IM test could provide a useful diagnostic check, as pointed out by Lee (1986).

The log-likelihood (ignoring constants) based on a sample of n independent observations on y_i is

$$\mathcal{L}(\beta) = \sum_{i=1}^{n}\left\{y_i x_i'\beta - \exp(x_i'\beta)\right\} \tag{22}$$

giving the following likelihood equations

$$\sum_{i=1}^{n}\{y_i - \hat{\lambda}_i\}x_i = 0,$$

where $\hat{\lambda}_i = \exp(x_i'\hat{\beta})$.

Differentiating (22) twice gives the matrix of second derivatives

$$\left.\begin{aligned} A_n(\mathbf{u};\beta) &= \frac{1}{n}[-X'\Lambda X] = \frac{1}{n}\sum_{i=1}^{n}\{-a(u_i;\beta)\}, \\ a(u_i;\beta) &= \lambda_i x_i x_i', \end{aligned}\right\} \tag{23}$$

where Λ is a diagonal matrix with non-zero elements equal to λ_i and X is a $(n \times k)$ matrix with rows x_i' and of rank $k < n$. Notice that in this model $A_n(\mathbf{u};\beta)$ is identical to $A(\beta)$. From the above equations it is seen that contributions to the IM test indicator vector are

$$d(u_i;\hat{\beta}) = \{(y_i-\hat{\lambda}_i)^2 - \hat{\lambda}_i\}z_i$$

where z_i is a column vector containing distinct elements of the symmetric matrix $x_i x_i'$.

For this model, also, it turns out that the asymptotically efficient variant of the IM test statistic can be obtained as the uncentred explained sum of squares from an artificial regression. The appropriate artificial regression can be established by first constructing the asymptotically efficient covariance matrix estimate using the formula given in equation (4). After some calculation this yields

$$V(\hat{\beta}) = \frac{1}{n}\left[Z'\hat{\Lambda}Z - Z'\hat{\Lambda}X(X'\hat{\Lambda}X)^{-1}X'\hat{\Lambda}Z + 2Z'\hat{\Lambda}^2Z\right] \tag{24}$$

where $\hat{\Lambda}^2$ is a diagonal matrix with non-zero elements equal to $\hat{\lambda}_i^2$ and Z is a $(n \times q)$ matrix with rows z_i'. If we now define the matrices $\hat{Z}^* = \hat{\Lambda}^{1/2}Z$, $\hat{X}^* = \hat{\Lambda}^{1/2}X$, where $\hat{\Lambda}^{1/2}$ is a diagonal matrix with non-zero elements equal to $\hat{\lambda}^{1/2}$, and $\hat{P}^* = I - \hat{X}^*(\hat{X}^{*\prime}\hat{X}^*)^{-1}\hat{X}^{*\prime}$ then $V(\hat{\beta})$ can be expressed as

$$V(\hat{\beta}) = \frac{1}{n}\left[\hat{Z}^{*\prime}\hat{P}^*\hat{Z}^* + 2\hat{Z}^{*\prime}\hat{\Lambda}\hat{Z}^*\right]$$

where $\hat{Z}^{*\prime}\hat{P}^*\hat{Z}^*$ is the "residual sum of squares" following a least squares regression of $\hat{Z}^*$ on $\hat{X}^*$. It will be useful, for the moment, to denote the residual matrix from this regression as $\tilde{Z}^* = \hat{P}^*\hat{Z}^*$, which has rows $\tilde{z}_i^{*\prime}$. Then $V(\hat{\beta})$ can be written as $n^{-1}\hat{W}'\hat{W}$ where $\hat{W}$ is a $(2n \times q)$ matrix defined in block partitioned form as

$$\hat{W} = \begin{bmatrix} \tilde{Z}^* \\ \sqrt{2}\hat{\Lambda}Z \end{bmatrix}. \qquad (25)$$

The matrix $\hat{W}$ has rows of the form

$$\begin{aligned} \hat{w}_i' &= \tilde{z}_i^{*\prime}, \ i = 1,...,n \\ &= \sqrt{2}\hat{\lambda}_{i-n} z_{i-n}', \ i = n+1,...,2n. \end{aligned}$$

Next define the $(n \times 1)$ vector, $\hat{g}$, to have typical element

$$\hat{g}_i = \left\{((y_i - \hat{\lambda}_i)^2/\hat{\lambda}_i) - 1)\right\} /\sqrt{2}$$

and the $(2n \times 1)$ vector $\hat{r}$ as having typical element $\hat{r}_i$ where

$$\left.\begin{aligned} \hat{r}_i &= 0, \ i=1,\ldots,n, \\ &= \hat{g}_{i-n} \ , \ i=n+1,...,2n. \end{aligned}\right\} \qquad (26)$$

Using the above notation the IM test indicator vector, $D_n(\mathbf{u};\hat{\beta})$, can be written as $n^{-1}\hat{W}'\hat{r}$. This means that the asymptotically efficient variant of the IM test statistic can be calculated as the (uncentred) explained sum of squares from the following (double length) regression

$$\begin{bmatrix} 0 \\ \hat{g} \end{bmatrix} = \begin{bmatrix} \tilde{Z}^* \\ \sqrt{2}\hat{\Lambda}Z \end{bmatrix} b \ + \ \text{errors}. \qquad (27)$$

Notice that the above formulation requires an initial (multivariate) artificial regression to be carried out in order to obtain the matrix of "residuals" $\tilde{Z}^*$. This can be circumvented by using the "residualing-out" interpretation of ordinary least squares. Thus, the same test statistic obtains if we calculate the (uncentred) explained sum of squares from

$$\begin{bmatrix} 0 \\ \hat{g} \end{bmatrix} = \begin{bmatrix} \hat{Z}^* & \hat{X}^* \\ \sqrt{2}\hat{\Lambda}Z & 0 \end{bmatrix} \begin{bmatrix} b_1 \\ b_2 \end{bmatrix} + \text{errors.} \tag{27a}$$

Alternatively, the test statistic could be obtained as the difference between the residual sums of squares from imposing $b_1 = 0$ in (27a) and unrestricted estimation.

(b) The exponential regression model

In an exponential model the random variable, y_i, has probability density function

$$f(y_i;\beta) = \begin{cases} \theta_i \exp(-\theta_i y_i), & y_i > 0 \\ 0, & \text{otherwise} \end{cases} \tag{28}$$

where $\log(\theta_i) = x_i'\beta$, which contains an intercept term.

Like the Poisson model, exponential models are relatively simple to estimate (again, see Maddala (1983)) and provide a first approach to empirical studies of duration data (see, for example, Lancaster (1979)). However, this specification is restrictive in that it imposes a constant hazard assumption. Since the specification of the hazard determines the distributional assumption, failure of the constant hazard restriction results in specification error. As pointed out by Lancaster (1983) (see also Kiefer (1985)) the IM test could be useful as a check for model misspecification.

The log-likelihood (ignoring constants) based on a sample of n observations of completed durations y_i is

$$\mathcal{L}(\beta) = \sum_{i=1}^{n} \left\{ x_i'\beta - \theta_i y_i \right\} \tag{29}$$

giving the following likelihood equations

$$\sum_{i=1}^{n} \{1 - \hat{\theta}_i y_i\} x_i = 0, \tag{30}$$

where $\hat{\theta}_i = \exp(x_i'\hat{\beta})$. Alternatively, (30) can be written

$$\sum_{i=1}^{n} \{1 - \hat{\varepsilon}_i\} x_i = 0,$$

where $\hat{\varepsilon}_i = \hat{\theta}_i y_i$, which Lancaster (1983) calls a generalised (ML) residual, in the sense of Cox and Snell (1968).

Differentiating (29) twice gives the matrix of second derivatives

$$\left.\begin{aligned} A_n(\mathbf{u};\beta) &= \frac{1}{n}\,[-X'EX] = \frac{1}{n}\sum_{i=1}^{n} \{-a(u_i;\beta)\}, \\ a(u_i\,;\beta) &= \varepsilon_i x_i x_i{}', \end{aligned}\right\} \tag{31}$$

where E is a diagonal matrix with non-zero elements equal to $\varepsilon_i = y_i \exp(x_i'\beta)$ and X is a $(n \times k)$ matrix with rows $x_i{}'$ and of rank $k < n$. Notice that in this model $A_n(\mathbf{u};\hat{\beta})$ is not identical to $A(\hat{\beta}) = -\,n^{-1}X'X$. From (30) and (31) it is seen that contributions to the IM test indicator vector are given by

$$d(u_i;\hat{\beta}) = \{(1 - \hat{\varepsilon}_i)^2 - \hat{\varepsilon}_i\} z_i,$$

where z_i is a column vector containing distinct elements of the symmetric matrix $x_i x_i'$.

Again, the asymptotically efficient IM test statistic can be obtained as the uncentred explained sum of squares from an artificial regression. The appropriate artificial regression can be established by first constructing the asymptotically efficient covariance matrix estimate using the formula given in (4). After some calculation this yields

$$V(\hat{\beta}) = \frac{1}{n}\left[5Z'Z - Z'X(X'X)^{-1}X'Z\right] \tag{32}$$

which like $A(\hat{\beta})$, although expressed as such, is not a function of $\hat{\beta}$. As before, Z is a $(n \times q)$ matrix with rows z_i'. If we now define the projection matrix $P = I - X(X'X)^{-1}X'$ and the residual matrix $\tilde{Z} = PZ$, obtained following a regression of Z on X, then $V(\hat{\beta})$ can be expressed as

$$V(\hat{\beta}) = \frac{1}{n}\left[\tilde{Z}'\tilde{Z} + 4Z'Z\right]$$

where $\tilde{Z}'\tilde{Z}$ is the "residual sum of squares" following a least squares regression of Z on X. Then $V(\hat{\beta})$ can be written as $n^{-1}W'W$ where W is a $(2n \times q)$ matrix defined in block partitioned form as

$$W = \begin{bmatrix} \tilde{Z} \\ 2Z \end{bmatrix}. \tag{33}$$

The matrix W has rows of the form

$$\begin{aligned} w_i &= \tilde{z}_i', \; i = 1,\ldots,n \\ &= 2z_{i-n}', \; i = n+1,\ldots,2n. \end{aligned}$$

Next define the $(n \times 1)$ vector $\hat{g}$ as having typical element

$$\hat{g}_i = \{(1 - \hat{\varepsilon}_i)^2 - \hat{\varepsilon}_i\}/2$$

and the $(2n \times 1)$ vector r as having typical element $\hat{r}_i$ where

$$\begin{aligned} \hat{r}_i &= 0, \; i = 1,\ldots,n \\ &= \hat{g}_{i-n}, \; i = n+1,\ldots,2n. \end{aligned}$$

Using the above notation the IM test indicator vector, $D_n(\mathbf{u};\hat{\beta})$, can be written as $n^{-1}W'\hat{r}$. As with the Poisson model, we can utilise the structure of the $\hat{W}$ matrix, and in particular the fact that $\tilde{Z}$ is the residual matrix following a regression of Z on X, in order to simplify the calculation procedure. This means that the asymptotically efficient variant of the IM test statistic can be calculated as the (uncentred) explained sum of squares from the following (double length) regression

$$\begin{bmatrix} 0 \\ \hat{g} \end{bmatrix} = \begin{bmatrix} Z & X \\ 2Z & 0 \end{bmatrix} \begin{bmatrix} b_1 \\ b_2 \end{bmatrix} + \text{errors} \qquad (34)$$

As with the Poisson model, the same test statistic obtains as the difference between the respective restricted ($b_1 = 0$) and unrestricted estimation of (34). Also, the estimated least squares coefficient of b_2, when added to the maximum likelihood estimate of β, gives the maximum likelihood corrected estimate. Such procedures are given a fuller discussion below.

(c) General conditional moment tests

The IM test is an example of a conditional moment test, as described by Newey (1985), and, in particular, the derivation of the asymptotic distribution of $\sqrt{n}D_n(\mathbf{U};\hat{\theta}) = \sqrt{n}n^{-1}\sum d(U_i;\hat{\theta})$ is applicable for any conditional moment test which is based on a statistic $d(U_i;\theta)$ where $E\{d(U_i;\theta^\circ)|x_i\} = 0$. The above algebra suggests that the construction of any asymptotically efficient conditional moment test, as applied in a Poisson model, an exponential model or any model where the conditional density of y_i given x_i can be expressed simply as a function of $x_i'\beta$ and y_i (with no other parameters), can be interpreted as an "explained sum of squares statistic" obtained from an artificial double length regression of the type arrived at in parts (a) and (b) above. This conjecture can be established as follows.

Assume that the conditional density of y_i given x_i can be expressed simply as a function of $x_i'\beta$ and y_i, i.e., $f(y_i|x_i;\beta) = f(y_i;m_i)$, where $m_i = x_i'\beta$.[18] Now, for this model any $(q \times 1)$ conditional moment test statistic indicator vector can be written

---

[18] For example, any member of the exponential family of densities $f(y;m) = \exp\{A(\theta)+B(y)+\theta T(y)\}$, where $\theta = \theta(x'\beta)$.

$$D_n(\mathbf{u};\hat{\beta}) = \frac{1}{n}\sum_{i=1}^{n} \mu(u_i;\hat{\beta})z(x_i;\hat{\beta}) \tag{35}$$

where $\mu(u_i;\beta)$ is a function of the data and parameters and the function $z(x_i;\beta) = z_i$ does not involve y_i. In (35), $\mu(u_i;\beta)$ and $z(x_i;\beta)$ are evaluated at $\beta = \hat{\beta}$, and in shorthand $z(x_i;\hat{\beta}) = \hat{z}_i$. Further, conditional on x_i, $E\{\mu(U_i;\beta°)\} = 0$ where expectations are taken with respect to $f(y_i|x_i;\beta°)$. The MLE $\hat{\beta}$ satisfies the likelihood equations $\Sigma\hat{\varepsilon}_i x_i = 0$, where $\hat{\varepsilon}_i$ is shorthand for $\varepsilon(u_i;\hat{\beta}) = \partial\log\{f(y_i;\hat{m}_i)\}/\partial m_i$ and $\hat{m}_i = x_i'\hat{\beta}$. Note that $E\{\varepsilon(U_i;\beta°)\} = 0$ from standard score theory.

We now define the following expectations and diagonal matrices, when they exist:

$$E\{\varepsilon(U_i;\beta°)\}^2 = \lambda(x_i;\beta°) = \lambda_i > 0; \quad \Lambda = \text{diag}\{\lambda_i\} \ (n \times n), \tag{36a}$$

$$E\{\varepsilon(U_i;\beta°).\mu(U_i;\beta°)\} = \gamma(x_i;\beta°) = \gamma_i; \quad \Gamma = \text{diag}\{\gamma_i\} \ (n \times n), \tag{36b}$$

$$E\{\mu(U_i;\beta°)\}^2 = \psi(x_i;\beta°) = \psi_i > 0; \quad \Psi = \text{diag}\{\psi_i\} \ (n \times n). \tag{36c}$$

Then applying the results of Newey (1985) (which essentially means evaluating (4) for the conditional moment test statistic $\sqrt{n}D_n(U;\hat{\beta})$ as written in (35)) the asymptotically efficient covariance matrix estimate is given by

$$V(\hat{\beta}) = \frac{1}{n}\left[\hat{Z}'\hat{\Psi}\hat{Z} - \hat{Z}'\hat{\Gamma}X(X'\hat{\Lambda}X)^{-1}X'\hat{\Gamma}\hat{Z}\right],$$

where $\hat{Z}$ is a $(n \times q)$ matrix having rows $\hat{z}_i'$ and X has rows x_i'. Since $\lambda_i > 0$ we can define $\hat{X}^* = \hat{\Lambda}^{1/2}X$, where $\hat{\Lambda}^{1/2} = \text{diag}\{\hat{\lambda}_i^{1/2}\}$. Further, we can always write $\gamma_i = \delta_i\lambda_i^{1/2}$, $\lambda_i^{1/2} > 0$. Then by defining the matrices $\hat{Z}^* = \hat{\Delta}\hat{Z}$, $\hat{\Delta} = \text{diag}\{\hat{\delta}_i\}$, $V(\hat{\beta})$ can be written

$$V(\hat{\beta}) = \frac{1}{n}\left[\hat{Z}^{*\prime}\hat{P}^*\hat{Z}^* + \hat{Z}'\hat{\Pi}\hat{Z}\right]$$

where $\hat{P}^* = I - \hat{X}^*(\hat{X}^{*\prime}\hat{X}^*)^{-1}\hat{X}^{*\prime}$ and $\hat{\Pi} = \text{diag}\{\hat{\pi}_i\} = \text{diag}\{\hat{\psi}_i - \hat{\delta}_i^2\}$.

Now, in order for the "double length regression" approach to be valid we must next establish that $\hat{\pi}_i > 0$. If $\hat{\pi}_i = 0$ then the calculation procedure is somewhat simpler and we shall detail this case shortly. In shorthand notation we write $\mu(U_i;\beta^\circ) = \mu_i$ and $\varepsilon(U_i;\beta^\circ) = \varepsilon_i$. Then

$$\pi_i = \psi_i - \delta_i^2 = E\{\mu_i^2\} - \frac{[E\{\varepsilon_i\mu_i\}]^2}{E\{\varepsilon_i^2\}} \geq 0 \tag{37}$$

by Cauchy-Schwartz. Assuming for the present that $\hat{\pi}_i > 0$, the matrix $\hat{V}(\beta)$ can be expressed as $n^{-1}\hat{W}'\hat{W}$ where $\hat{W}$ is a $(2n \times q)$ matrix defined in block partitioned form as

$$\hat{W} = \begin{bmatrix} \tilde{Z}^* \\ \hat{\Pi}^{1/2}\hat{Z} \end{bmatrix}$$

where $\tilde{Z}^*$ is the residual matrix, having rows $\tilde{z}_i^{*\prime}$, following a regression of $\hat{Z}^*$ on $\hat{X}^*$. The matrix $\hat{W}$ has rows

$$\left.\begin{aligned} \hat{w}_i' &= \tilde{z}_i^{*\prime},\ i=1,\ldots n \\ &= \sqrt{\hat{\pi}_{i-n}}\,\hat{z}_{i-n}',\ i=n+1,\ldots,2n \end{aligned}\right\}. \tag{38}$$

Next define the $(n \times 1)$ vector $\hat{g}$ as having typical element

$$\hat{g}_i = \{\mu_i(u_i;\hat{\beta})/\sqrt{\hat{\pi}_i}\}$$

and the $(2n \times 1)$ vector $\hat{r}$ as having typical element $\hat{r}_i$ where

$$\left.\begin{aligned} \hat{r}_i &= 0,\ i=1,\ldots,n \\ &= \hat{g}_{i-n},\ i=n+1,\ldots,2n \end{aligned}\right\}. \tag{39}$$

Then the conditional moment test indicator vector, $D_n(\mathbf{u};\hat{\beta})$, can be written as $n^{-1}\hat{W}'\hat{r}$ so that the asymptotically efficient conditional moment test statistic takes the form $\hat{r}'\hat{W}(\hat{W}'\hat{W})^{-1}\hat{W}'\hat{r}$ which is the (uncentred) explained sum of squares from the following (double length) regression

$$\begin{bmatrix} 0 \\ \hat{g} \end{bmatrix} = \begin{bmatrix} \tilde{Z}^* \\ \hat{\Pi}^{1/2}\hat{Z} \end{bmatrix} b \ + \ \text{errors}.$$

The procedure can be summarised as follows:

(1) Estimate the model to obtain $\hat{\beta}$ and generate the variables $\hat{\mu}_i = \mu(u_i;\hat{\beta})$ and $\hat{z}_i = z(x_i;\hat{\beta})$.

(2) Calculate the expectations given in (36a)-(36c), under the assumption of correct model specification and use these to calculate π_i given in (37). Then generate the variables $\hat{\lambda}_i = \lambda(x_i;\hat{\beta})$, $\hat{\gamma}_i = \gamma(x_i;\hat{\beta})$, $\hat{\psi}_i = \psi(x_i;\hat{\beta})$ and $\hat{\pi}_i = \pi(x_i;\hat{\beta})$.

(3) Generate the variables $\hat{z}_i^{*\prime} = \{\hat{\gamma}_i/\surd\hat{\lambda}_i\}\hat{z}_i'$ and $\hat{x}_i^{*\prime} = \{\surd\hat{\lambda}_i\}x_i'$, regress $\hat{z}_i^{*\prime}$ on $\hat{x}_i^{*\prime}$ and retain the residuals $\tilde{z}_i^{*\prime}$.

(4) Generate the variables $\{\surd\hat{\pi}_i\}\hat{z}_i'$ and $\hat{g}_i = \hat{\mu}_i/\surd\hat{\pi}_i$. Then form the $(2n \times q)$ regressor matrix W having rows w_i' as defined in (38) and the $(2n \times 1)$ vector of left hand side variables, $\hat{r}$, as defined in (39).

(5) Regress $\hat{r}_i$ on $\hat{w}_i'$, and obtain the (uncentred) explained sum of squares statistic: $\hat{r}'\hat{W}(\hat{W}'\hat{W})^{-1}\hat{W}'\hat{r}$. This is the asymptotically efficient conditional moment test statistic which has an asymptotic χ^2 distribution on q degrees of freedom (the dimension of $\hat{w}_i'$) when the model is correctly specified.[19]

Notice that the procedure as described above requires an initial artificial regression to be carried out at stage (3) in order to construct the residuals $\tilde{z}_i^{*\prime}$. To save time, this regression can be avoided by appealing to the familiar "residualing-out" interpretation of ordinary least squares, as intimated previously in the special cases of the Poisson and exponential models. This implies that exactly the same test

---

[19] These double length regressions are different to those proposed by Davidson and MacKinnon (1988).

statistic can be obtained as the uncentred explained sum of squares from the following double length regression

$$\begin{bmatrix} 0 \\ \hat{g} \end{bmatrix} = \begin{bmatrix} \hat{Z}^* \\ \hat{\Pi}^{1/2}\hat{Z} \end{bmatrix} b_1 + \begin{bmatrix} \hat{X}^* \\ 0 \end{bmatrix} b_2 + \text{errors.}$$

As pointed out before, this sort of artificial regression has a number of interesting features, and we discuss them more fully here. The last three of them relate to the idea of "local" corrections, discussed in Chapter 6 of this volume. Firstly, observe that the explained sum of squares obtained from this regression but when the restriction $b_1 = 0$ is imposed is identically zero, so that the residual sum of squares from this restricted regression is identical to $\hat{r}'\hat{r}$, the total sum of squares. This means that the required test statistic can be calculated as the difference between the residual sums of squares from imposing $b_1 = 0$ and not. As such, the test can be viewed as a test of whether $b_1 = 0$.

Secondly, the estimated ordinary least squares coefficient of b_2 can be thought of as indicating possible inconsistency in the maximum likelihood estimator, $\hat{\beta}$, when $b_1 \neq 0$; i.e., when the model is misspecified. Davidson and MacKinnon (1989) have discussed such an interpretation which leads to the construction of a Hausman test. This test investigates the possible inconsistency of the MLE by looking at the statistical significance of the least squares estimate of b_2; the form of this sort of Hausman test statistic is given in Chapter 6.

Thirdly, if the test is an IM test (as was the case in the Poisson and exponential models considered above), then the estimate of b_2 when added to the original MLE, $\hat{\beta}$, gives an estimator $\tilde{\beta}$ which is designed to "correct" the MLE for potential asymptotic bias caused by (local) neglected heterogeneity of unknown form. The idea of using such a procedure was introduced by Chesher, Lancaster and Irish (1985) and developed further by Orme (1989a,b); this correction procedure provided the theme of Chapter 6. Fourthly, and carrying on the "correction" theme, if the left hand side vector is replaced by $\left[\hat{\beta}'\hat{X}^{*\prime}, \hat{g} \right]'$, then the required test statistic is still the difference between the residual sums of squares from imposing $b_1 = 0$, and unrestricted least squares estimation, but the estimated coefficient for b_2 is now the exact maximum likelihood correction, $\tilde{\beta}$.

A much simpler, single length regression, procedure obtains when $\pi_i = 0$. From the Cauchy-Schwartz inequality given in equation (37) it can be seen that this arises when the zero mean random variables μ_i and ε_i satisfy $\mu_i = h_i\varepsilon_i$, where $h_i = \gamma_i/\lambda_i$ is not a function of y_i. This is the case for the IM test in the probit and logit models. Note that both models are members of the class of models considered here (see Orme (1988)). If it is established at stage (2) that $\pi_i = 0$ then the test indicator vector can be expressed as $n^{-1}\Sigma\varepsilon(u_i;\hat{\beta})\hat{h}_i\hat{z}_i$, where $\hat{h}_i$ and $\hat{z}_i$ may be functions of x_i and $\hat{\beta}$, but not y_i. Correspondingly, $V(\hat{\beta})$ can be written $n^{-1}[\hat{Z}^{*\prime}\hat{P}^*\hat{Z}^*]$. The procedure then changes at (3) to become:

(3)* Generate the variables $\hat{z}_i^{*\prime} = \{\hat{\gamma}_i/\surd\hat{\lambda}_i\}\hat{z}_i'$ and $\hat{x}_i^{*\prime} = \{\surd\hat{\lambda}_i\}x_i'$. Then form the regressors $\hat{w}_i^{*\prime} = (\hat{z}_i^{*\prime}, \hat{x}_i^{*\prime})$.

(4)* Generate the left hand side variable $\hat{r}_i = \{\surd\hat{\lambda}_i/\hat{\gamma}_i\}\hat{\mu}_i$.

(5)* Regress $\hat{r}_i$ on $\hat{w}_i^{*\prime}$ and obtain the (uncentred) explained sum of squares statistic or the asymptotically equivalent (in this case, since *plim* $n^{-1}\hat{r}'\hat{r} = 1$) (uncentred) nR^2 statistic, both of which are asymptotically distributed as a χ^2 variate on q degrees of freedom (the dimension of $\hat{w}_i^{*\prime}$) when the model is correctly specified.

Such regressions, generation of variables and retention of residuals for use in further regressions are easily accomplished on most standard econometric packages; for example, TSP or alternatively LIMDEP where the manipulation of data matrices and direct calculation of the relevant quadratic forms are easily achieved.

Also observe that in the special case considered above, where $\pi_i = 0$, the test statistic examines the sample covariance between the variables $\hat{\varepsilon}_i$ and $\hat{h}_i\hat{z}_i' = \{\hat{\gamma}_i/\hat{\lambda}_i\}\hat{z}_i'$. Since these latter variables are not directly functions of y_i this suggests that an asymptotically equivalent procedure would be to construct $\hat{h}_i\hat{z}_i'$ and regard these as "candidate omitted variables" and re-estimate the original model with these variables included as extra "regressors" and test their joint significance via an asymptotically valid likelihood ratio or Wald test procedure. This is pointed out because in many econometric packages (e.g., LIMDEP) it is often easier to simply re-estimate a non-linear model than to construct and save all the necessary variables required to implement the "explained sum of squares" procedure.

(d) An information matrix test for the Weibull duration model

A generalisation of the exponential duration model, discussed in part (b) above, is the Weibull specification. For this model, too, an asymptotically efficient information matrix test can be computed as an explained sum of squares statistic. Here the conditional density of completed durations is given by

$$f(y_i;\beta',\sigma) = \begin{cases} \sigma y_i^{\sigma-1}\lambda_i \exp(-\lambda_i y_i^{\sigma}), \ \sigma > 0, \ y_i > 0 \\ \\ 0, \ \text{otherwise} \end{cases} \tag{40}$$

where $\log(\lambda_i) = x_i'\beta$, which contains an intercept term. This model is not in the class previously considered, as the conditional density is defined not only in terms of the regression function, $m_i = x_i'\beta$, but also an ancillary parameter $\sigma > 0$. The exponential model obtains when $\sigma = 1$.

The log-likelihood (ignoring constants) based on a sample of n observations of completed durations is

$$\mathcal{L}(\beta',\sigma) = n\log(\sigma) + \sum_{i=1}^{n} \left\{\log(\varepsilon_i) - \varepsilon_i\right\} \tag{41}$$

where $\varepsilon_i = \lambda_i y_i^{\sigma}$ are unit exponential (generalised) errors in the sense of Cox and Snell (1968) and as discussed by Lancaster (1985). As Lancaster points out, not only do these errors simplify the algebra but they also facilitate the calculation of the variance estimator for the asymptotically efficient information matrix test.

From (41) we readily obtain the following likelihood equations (differentiating (41) with respect to β' and σ respectively),

$$\sum_{i=1}^{n} (1 - \hat{\varepsilon}_i)x_i = 0 , \qquad \frac{n}{\hat{\sigma}} + \sum_{i=1}^{n} (1 - \hat{\varepsilon}_i)\log(y_i) = 0 \tag{42}$$

where $\hat{\varepsilon}_i$ is simply ε_i evaluated at the MLE and which Lancaster (1985) calls a "generalised (ML) residual".

Writing $\theta' = (\beta', \sigma)$ and differentiating (41) twice gives the following hessian matrix (partitioned conformably with θ)

$$A_n(\mathbf{u};\theta) = -\frac{1}{n}\sum_{i=1}^{n} \begin{bmatrix} \varepsilon_i x_i x_i' & \varepsilon_i x_i \log(y_i) \\ \varepsilon_i \log(y_i) x_i' & \varepsilon_i \{\log(y_i)\}^2 + \sigma^{-2} \end{bmatrix} \tag{43}$$

which is evidently negative definite (assuming no collinear regressors in x_i'), so that any solution to (42) will be the unique MLE. Consider, now, that part of the information matrix test which only examines the "top left block" of the information matrix equality. From (42) and (43) it is seen that this information matrix test indicator vector has contributions given by $d(u_i;\hat{\theta}) = \{(1 - \hat{\varepsilon}_i)^2 - \hat{\varepsilon}_i\}z_i$, where z_i is a column vector containing distinct elements of the symmetric matrix $x_i x_i'$. This test is of interest because, as before, the asymptotically efficient variant of the test statistic can be calculated as an explained sum of squares statistic, but from a triple length regression rather than a double length regression.

The calculation of the asymptotically efficient variance estimator is slightly more complicated in this case because of the presence of the ancillary parameter σ, which is also estimated. To establish the variance in the limiting distribution of $\sqrt{n}n^{-1}\Sigma d(u_i;\hat{\theta})$ we could, for example, use the formula given by Lancaster (1984; halfway down page 1052) and repeated in this chapter as equation (4) (see also White (1982) and Newey (1985)). This requires some intermediary calculations which we shall detail for completeness. In what follows below, expectations are taken conditional on x_i taken with respect to $f(y_i;\theta°)$:

$$E[\{d(U_i;\theta°)\}\{d(U_i;\theta°)\}'] = 5z_iz_i',$$

$$E[\{d(U_i;\theta°)\}\{\partial \log f(y_i;\theta°)/\partial\theta'\}] = -z_i(x_i', \sigma°^{-1}\{\eta_i° + 2\})$$

where $\eta_i° = \psi(2) - x_i'\beta°$ and $\psi(.)$ denotes the logarithmic derivative of the complete gamma integral so that $\psi(2) = \psi(1) + 1 = 0.4228$. Finally, the average information matrix is given by

$$-A(\theta°) = - E[A_n(\mathbf{U};\theta°)] = \frac{1}{n}\sum_{i=1}^{n}\begin{bmatrix} x_ix_i' & x_i\eta_i°/\sigma° \\ \eta_i°x_i'/\sigma° & \{\dot{\psi}(1)+\eta_i°^2\}/\sigma°^2 \end{bmatrix}. \qquad (44)$$

where $\dot{\psi}(.)$ denotes the first derivative of $\psi(.)$ and $\dot{\psi}(1) = 1.6449$. It is not too difficult to establish the inverse of (44) and this turns out to be

$$[-A(\theta°)]^{-1} = \begin{bmatrix} (X'X/n)^{-1} + b°b°'/\dot{\psi}(1) & \sigma°b°/\dot{\psi}(1) \\ \sigma°b°'/\dot{\psi}(1) & \sigma°^2/\dot{\psi}(1) \end{bmatrix} \qquad (45)$$

where $b°' = (\beta_0° - \psi(2), \beta_1°')$ and β_0 is the intercept term in the regression function with $\beta' = (\beta_0, \beta_1')$; X is the familiar regression matrix with rows x_i'. Applying these results in equation (4) and then replacing any unknown parameters by their MLE gives the asymptotically efficient covariance matrix estimator. However, in this case, as with the exponential model, $V(\theta°)$ does not involve $\theta°$ and the variance is estimated by

$$V = \frac{5Z'Z}{n} - \frac{Z'X(X'X)^{-1}X'Z}{n} - \frac{4}{\dot{\psi}(1)}\bar{z}\,\bar{z}' \qquad (46)$$

where the matrix Z has rows z_i' and the vector $\bar{z}$ contains the sample means of the columns of Z, i.e., $\bar{z}' = n^{-1}\Sigma z_i'$.

Now, (46) is in a particularly simple form already, but can be re-expressed as

$$V = \frac{1}{n}\left[Z'PZ + 4\pi\bar{Z}'\bar{Z} + 4(1-\pi)Z'Z\right] \tag{47}$$

where the matrix $\bar{Z}$ expresses the columns of Z in mean deviation form, i.e., $\bar{Z}$ has rows $z_i' - \bar{z}'$, $P = I - X(X'X)^{-1}X'$ and $\pi = 1/\dot{\psi}(1)$ $(0 < \pi < 1)$. We can then define the matrix W to be

$$W = \begin{bmatrix} \tilde{Z} \\ 2\sqrt{\pi}\ \bar{Z} \\ 2\sqrt{1-\pi}\ Z \end{bmatrix} \tag{48}$$

where $\tilde{Z}$ is the residual matrix following a regression of Z on X. In a similar fashion to before we now define the $(n \times 1)$ column vector to $\hat{g}$ to have typical element $\hat{g}_i = \{(1 - \hat{\varepsilon}_i)^2 - \hat{\varepsilon}_i\}/2\sqrt{1-\pi}$ and the $(3n \times 1)$ matrix $\hat{r}$ to have typical elements

$$\hat{r}_i = \begin{cases} 0\ , & i=1,\ldots,2n \\ \hat{g}_{i-2n}\ , & i=2n+1,\ldots,3n. \end{cases} \tag{49}$$

Then the asymptotically efficient variant of this information matrix test can be calculated as $\hat{r}'W(W'W)^{-1}W'\hat{r}$, the (uncentred) explained sum of squares obtained from a triple length regression of $\hat{r}$ on the W. As with the previous models considered the number of ordinary least squares computations can be reduced in order to calcualte the IM test statistic. Previous analysis correctly suggests that the following triple length regression can be employed

$$\begin{bmatrix} 0 \\ 0 \\ \hat{g} \end{bmatrix} = \begin{bmatrix} Z & X & 0 \\ 2\sqrt{\pi}\ Z & 0 & s \\ 2\sqrt{1-\pi}\ Z & 0 & 0 \end{bmatrix} \begin{bmatrix} b_1 \\ b_2 \\ b_3 \end{bmatrix} + \text{errors}$$

where s denotes the vector of ones (the sum vector). The IM test statistic can be calculated as the difference between the restricted $(b_1 = 0)$ and unrestricted residual sums of squares.

(Note, however, that corrections for neglected parameter variation are not directly obtainable from the output of this estimated regression due to the construction of the "regressor" matrix employed.)

Some Monte Carlo results, comparing the sampling performances of the OPG variant and efficient variant of the above IM test statistics, confirms the conclusions drawn in Section 4 of this chapter that the OPG variant is disastrously over-sized. The efficient variant, in all three models, is vastly superior but still not quite perfect. Its sampling mean appears smaller than it should be, resulting in slightly fewer rejections at the 10% nominal level. Conversely, there are a few large realisations of the test statistic which cause it to be over-sized at the 1% level. However, the Monte Carlo experiments provided acceptable rejection rates at the 5% nominal level for all sample sizes considered across all models (see Orme (1990b) for further details).

(e) An efficient information matrix test for the truncated regression model.

The structure of the artificial regressions used to construct the asymptotically efficient IM test statistics discussed in (a) to (d) above suggests that a general procedure of this type is available in the context of regression based econometric models. This general procedure is explored by Orme (1991a). For the present illustrative purposes, we return to the information matrix test statistic for the truncated regression model. The paper by Orme (1990a), as reviewed in this chapter, provided a class of asymptotically equivalent, regression based, IM test statistics. However, the work did not indicate that such a form could be derived for the efficient version. To complete this chapter we update that work by providing such a form, based on a double length regression. (Further details of the calculation procedure are given in Orme (1991a).)

As outlined in Section 4 above, in the truncated regression model (see equations (9) and (10)) the conditional density of y_i

given x_i is defined by

$$f(y_i;a_i,\sigma) = \left[\frac{1}{\sigma}\phi(\{y_i-a_i\}/\sigma)\right]\left[\Phi(a_i/\sigma)\right]^{-1}, \quad y_i > 0$$

where $a_i = x_i'\beta$, $\varepsilon_i = y_i - a_i$. Here (unlike Orme (1990a)) we shall consider the model in its "natural" parameterisation. Thus, adapting the algebra of Section 4 accordingly, the maximum likelihood estimators, $\hat{\beta}$ and $\hat{\sigma}$, solve the following likelihood equations

$$\sum_i \hat{\mu}_i x_i = 0, \quad \sum_i \hat{\gamma}_i = 0,$$

where

$$\mu_i = \sigma^{-1}\{v_i - h_i\}$$

and

$$\gamma_i = \sigma^{-1}\{(v_i^2 - 1) + c_i h_i\},$$

in which the following definitions apply: $v_i = \varepsilon_i/\sigma$, $c_i = a_i/\sigma$ and $h_i = \phi(c_i)/\Phi(c_i)$.

The following second order moments are defined conditional on x_i:

$$\text{var}(\mu_i) = \sigma^{-2}(1-h_i(h_i+c_i)) > 0;$$

$$\text{cov}(\mu_i,\gamma_i) = \sigma^{-2}h_i(1+c_i(h_i+c_i)) > 0;$$

$$\text{var}(\gamma_i) = \sigma^{-2}(2-c_i h_i(1+c_i(h_i+c_i))) > 0.$$[20]

Then, following the method of Orme (1991a, 1991b) various variables need to be constructed from the above (evaluated at

---

[20] These results and associated inequalities are well established.

maximum likelihood estimates). These are:

$$\upsilon_{1i} = \{\text{var}(\mu_i) - \{\text{cov}(\mu_i,\gamma_i)\}^2/\text{var}(\gamma_i)\}^{1/2};$$

$$\upsilon_{2i} = \text{cov}(\mu_i,\gamma_i)/\{\text{var}(\gamma_i)\}^{1/2}$$

and

$$\upsilon_{3i} = \{\text{var}(\gamma_i)\}^{1/2}.$$

From these expressions we obtain $\hat{\upsilon}_{1i}$, $\hat{\upsilon}_{2i}$ and $\hat{\upsilon}_{3i}$ where the hats indicate that unknown parameters have been replaced by maximum likelihood estimates. In particular, it can be shown that the variable $\hat{\upsilon}_{1i}$ is strictly positive. As before the $(q \times 1)$ vector z_i is defined to contain distinct non-constant terms of $\text{vech}(x_i x_i')$.[21]

For the truncated regression model the information matrix test indicator, corresponding to the $\beta\beta'$ partition of the information matrix equality is given by

$$D_n(\mathbf{u};\hat{\theta}) = \frac{1}{n}\sum_{i=1}^{n} \hat{\sigma}^{-2}\left\{(\hat{v}_i - \hat{h}_i)^2 - (1 - \hat{h}_i(\hat{h}_i + \hat{c}_i))\right\} z_i. \quad (50)$$

A closer inspection of the expression in (50) reveals that it is identical to

$$D_n(\mathbf{u};\hat{\theta}) = \frac{1}{n}\sum_{i=1}^{n} \hat{\sigma}^{-1}\left\{\hat{\gamma}_i - 2\hat{h}_i\hat{\mu}_i\right\} z_i.$$

It is the ability to express the test indicator in this form which facilitates the following double length regression interpretation. Firstly, the efficient estimator of the variance

---

[21] As in the probit model, the first element in the IM test indicator vector is identically zero.

matrix of $\sqrt{n}D_n(u;\hat{\theta})$ is given by the inverse of the bottom right $(q \times q)$ partition of $\hat{\mathcal{I}}^{-1}$, where the $(k+1+q \times k+1+q)$ matrix $\hat{\mathcal{I}}$ is defined as

$$\hat{\mathcal{I}} = \frac{1}{n}\begin{bmatrix} X' & 0' \\ 0' & s' \\ \hat{Z}_1' & \hat{Z}_2' \end{bmatrix}\begin{bmatrix} \hat{A} & \hat{C} \\ \hat{C} & \hat{B} \end{bmatrix}\begin{bmatrix} X & 0 & \hat{Z}_1 \\ 0 & s & \hat{Z}_2 \end{bmatrix} = \frac{1}{n}\hat{H}'\hat{\Omega}\hat{H}$$

where X is the $(n \times k)$ regressor matrix; $\hat{Z}_1$ is an $(n \times q)$ matrix with rows $-2\hat{\sigma}^{-1}\hat{h}_i z_i'$; $\hat{Z}_2$ is an $(n \times q)$ matrix with rows $\hat{\sigma}^{-1}z_i'$; $\hat{A}$, $\hat{B}$ and $\hat{C}$ are diagonal matrices defined by $A = \text{diag}(\text{var}(\mu_i))$, $B = \text{diag}(\text{var}(\gamma_i))$ and $C = \text{diag}(\text{cov}(\mu_i,\gamma_i))$. This is a fairly straightforward result to derive if one regards the IM test indicator as the sum of two score tests for omission of variables from the variance and regression specification respectively. Now, the matrix $\hat{\Omega}$ can be expressed as $\hat{P}'\hat{P}$ where

$$\hat{P} = \begin{bmatrix} (\hat{A} - \hat{C}\hat{B}^{-1}\hat{C})^{1/2} & 0 \\ \hat{C}\hat{B}^{-1/2} & \hat{B}^{1/2} \end{bmatrix} = \begin{bmatrix} \hat{V}_1 & 0 \\ \hat{V}_2 & \hat{V}_3 \end{bmatrix}.$$

Thus, the matrix $\hat{\mathcal{I}}$ is expressible as $\hat{\mathcal{I}} = n^{-1}\hat{W}'\hat{W}$, where the $(2n \times k+1+q)$ matrix $\hat{W}$ is defined by

$$\hat{W} = \hat{P}\hat{H} = \begin{bmatrix} \hat{V}_1X & 0 & \hat{V}_1\hat{Z}_1 \\ \hat{V}_2X & \hat{V}_3s & \hat{V}_2\hat{Z}_1 + \hat{V}_3\hat{Z}_2 \end{bmatrix}.$$

The artificial regression employed to calculate the efficient IM test statistic (and associated corrections) is of "double length" form. The first set of n observations on the left hand side variable are set equal to

$$\hat{r}_{1i} = (\hat{\mu}_i - \{\hat{\gamma}_i/\hat{\upsilon}_{3i}\}\hat{\upsilon}_{2i})/\hat{\upsilon}_{1i},\ i = 1,\ldots,n$$

and the second set of n elements are

$$\hat{r}_{2i} = \hat{\gamma}_i/\hat{\upsilon}_{3i},\ i = 1,\ldots,n.$$

Correspondingly, from the above definition of $\hat{W}$, the right hand side variables come in two sets of n observations. The first set is typically

$$\hat{w}_{1i}' = \{\hat{\upsilon}_{1i}x_i', 0, -2\hat{\sigma}^{-1}\hat{\upsilon}_{1i}\hat{h}_i z_i'\} \quad (1 \times k+1+q)$$

and the second set is typically

$$\hat{w}_{2i}' = \{\hat{\upsilon}_{2i}x_i', \hat{\upsilon}_{3i}, \hat{\sigma}^{-1}(\hat{\upsilon}_{3i} - 2\hat{\upsilon}_{2i}\hat{h}_i)z_i'\} \quad (1 \times k+1+q).$$

Then, stacking the observations into appropriate vectors and matrices, the double length artificial regression can be expressed as

$$\underset{(2n \times 1)}{\begin{bmatrix} \hat{r}_1 \\ \hat{r}_2 \end{bmatrix}} = \underset{(2n \times k+1+q)}{\begin{bmatrix} \hat{W}_1 \\ \hat{W}_2 \end{bmatrix}} \begin{bmatrix} b_1 \\ b_2 \end{bmatrix} + \text{errors}$$

where b_1 is $(k+1 \times 1)$ and b_2 is $(q \times 1)$. The efficient IM test statistic is simply the difference between the two residual sums of squares obtained from imposing the restriction that $b_2 = 0$ and unrestricted ordinary least squares estimation (see Orme (1991a)). The corrected regression parameter estimates (i.e., corrected estimates of β, $(k \times 1)$) are obtained by taking the first k elements of the ordinary least squares estimate of b_1, from the above regression, and adding them to the elements of $\hat{\beta}$, the maximum likelihood estimate of β. The appropriate estimated covariance matrix of these corrections is the top left $(k \times k)$ partition of $(\hat{W}'\hat{W})^{-1}$.

For this IM test statistic we provide some evidence on its sampling performance under the null hypothesis of no model misspecification and compare it to that of the Chesher/Lancaster OPG variant. For this Monte Carlo experiment the regression function contained a constant term, a standard normal regressor and an independent uniform regressor distributed over the range

(0,4); the error terms in the latent regression model were independent standard normal random variates. The data generation process involved simulating the latent variable y_i^* and rejecting all simulated data if $y_i^* \leq 0$. The results reported in Table 4 are based on 2000 replications. The efficient variant described above is denoted EFF and the Chesher/Lancaster variant is denoted OPG.

TABLE 4

Truncated Regression Model: IM Test Statistics
(first k(k+1)/2 set of indicators)

Number of regressors: k = 3

sample size	statistic	mean	st.dev	rej.rate [a]
		6.00	3.46	5%
50	EFF	5.67	3.96	5.0
50	OPG	9.86	4.29	24.7
100	EFF	5.65	3.85	5.0
100	OPG	9.04	4.84	21.0
200	EFF	5.89	3.95	6.0
200	OPG	8.41	4.98	17.5

[a] The standard error of the rejection rate is 0.487

These results again illustrate the superior sampling performance of the efficient variant of the IM test statistic over the Chesher/Lancaster variant. Occasionally the efficient variant produces a large value which gives it a larger standard deviation than anticipated; this phenomenon has been noted before by Chesher and Spady (1991). However, the empirical rejection rate of this variant, under the null hypothesis, appears to be adequately predicted by the appropriate χ_6^2 sampling distribution (in all cases a 99% confidence interval contains the nominal (correct) rejection rate of 5%). The empirical rejection rates of the OPG variant are appalling, as previously encountered.

6. Conclusion

This chapter has reviewed the IM test procedure, originally proposed by White (1982), and has investigated the size performance of several variants of the IM test statistic for the truncated normal regression model and binary probit model. A set, Ω, of IM test statistics can be defined in which each member has an nR^2 interpretation and thus, for the applied worker, is relatively simple to compute. As pointed out, this set is by no means exhaustive and does not, for instance, contain the so-called asymptotically efficient variant. The Lancaster/Chesher variant is a member of Ω and has been shown, in Section 3, to be always the largest numerically. The sampling experiments show that the Lancaster/Chesher variant exhibits appalling size bias even in quite large samples and this appears to get worse as the number of indicators used in the test increases. This last observation is also true for variant ω_4 which, like the OPG variant, does not employ third derivatives of the log-density in its construction. The evidence presented here together with that reported by Taylor (1987) and Kennan and Neumann (1987) indicates that these two variants can produce very misleading inferences.

The IM test is an example of a conditional moment test and the Lancaster/Chesher variant is exactly the type of construction that Newey (1985) proposed for general conditional moment test statistics. Thus the appalling size bias exhibited by the Lancaster/Chesher variant of the IM test may well be indicative of the sampling performance to be expected from Newey's conditional moment test statistics. In particular, the analysis presented in Section 3 is directly applicable to conditional moment tests: we can define a set Ω of conditional moment test statistics in exactly the same way and Newey's variant (the OPG version) will always be numerically the largest in that set. Thus the comments and conclusions presented concerning the IM test procedure may well be more generally applicable, or at least should not be ignored when implementing a conditional moment test procedure.

A reduction in the upward size bias exhibited by the Lancaster/Chesher variant can always be achieved by employing another member of Ω, of which one is White's original version of the IM test statistic. For the binary probit model this also performed badly, being only marginally superior to the OPG variant. However, quite substantial reductions in the size bias exhibited by the OPG variant were achieved by employing expected values of third derivatives in the nR^2 formulation, i.e. variants ω_3, ω_6, ω_8 (which are members of Ω), the asymptotically efficient variant and its modified version (ω_A and ω_B) (which are not members of Ω). (Substantial reductions in size bias were also achieved by variants employing third derivatives, or expected values thereof, of the log-density in the truncated regression model experiments.) Interestingly these are exactly the sort of variants that Chesher and Lancaster were trying to avoid by proposing the OPG variant. Of the rest, the performance of ω_3 was quite encouraging for both the models considered in the Monte Carlo experiments reported in Section 4 of this chapter. In particular, its performance did not deteriorate like the OPG variant as the number of indicators increased.

Variants ω_A and ω_B performed well for the binary probit model and their use, over other variants, may well be recommended in other models (where available) in order to obtain a test procedure with reasonable size properties. In order to facilitate this, in Section 5 asymptotically efficient variants of the IM test, and other conditional moment tests, were provided for any model where the conditional density of y_i given x_i can be simply expressed in terms of a linear regression function ($x_i'\beta$) and y_i; for example the Poisson and exponential models. Moreover, these were given a convenient (and previously unreported) "explained sum of squares" interpretation to further aid calculation. The relevant explained sum of squares is obtained from a double length artificial regression which is run after model estimation. This double length regression, however, is not of the sort proposed by Davidson and MacKinnon (1988). Interestingly, for the Weibull duration model an asymptotically efficient information matrix test can be calculated as the explained sum of squares from a triple length regression.

The analysis of Orme (1990a) was extended to provide the

efficient variant of the IM test for the truncated regression model, also obtainable from a double length artificial regression. Monte Carlo evidence indicates that the efficient variant is vastly superior to the OPG variant. Further, the tail of the sampling distribution, for the efficient IM test statistic, approximated that of the appropriate asymptotic χ^2 distribution adequately, if one judges it on the basis of observed rejections at the 5% nominal level (see also Orme (1990b)). However, further refinements could be made to the efficient IM test statistic along the lines of Chesher and Spady (1991).

The development of artificial regressions (triple and double length) for the purposes of calculating information matrix tests for the Weibull and truncated regression models suggests that a more general procedure along these lines can be employed to construct conditional moment test statistics in the context of econometric models specified solely in terms of a regression function ($x_i'\beta$) and an ancillary parameter (σ). Such a general approach provides the focus of the paper by Orme (1991a).

In the sorts of microeconometric models considered in this chapter, sample sizes of more than one thousand are common. At the same time, however, "real-life" models contain a substantially greater number of parameters compared with the numbers used in Monte Carlo studies. With these thoughts in mind, it is hoped that the results presented here (smaller sample sizes but with fewer parameters) provide, at least, some guidance as to the relative sampling performances of differing variants of the IM test statistics in practical situations (larger sample sizes and more parameters). In particular, it draws attention to the, potentially, highly misleading OPG variant.

Appendix A

The derivation of the matrix $\nabla D(\theta)$ for the truncated regression model

Firstly, consider the standard normal random variable, Z, whose distribution is truncated to the left of -c. The density function for this truncated variate, at the point ζ, is thus $f(\zeta)=\phi(\zeta)/\Phi(c)$, where the usual notation applies. The moment generating function is easily shown to be

$$m(t) = \exp(t^2/2)\left\{\frac{\Phi(c+t)}{\Phi(c)}\right\}$$

from which the following first four moments of Z are obtained:

$$\begin{aligned}
E(Z) &= \phi(c)/\Phi(c) = h(c),\\
E(Z^2) &= 1 - ch,\ \text{var}(Z) = 1 - h(h+c) = 1 - \lambda,\\
E(Z^3) &= 2h + c^2h,\\
E(Z^4) &= 3 - 3ch - c^3h.
\end{aligned}$$

The above will be of use in the construction of $\nabla D(\theta)$.

Contributions to the IM test indicator vector are given in equation (15). In what follows, for notational convenience, the subscript i will be dropped and we shall write $d(u;\theta)' = (d_1', d_2', d_3)$. We now calculate the partial derivatives of d_1, d_2 and d_3 with respect to $\theta' = (\delta', \tau)$, where the notation is as presented in the main text.

$$\frac{\partial d_1}{\partial \delta'} = \{-2(v - h)(1 - \lambda) + h - \lambda(2h + x'\delta)\}zx',$$

$$\frac{\partial d_2}{\partial \delta'} = \{-(1 - \lambda)(\frac{1}{\tau} - vy) + (v - h)y\}xx',$$

$$\frac{\partial d_3}{\partial \delta'} = 2(\frac{1}{\tau} - vy)yx',$$

$$\frac{\partial d_1}{\partial \tau} = 2(v - h)yz,$$

$$\frac{\partial d_2}{\partial \tau} = \{(\tfrac{1}{\tau} - vy)y - (v - h)\left(\frac{1}{\tau^2} + y^2\right)\}x,$$

$$\frac{\partial d_3}{\partial \tau} = - 2(\tfrac{1}{\tau} - vy)\left(\frac{1}{\tau^2} + y^2\right) + \frac{2}{\tau^3} .$$

Observe that the sample average of the above six expressions, evaluated at the maximum likelihood estimate, and arranged in the appropriate ($q \times k+1$) matrix gives $\nabla D_n(\mathbf{u};\theta)$. Writing $\tau y = x'\delta + v$ and recalling that v is a standard normal variate truncated to the left of $-x'\delta$, expectations of the above derivatives (conditional on x) can easily be obtained using the moments given previously. We thus obtain the following expressions:

$$E\left(\frac{\partial d_1}{\partial \delta'} |x\right) = \{h - \lambda(2h + x'\delta)\}zx',$$

$$E\left(\frac{\partial d_2}{\partial \delta'} |x\right) = \frac{(1-\lambda)}{\tau} xx',$$

$$E\left(\frac{\partial d_3}{\partial \delta'} |x\right) = -2 \frac{(h+x'\delta)}{\tau^2} x',$$

$$E\left(\frac{\partial d_1}{\partial \tau} |x\right) = 2 \frac{(1-\lambda)}{\tau} z,$$

$$E\left(\frac{\partial d_2}{\partial \tau} |x\right) = - \frac{1}{\tau^2} \{2(h + x'\delta) + x'\delta(1 - \lambda)\}x,$$

$$E\left(\frac{\partial d_3}{\partial \tau} |x\right) = \frac{4}{\tau^3} \{1 + (x'\delta)^2 + (x'\delta)h\} + \frac{2}{\tau^3} .$$

Taking the sample average of the above expressions, evaluated at the maximum likelihood estimate, and arranged in the appropriate ($q \times k+1$) matrix gives the expression for $\nabla D(\theta)$. Finally, observe that the likelihood equations, (12a and 12b) afford some simplifications when constructing this matrix since we can write

$$\sum yx' = \hat{\tau}^{-1} \sum \{\hat{h}+x'\hat{\delta}\}x', \text{ and } \sum y^2 = \hat{\tau}^{-2} \sum \{1+(x'\hat{\delta})^2+(x'\hat{\delta})\hat{h}\}.$$

References

Arabmazar, A. and P. Schmidt (1981), 'Further Evidence on the Robustness of the Tobit Estimator to Heteroscedasticity', *Journal of Econometrics,* 17, 253-8.

Arabmazar, A. and P. Schmidt (1982), 'An Investigation into the Robustness of the Tobit Estimator to Non-Normality', *Econometrica,* 50, 1055-63.

Berndt, E.R, B.H. Hall, R.E. Hall and J.A.Hausman (1974), 'Estimation and Inference in Non-Linear Structural Models', *Annals of Economic and Social Measurement,* 3, 563-5.

Cameron, A. and P. Trivedi (1986), 'Econometric Models Based on Count Data : Comparisons and Applications of Some Estimators', *Journal of Applied Econometrics,* 1, 29-54.

Chesher, A.D. (1983), 'The Information Matrix Test: A Simplified Calculation via a Score Test Interpretation', *Economics Letters,* 13, 45-8.

Chesher, A.D. (1984), 'Testing for Neglected Heterogeneity', *Econometrica,* 52, 865-72.

Chesher, A.D., T. Lancaster and M. Irish (1985), 'On Detecting the Failure of Distributional Assumptions', *Annals de l'Insee,* 59/60, 7-44.

Chesher, A.D. and R. Spady (1991), 'Asymptotic Expansions of the Information Matrix Test', *Econometrica,* 59, 787-817.

Cox, D.R. and E.J. Snell (1968), 'A General Definition of Residuals (with discussion)', *Journal of the Royal Statistical Society (Series B),* 30, 248-75.

Davidson, R. and J.G. MacKinnon (1983), 'The Small Sample Performance of the Lagrange Multiplier Test', *Economics Letters,* 269-75.

Davidson, R. and J.G. MacKinnon (1984), 'Convenient Specification Tests for Logit and Probit Models', *Journal of Econometrics,* 25, 241-62.

Davidson, R. and J.G. MacKinnon (1988), 'Double-Length Artificial Regressions', *Oxford Bulletin of Economics and Statistics,* 50, 203-17.

Davidson, R. and J.G. MacKinnon (1989), 'Testing for Consistency

Using Artificial Regressions', *Econometric Theory,* 5, 363-84.

Efron, B. and D.V. Hinkley (1978), 'Assesing the Accuracy of the Maximum Likelihood Estimator: Observed versus Expected Information', *Biometrika,* 65, 457-87.

Godfrey, L.G. (1978a), 'Testing against General Autoregressive and Moving Average Error Models when the Regressors Include Lagged Dependent Variables', *Econometrica,* 46, 1293-1302.

Godfrey, L.G. (1978b), 'Testing for Higher Order Serial Correlation in Regression Models when the Regressors Include Lagged Dependent Variables', *Econometrica,* 46, 1303-10.

Godfrey, L.G. and M.R. Wickens (1982), 'Tests of Misspecification Using Locally Equivalent Alternative Models' in: G.C. Chow and P.Corsi, *eds., Evaluating the Reliability of Macro-economic Models,* Wiley, New York.

Hall, A. (1987), 'The Information Matrix Test for the Linear Model', *Review of Economic Studies,* 54, 257-63.

Harris, P. (1985), 'An Asymptotic Expansion for the Null Distribution of the Efficient Score Test Statistic', *Biometrika,* 72, 653-9.

Hausman, J.A. (1984), 'Specification and Estimation of Simultaneous Equation Models', Ch.7 in: Z.Griliches and M.D.Intriligator, eds., *Handbook of Econometrics, Vol. 1,* North-Holland, Amsterdam.

Hausman, J.A., B.H. Hall and Z. Griliches (1984), 'Econometric Models for Count Data with an Application to Patents - R & D Relationships', *Econometrica,* 52, 909-38.

Kennan, J. and G.R. Neumann (1987), 'Why Does the Information Matrix Test Reject So Often ? A Diagnosis with Some Monte Carlo Evidence', *University of Iowa, mimeo.*

Kiefer, N.M. (1985), 'Specification Diagnostics Based on Laguerre Alternatives for Econometric Models of Duration', *Journal of Econometrics,* 28, 135-54.

Lancaster, T. (1979), 'Econometric Models for the Duration of Unemployment', *Econometrica,* 47, 939-56.

Lancaster, T. (1983), 'Generalised Residuals and Heterogeneous Duration Models: The Exponential Case', *Bulletin of Economic*

Research, 35, 71-85.

Lancaster, T. (1984), 'The Covariance Matrix of the Information Matrix Test', *Econometrica,* 52, 1051-3.

Lancaster, T. (1985), 'Generalised Residuals and Heterogeneous Duration Models: The Weibull Case', *Journal of Econometrics,* 28, 155-69.

Lee, L - F. (1986), 'Specification Tests for Poisson Regression Models', *International Economic Review,* 27, 689-706.

Maddala, G.S. (1983), *Limited Dependent and Qualitative Variables in Econometrics*, Cambridge University Press, Cambridge.

Mullahy, J. (1986), 'Specification and Testing of Some Modified Count Data Models', *Journal of Econometrics,* 33, 341-65.

Newey, W.K. (1985), 'Maximum Likelihood Specification Testing and Conditional Moment Tests', *Econometrica,* 53, 1047-70.

Orme, Chris (1987), 'Specification Tests for Binary Data Models', *Discussion Paper No. 65, Department of Economics, University of Nottingham.*

Orme, Chris (1988), 'The Calculation of the Information Matrix Test for Binary Data Models', *The Manchester School,* 54, 370-6.

Orme, Chris (1989a), 'Evaluating the Performance of Maximum Likelihood Corrections in the Face of Local Misspecification', *Bulletin of Economic Research,* 41, 29-44.

Orme, C.D. (1989b), *Misspecification and Inference in Micro-Econometrics,* Unpublished Ph.D. Thesis, University of York.

Orme, Chris (1990a), 'The Small Sample Performance of the Information Matrix Test', *Journal of Econometrics,* 46, 309-31.

Orme, Chris (1990b), 'Double and Triple Length Regressions for the Information Matrix Test and other Conditional Moments Tests', *mimeo. Department of Economics and Related Studies, University of York.*

Orme, Chris (1991a), 'On the Use of Artificial Regressions in Certain Microeconometric Models', *mimeo. Department of Economics and Related Studies, University of York.*

Orme, Chris (1992), 'Efficient Score Tests for Heteroskedasticity in Microeconometrics', *Econometric Reviews* (forthcoming).

Tauchen, G. (1985), 'Diagnostic Testing and Evaluation of Maximum Likelihood Models', *Journal of Econometrics,* 30, 415-43.

Taylor, L. (1987), 'The Size Bias of White's Information Matrix Test', *Economics Letters,* 24, 63-8.

White, H. (1982), 'Maximum likelihood Estimation of Misspecified Models', *Econometrica,* 50, 1-25.

Yatchew, A. and Z. Griliches (1985), 'Specification Error in Probit Models', *Review of Economics and Statistics,* 67, 134-9.

Index